MW00844318

A GUIDE TO FIRST-PASSAGE PROCESSES

First-passage properties underlie a wide range of stochastic processes, such as diffusion-limited growth, neuron firing, and the triggering of stock options. This book provides a unified presentation of first-passage processes, which highlights their interrelations with electrostatics and the resulting powerful consequences. The goal of this book is to help those with modest backgrounds learn essential results quickly.

The author begins with a modern presentation of fundamental theory including the connection between the occupation and first-passage probabilities of a random walk and the connection to electrostatics and current flows in resistor networks. The consequences of this theory are then developed for simple, illustrative geometries including finite and semi-infinite intervals, fractal networks, spherical geometries, and the wedge. Various applications are presented including neuron dynamics, self-organized criticality, stochastic resonance diffusion-limited aggregation, the dynamics of spin systems, and the kinetics of diffusion-controlled reactions.

First-passage processes provide an appealing way for graduate students and researchers in physics, chemistry, theoretical biology, electrical engineering, chemical engineering, operations research, and finance to understand all of these systems.

Sidney Redner is a Professor of Physics at Boston University and both a Member and a Fellow of the American Physical Society. He is interested in nonequilibrium statistical mechanics and continues to research reaction kinetics, diffusive transport in random media, the structure of percolation, coarsening kinetics, filtration processes, and the structure of random growing networks. He has published 150 papers in these areas.

A GUIDE TO FIRST-PASSAGE PROCESSES

SIDNEY REDNER
Boston University

CAMBRIDGE
UNIVERSITY PRESS

CAMBRIDGE UNIVERSITY PRESS
Cambridge, New York, Melbourne, Madrid, Cape Town, Singapore, São Paulo

Cambridge University Press
The Edinburgh Building, Cambridge CB2 8RU, UK

Published in the United States of America by Cambridge University Press, New York

www.cambridge.org
Information on this title: www.cambridge.org/9780521652483

© Cambridge University Press 2001

This publication is in copyright. Subject to statutory exception
and to the provisions of relevant collective licensing agreements,
no reproduction of any part may take place without the written
permission of Cambridge University Press.

First published 2001
This digitally printed version (with corrections) 2007

A catalogue record for this publication is available from the British Library

Library of Congress Cataloguing in Publication data
Redner, Sidney, 1951–
A guide to first passage processes / Sidney Redner.
p. cm.
Includes bibliographical references and index.
ISBN 0-521-65248-0
1. Stochastic processes. I. Title.
QA274.2 .R44 2001
519.2'3 dc21   00-067493

ISBN 978-0-521-65248-3 hardback
ISBN 978-0-521-03691-7 paperback

Cambridge University Press has no responsibility for the persistence or
accuracy of URLs for external or third-party Internet websites referred to in
this publication, and does not guarantee that any content on such websites is,
or will remain, accurate or appropriate.

Contents

v

Preface

You arrange a 7 P.M. date at a local bistro. Your punctual date arrives at 6:55, waits until 7:05, concludes that you will not show up, and leaves. At 7:06, you saunter in – "just a few minutes" after 7 (see Cover). You assume that you arrived first and wait for your date. The wait drags on and on. "What's going on?" you think to yourself. By 9 P.M., you conclude that you were stood up, return home, and call to make amends. You explain, "I arrived around 7 and waited 2 hours! My probability of being at the bistro between 7 and 9 P.M., $P(\text{bistro}, t)$, was nearly one! How did we miss each other?" Your date replies, "I don't care about your *occupation* probability. What mattered was your *first-passage* probability, $F(\text{bistro}, t)$, which was zero at 7 P.M. GOOD BYE!" Click!

The moral of this juvenile parable is that first passage underlies many stochastic processes in which the event, such as a dinner date, a chemical reaction, the firing of a neuron, or the triggering of a stock option, relies on a variable reaching a specified value *for the first time*. In spite of the wide applicability of first-passage phenomena (or perhaps because of it), there does not seem to be a pedagogical source on this topic. For those with a serious interest, essential information is scattered and presented at diverse technical levels. In my attempts to learn the subject, I also encountered the proverbial conundrum that a fundamental result is "well known to (the vanishingly small subset of) those who know it well."

In response to this frustration, I attempt to give a unified presentation of first-passage processes and illustrate some of their beautiful and fundamental consequences. My goal is to help those with modest backgrounds learn essential results quickly. The intended audience is physicists, chemists, mathematicians, engineers, and other quantitative scientists. The technical level should be accessible to the motivated graduate student.

My literary inspirations for this book include *Random Walks and Electric Networks*, by P. G. Doyle and J. L. Snell (Carus Mathematical Monographs #22, Mathematical Association of America, Washington, D.C., 1984), which cogently describes the relation between random walks and electrical networks,

vii

and *A Primer on Diffusion Problems*, by R. Ghez (Wiley, New York, 1988), which gives a nice exposition of solutions to physically motivated diffusion problems. This book is meant to complement classic monographs, such as *An Introduction to Probability Theory and Its Applications*, by W. Feller (Wiley, New York, 1968), *Aspects and Application of the Random Walks*, by G. H. Weiss (North-Holland, Amsterdam, 1996), and *Stochastic Processes in Physics and Chemistry*, by N. G. van Kampen (North-Holland, Amsterdam, 1997). Each of these very worthwhile books discusses first-passage phenomena, but secondarily rather than as a comprehensive overview.

I begin with fundamental background in Chapter 1 and outline the relation between occupation and first-passage probabilities, as well as the connection between first passage and electrostatics. Many familiar results from electrostatics can be easily adapted to give first-passage properties in the same geometry. In Chapter 2, I discuss first passage in a one-dimensional interval. This provides a simple laboratory for answering basic questions, such as: What is the probability that a diffusing particle eventually exits at either end? How long does it take to exit? These problems are solved by both direct approaches and developing the electrostatic equivalence. Chapter 3 treats first passage in a semi-infinite interval by both standard approaches and the familiar image method. I also discuss surprising consequences of the basic dichotomy between certain return to the starting point and infinite mean return time.

Chapter 4 is devoted to illustrations of the basic theory. I discuss neuron dynamics, realizations of self-organized criticality, and the dynamics of spin systems. These all have the feature that they can be viewed as first-passage processes in one dimension. I also treat stochastic resonant escape from fluctuating and inhomogeneous media, for which the time-independent electrostatic formalism provides a relatively easy way to solve for mean first-passage times. Finally, I discuss the survival of a diffusing particle in a growing "cage" and near a moving "cliff," where particularly rich behavior arises when diffusion and the motion of the boundary have the same time dependence.

In Chapter 5, I turn to first passage on branched, self-similar structures. I emphasize self-similar systems because this feature allows us to solve for the first-passage probability by renormalization. Another essential feature of branched systems is the competition between transport along the "backbone" from source to sink and detours along sidebranches. I give examples that illustrate this basic competition and the transition from scaling, in which a single time scale accounts for all moments of the first-passage time, to multiscaling, in which each moment is governed by a different time scale.

I then treat spherically symmetric geometries in Chapter 6 and discuss basic applications, such as efficient simulations of diffusion-limited aggregation and the Smoluchowski chemical reaction rate. First passage in wedge and conical domains are presented in Chapter 7. I discuss how the wedge geometry can be solved elegantly by mapping to electrostatics and conformal transformations. These systems provide the kernel for understanding the main topic of Chapter 8, namely, the kinetics of one-dimensional diffusion-controlled reactions. This includes trapping, the reactions among three diffusing particles on the line, as well as basic bimolecular reactions, including capture $p + P \rightarrow P$, annihilation $A + A \rightarrow 0$, coalescence $A + A \rightarrow A$, and aggregation $A_i + A_j \rightarrow A_{i+j}$. The chapter ends with a brief treatment of ballistic annihilation.

A large fraction of this book discusses either classical first-passage properties or results about first passage from contemporary literature, but with some snippets of new results sprinkled throughout. However, several topics are either significant extensions of published results or are original. This includes the time-integrated formalism to compute the first-passage time in fluctuating systems (Section 4.5), aspects of survival in an expanding interval (Section 4.7), return probabilities on the hierarchical tree and homogeneous comb (Section 5.3 and Subsection 5.4.2), the first-passage probability on the hierarchical blob (Section 5.5), and reactions of three diffusing particles on the line (Section 8.3).

This book has been influenced by discussions or collaborations with Dani ben-Avraham, Eli Ben-Naim, Charlie Doering, Laurent Frachebourg, Slava Ispolatov, Joel Koplik, Paul Krapivsky, Satya Majumdar, Francois Leyvraz, Michael Stephen, George Weiss, David Wilkinson, and Bob Ziff, to whom I am grateful for their friendship and insights. I thank Bruce Taggart of the U.S. National Science Foundation for providing financial support at a crucial juncture in the writing, as well as Murad Taqqu and Mal Teich for initial encouragement. Elizabeth Sheld helped me get this project started with her invaluable organizational assistance. I also thank Satya Majumdar for advice on a preliminary manuscript and Erkki Hellén for a critical reading of a nearly final version. I am especially indebted to Paul Krapivsky, my next-door neighbor for most of the past 6 years, for many pleasant collaborations, and for much helpful advice. While it is a pleasure to acknowledge the contributions of my colleagues, errors in presentation are mine alone.

Even in the final stages of writing, I am acutely aware of many shortcomings in my presentation. If I were to repair them all, I might never finish. This book is still work "in progress" and I look forward to receiving your corrections, criticisms, and suggestions for improvements (redner@bu.edu).

Finally and most importantly, I thank my family for their love and constant support and for affectionately tolerating me while I was writing this book.

List of Errata

(updated February 23, 2007)

Chapter 1:

1. Page 2. The second paragraph should read: "To make these questions precise and then answer them, note that the price performs a symmetric random walk in n, where $n = N_- - N_+$ is the difference in the number of "down" and "up" days, N_- and N_+, respectively.

   I thank Ronny Straube for pointing out this error.

2. Page 5. The second sentence of Sec. 1.2 should read: "An important starting fact is that the survival probability $S(t)$ in an absorbing domain is closely related to the time integral of the first-passage probability up to time t over the spatial extent of the boundary (see Eq. (1.5.7))".

3. Page 8. The integral in Eq. (1.3.5) is over the range $-\pi \leq k \leq \pi$. This is not a contour integral.

4. Page 9. In Eq. (1.3.8), the factor Np in the numerator of the exponential should be $N(p - q)$.

   I thank David Waxman for pointing out these three errors.

5. Page 9. The second line of the un-numbered formula after Eq. (1.3.8) should read:

$$\sim e^{ik\langle x\rangle - \frac{1}{2}k^2(\langle x^2\rangle - \langle x\rangle^2)}, \qquad k \to 0.$$

6. Page 9. Eq. (1.3.9) should read:

$$P(x,N) \to \frac{1}{\sqrt{2\pi N\left(\langle x^2\rangle - \langle x\rangle^2\right)}} e^{(x-\langle x\rangle)^2 / \left[2N(\langle x^2\rangle - \langle x\rangle^2)\right]}.$$

   I thank Paul Krapivsky for pointing out these two errors.

7. Page 10. In Eq. (1.3.11), the denominator inside the square brackets should be $2zq$ for $x > 0$ and $2zp$ for $x < 0$.

   I thank David Waxman for pointing out this error.

8. Page 11. In Eq. (1.3.13), the left-hand side should read $\frac{\partial P(n,t)}{\partial t}$.

9. Page 22. In Eq. (1.5.6), the factor $(1 - R)$ should read $(1 - \mathcal{R})$.

10. Page 31. In Eq. (1.6.22), the argument on the left-hand side should be $\vec{r}$, not $\vec{r}'$.

Chapter 2:

1. Page 46. The statement of "complete parallelism" in the 4th line is a bit misleading. While the splitting probabilities and the exit times for the discrete random walk and continuum diffusion agree, this correspondence does not extend to all higher moments.

   I thank Tibor Antal for clarifying this point.

2. Page 47. The factor in the second of Eqs. (2.2.16) should be $(1 - u_0^2)$ not $(1 - u_0)^2$.

3. Page 57. The signs of the v and D terms are *not* opposite to the sense of the corresponding terms in the convection-diffusion equation.

4. Page 63. Just above (2.3.23), the reference is to (1.6.27) not (1.6.29).

5. Page 75. In the line after (2.4.12), the word "completely" should be replaced by "asymptotically".

   I thank Tibor Antal for clarifying this point.

Chapter 3:

1. Page 83. In Eq. (3.2.4) the derivative should be with respect to x not t.

   I thank Carl Gold for pointing out this error.

2. Page 84. In Eq. (3.2.6) there is no factor of t in the square root.

   I thank Robin Groenevelt for pointing out this error.

3. Page 87. In Eq. (3.2.12) the prefactor in the second term should be $e^{-vx_0/D}$. Similarly, in the first and also the third line of text below this equation the factor should read $e^{-vx_0/D}$.

4. Page 88. Replace x by x_0 in the line after (3.2.14) and in the first line of (3.2.15). Also, the fontsize for the first factor of Pe in the 3rd line of this formula should be larger.

5. Page 89. The second half of Eq. (3.2.16) is incorrect. The correct result is

$$\left(\frac{x_0}{\sqrt{Dt}}\right)^3 \frac{1}{\sqrt{4\pi}} \frac{1}{Pe^2} e^{-Pe^2 Dt/x_0^2} = \sqrt{\frac{4}{\pi}} \frac{x_0 \sqrt{Dt}}{(vt)^2} e^{-v^2 t/4D}$$

   I thank David Mukamel for pointing out the above two errors.

6. Page 93. The prefactor in Eq. (3.3.1) should be $\frac{1}{\sqrt{4Ds}}$.

7. Page 99. In the last two lines of Eq. (3.4.4), the factor in the denominator should be $2(t - 1)$ not $2t - 1$.

8. Page 108. Add the word "with" in the 9th line. It should read "... returned to site 1 with probability r ...". I thank Ronny Straube for pointing out this error.

Chapter 4:

1. Page 123. Sec. 4.3.2, third line: "Flyvbjerg", not "Flyvbjery".

2. Page 130. The last term in the equation at the top of the page should read
$\frac{1}{2}\left(r_{in}+r_{out}\right)\frac{\partial^2 P}{\partial n^2}$.
I thank Paul Krapivsky for pointing out this error.

3. Page 151. The sign of the third term on the right-hand side of Eq. (4.6.1) should be plus, not minus.
I thank Alex Petersen for pointing out this error.

4. Page 151/152. The prefactor on the right side of Eq. (4.6.5) should be 1 not 4, and the prefactor on the right side of Eq. (4.6.6) should be $\frac{1}{2}$ not 2.
I thank Alex Petersen for pointing out these errors.

5. Page 152. Remove the word "a" four lines after Eq. (4.6.6).

6. Page 166. The double subscript y_{y1} should simply be y_1.
I thank Robin Groenevelt for pointing out this error.

Chapter 5:

1. Page 202. Fifth line in section 5.5.3.1, a factor should read $1/(3 + 2\epsilon)$, not $1(3 + 2\epsilon)$.

Chapter 6:

1. Page 209. The result in the second of Eq. (6.2.2) actually gives $\mathcal{E}_-(r)$. The correct result is:
$$\mathcal{E}_+(r)=\frac{\ln(r/R_-)}{\ln(R_+/R_-)}.$$
I thank Ronny Straube for pointing out this error.

2. Page 215. In the line after Eq. (6.2.3b), the factor should read $\left(a/r_0\right)^{d\,2}$.

3. Pages 226–227. In the first formula in subsection 6.5.2.2 and in the first un-numbered formula after Eq. (6.5.6), the prefactor should be $\sqrt{\frac{D}{s}}$ rather than $\frac{D}{s}$.
I thank Ronny Straube for pointing out these two errors.

Chapter 7:

1. Page 236. In the first sentence of the third paragraph, the phrase "We thus define $c(r, \theta, t = 0)\ldots$" is missing an equal sign.

2. Page 243. In the first line of Eq. (7.4.3), the fraction $\frac{1}{2\pi D}$ should be replaced by D.

   I thank Robin Groenevelt for pointing out these two errors.

Chapter 8:

1. Page 261. In Eq. (8.2.10) the factor $\ln w$ in the exponential should read $\ln q$ and not $\ln w$.

2. Page 266. The expression for Θ_{end} should read:

$$\Theta_{\text{end}} = \pi - \cos^{-1} \frac{D_3}{\sqrt{(D_1 + D_3)(D_2 + D_3)}},$$

   while the final expression for β_{end} should read

$$\left[2 - \frac{2}{\pi} \cos^{-1} \frac{D_3}{\sqrt{(D_1 + D_3)(D_2 + D_3)}} \right]^{-1}.$$

   I thank Alan Bray for pointing out this error.

3. Page 269. In figure 8.6(a) the labels $1 = 3$ and $2 = 3$ should be transposed.

4. Page 284. In Eq. (8.4.24), the expression for $c_k(t)$ should have a $t^{-3/2}$ time dependence in the prefactor, not $t^{-1/2}$.

   I thank Pu Chen for pointing out this error.

1

First-Passage Fundamentals

1.1. What Is a First-Passage Process?

This book is concerned with the first-passage properties of random walks and diffusion, and the basic consequences of first-passage phenomena. Our starting point is the *first-passage probability*; this is the probability that a diffusing particle or a random-walk *first* reaches a specified site (or set of sites) at a specified time. The importance of first-passage phenomena stems from its fundamental role in stochastic processes that are triggered by a first-passage event. Typical examples include fluorescence quenching, in which light emission by a fluorescent molecule stops when it reacts with a quencher; integrate-and-fire neurons, in which a neuron fires only when a fluctuating voltage level first reaches a specified level; and the execution of buy/sell orders when a stock price first reaches a threshold. Further illustrations are provided throughout this book.

1.1.1. A Simple Illustration

To appreciate the essential features of first-passage phenomena, we begin with a simple example. Suppose that you are a nervous investor who buys stock in a company at a price of \$100. Suppose also that this price fluctuates daily in a random multiplicative fashion. That is, at the end of each day the stock price changes by a multiplicative factor $f < 1$ or by f^{-1} compared with the previous day's price, with each possibility occurring with probability $1/2$ (Fig. 1.1). The multiplicative change ensures that the price remains positive. To be concrete, let's take $f = 90\%$ and suppose that there is a loss on the first day so that the stock price drops to \$90. Being a nervous investor, you realize that you don't have the fortitude to deal with such a loss and wish to sell your stock. However, because the price fluctuates randomly you also think that it might be reasonable to wait until the stock recovers to its initial price before selling out.

Some of the basic questions you, as an skittish investor, will likely be asking yourself are: (1) Will I eventually break even? (2) How long do I have to wait

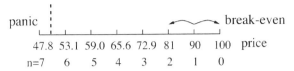

Fig. 1.1. Daily stock price when it changes by a multiplicative factor of $f = 0.9$ or $f^{-1} = 1.11\cdots$ with equal probabilities. The starting price is \$100. Panic selling occurs if the stock price drops to less than half its initial value.

until I break even? (3) While I am waiting to break even, how low might the stock price go? (4) Is it a good idea to place a limit order, i.e., automatically sell if the stock falls below a specified price? Clearly, the answers to these questions will help you to make an informed decision about how to invest in this toy market.

To make these questions precise and then answer them, note that the price performs a symmetric random walk in n, where $n = N_+ - N_-$ is the difference in the number of "up" and "down" days, N_+ and N_-, respectively. After $N = N_+ + N_-$ trading days, the stock price will be $f^n \times \$100$. In our example, the random walk started at $n = 0$ (price \$100) and after one day it jumped to $n = 1$, corresponding to a stock price of \$90. The question of break even after a first-day loss can be rephrased as, What is the probability that a random walk that starts at $n = 1$ *eventually* returns to, or equivalently, eventually hits $n = 0$? As discussed in Section 1.5, this eventual return probability equals one; you are sure to recoup your initial loss. However, the time required for recouping this loss, averaged over *all* possible histories of the stock price, is infinite! Just as disconcerting, while you are waiting forever to ensure recovery of your initial investment, your capital can achieve a vanishingly small value. You would need a strong stomach to survive this stock market! These seemingly paradoxical features have a natural explanation in terms of the first-passage probability of a random walk. Our goal will be to develop a general understanding of first-passage properties for both random walk and diffusion phenomena in a variety of physically relevant situations.

An important aspect of first-passage phenomena is the conditions by which a random-walk process terminates. Returning to our stock market with 10% daily price fluctuations, suppose that you would panic and sell out if the stock price were to sink to less than one-half of its initial value, in addition to selling if and when the price returns to its original value. Selling out at a loss would occur the *first* time that there are 7 more down days than up days. We can then ask, What it the probability of selling at break even or selling at a loss? How long will it take before one of these two events occurs? These are the types of questions that are answered by the study of first-passage phenomena.

1.1.2. Fundamental Issues

The basic questions of first-passage phenomena are the following:

- What is the time dependence of the first-passage probability $F(\vec{r}, t)$? This is the probability that a random walk or a diffusing particle hits a specified point $\vec{r}$ *for the first time* at time t. More generally, what is the first-passage probability for a diffusing particle to hit a *set* of points?

As we shall discuss in Section 1.2, the first-passage probability can be obtained directly from the more familiar *occupation probability* $P(\vec{r}, t)$. This is the probability that a diffusing particle is located at $\vec{r}$ at time t. Note that we are using the terms random walk and diffusing particle loosely and interchangeably. Although these two processes are very different microscopically, their long-time properties – including first-passage characteristics – are essentially the same. In Section 1.3 it is shown how diffusion is merely the continuum limit of any sufficiently well-behaved random-walk process. Thus in the rest of this book we will use the description that best suits the situation under discussion.

In many cases, the diffusing particle physically disappears or "dies" when the specified point or set of points is hit. We can therefore think of this set as an absorbing boundary for the particle. This picture then leads to several more fundamental first-passage-related questions:

- What is the survival probability $S(t)$, that is, the probability that a diffusing particle has not hit the absorbing boundary by time t?
- How long does it take for the diffusing particle to die, that is, hit the absorbing boundary? More precisely, what is the mean time until the particle hits a site on the absorbing boundary as a function of the particle's starting point? This is often referred to as the mean first-passage time or mean exit time.
- Where on the absorbing boundary does the particle get absorbed?

Another interesting feature arises in systems for which the boundary B consists of disjoint subsets, for example, $B = B_1 \cup B_2$, with B_1 and B_2 nonoverlapping. Then it is worthwhile to determine whether the diffusing particle is eventually absorbed in B_1 or in B_2 as a function of the initial particle position. This is often known as the *splitting probability*. For the nervous-investor example presented above, the splitting probabilities refer to the probabilities of ultimately selling at break even or at a 50% loss. More precisely:

- For an absorbing boundary that can be decomposed into disjoint subsets B_1 and B_2, what is the splitting probability, namely, the probability that a diffusing particle will eventually be trapped on B_1 or trapped on B_2 as a function of the initial particle position?

We conclude this introduction with a brief synopsis of the answers to the above questions. For a diffusing particle in a finite domain with an absorbing boundary, the survival probability $S(t)$ typically decays exponentially with time (Chap. 2). Roughly speaking, this exponential decay sets in once the displacement of the diffusing particle becomes comparable with the linear size of the domain. Correspondingly, the mean time $\langle t \rangle$ until the particle hits the boundary is finite. On the other hand, if the domain is unbounded (Chaps. 3 and 5–7) or if boundaries move (Chap. 4), then the geometry of the absorbing boundary is an essential feature that determines the time dependence of the survival probability. For such situations, $S(t)$ either asymptotically approaches a nonzero value or it decays as a power law in time, with the decay exponent a function of the boundary geometry. In our nervous-investor example, the techniques of Chap. 2 will tell us that (s)he will (fortunately) break even 6/7 of the time and will be in the market for 6 days, on average, before selling out.

To determine where on the boundary a diffusing particle is absorbed, we shall develop a powerful but simple analogy between first-passage and electrostatics (Section 1.6). From this classic approach, we will learn that the probability for a particle, which starts at $\vec{r}_0$, to eventually exit at a point $\vec{r}$ on the boundary equals the electric field at $\vec{r}$, when all of the boundaries are grounded conductors and a point charge is placed at $\vec{r}_0$. From this electrostatic analogy, the splitting probability between two points is simply the ratio of the corresponding electrostatic potentials at these points. Because the potential is a linear function of the distance between the source and the observer, we can immediately deduce the splitting probabilities of 6/7 (break even) and 1/7 (panic sell at a 50% loss) of the nervous-investor example.

In the rest of this chapter, basic ideas and techniques are introduced to aid in understanding the first-passage probability and the fundamental quantities that derive from it. In the remaining chapters, many features of first-passage phenomena are illustrated for diffusion processes in a variety of physically relevant geometries and boundary conditions. Many of these topics are covered in books that are either devoted entirely to first-passage processes [Kemperman (1961), Dynkin & Yushkevich (1969), Spitzer (1976), and Syski (1992)] or in books on stochastic processes that discuss first-passage processes as a major subtopic [Cox & Miller (1965), Feller (1968), Gardiner (1985), Risken (1988), Hughes (1995), and van Kampen (1997)].

1.2. Connection between First-Passage and Occupation Probabilities

We now discuss how to relate basic first-passage characteristics to the familiar occupation probability of the random walk or the probability distribution of diffusion. An important starting fact is that the survival probability $S(t)$ in an absorbing domain equals the integral of the first-passage probability over all time and over the spatial extent of the boundary. This is simply a consequence of the conservation of the total particle density. Thus all first-passage characteristics can be expressed in terms of the first-passage probability itself.

We now derive the relation between the first-passage probability and the probability distribution. Derivations in this spirit are given, e.g., in Montroll and Weiss (1965), Fisher (1984), and Weiss (1994). We may express this connection in equivalent ways, either in terms of discrete-space and -time or continuous-space and -time variables, and we will use the representation that is most convenient for the situation being considered. Let us start with a random walk in discrete space and time. We define $P(\vec{r}, t)$ as the occupation probability; this is the probability that a random walk is at site $\vec{r}$ at time t when it starts at the origin. Similarly, let $F(\vec{r}, t)$ be the first-passage probability, namely, the probability that the random walk visits $\vec{r}$ for the first time at time t with the same initial condition. Clearly $F(\vec{r}, t)$ asymptotically decays more rapidly in time than $P(\vec{r}, t)$ because once a random walk reaches $\vec{r}$, there can be no further contribution to $F(\vec{r}, t)$, although the same walk may still contribute to $P(\vec{r}, t)$.

Strategically, it is simplest to write $P(\vec{r}, t)$ in terms of $F(\vec{r}, t)$ and then invert this relation to find $F(\vec{r}, t)$. For a random walk to be at $\vec{r}$ at time t, the walk must first reach $\vec{r}$ at some earlier time step t' and then return to $\vec{r}$ after

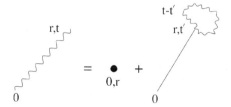

Fig. 1.2. Diagrammatic relation between the occupation probability of a random walk (propagation is represented by a wavy line) and the first-passage probability (straight line).

$t - t'$ additional steps (Fig. 1.2). This connection between $F(\vec{r}, t)$ and $P(\vec{r}, t)$ may therefore be expressed as the convolution relation

$$P(\vec{r}, t) = \delta_{\vec{r},0}\delta_{t,0} + \sum_{t' \le t} F(\vec{r}, t') P(0, t - t'), \qquad (1.2.1)$$

where $\delta_{t,0}$ is the Kronecker delta function. This delta function term accounts for the initial condition that the walk starts at $\vec{r} = 0$. Note that returns *before* time t are also permissible as long as there is also a return $t - t'$ steps after the first return. Because of the possibility of multiple visits to $\vec{r}$ between time t' and t, the return factor involves P rather than F.

This convolution equation is most conveniently solved by introducing the generating functions,

$$P(\vec{r}, z) = \sum_{t=0}^{\infty} P(\vec{r}, t)z^t, \quad F(\vec{r}, z) = \sum_{t=0}^{\infty} F(\vec{r}, t)z^t.$$

If we were dealing with a random walk in continuous time, we would first replace the sum over discrete time in Eq. (1.2.1) with an integral and then use the Laplace transform. However, the ensuing asymptotic results would be identical. To solve for the first-passage probability, we multiply Eq. (1.2.1) by z^t and sum over all t. We thereby find that the generating functions are related by

$$P(\vec{r}, z) = \delta_{\vec{r},0} + F(\vec{r}, z)P(0, z). \qquad (1.2.2)$$

Thus we obtain the fundamental connection

$$F(\vec{r}, z) = \begin{cases} \dfrac{P(\vec{r}, z)}{P(0, z)}, & \vec{r} \neq 0 \\[2ex] 1 - \dfrac{1}{P(0, z)}, & \vec{r} = 0 \end{cases}, \qquad (1.2.3)$$

in which the generating function, or equivalently, the Laplace transform of the first-passage probability, is determined by the corresponding transform of the probability distribution of diffusion $P(\vec{r}, t)$. This basic relation and its many consequences are also treated extensively in Montroll (1965), Montroll and Weiss (1965), Weiss and Rubin (1983), and Weiss (1994).

Because this probability distribution is basic to understanding diffusion and because it also determines the first-passage probability, several complementary approaches are presented to derive the first-passage probability in Section 1.3. In Section 1.4, it is shown how to relate basic probabilities in real time with their corresponding generating function or Laplace transform. Armed with this information, we then obtain the asymptotics of the first-passage probability in Section 1.5.

1.3. Probability Distribution of a One-Dimensional Random Walk

We now discuss basic features of the probability distribution of a random walk, as well as the relation between random walks and diffusion. Much of this discussion is standard but is included for completeness and to illustrate the universal asymptotic behavior of the various representations of random walks in terms of diffusion. Derivations of varying degrees of completeness and sophistication can be found, e.g., in Reif (1965) and van Kampen (1997). In deriving $P(\vec{r}, t)$, we treat in detail the nearest-neighbor random walk in one dimension and merely quote results for higher dimension as needed. We start with the easily visualized discrete-space and -time hopping process and then proceed to the continuum diffusion limit. The order of presentation is chosen to help advertise the relative simplicity of the continuum representation. These derivations also serve to introduce the mathematical tools that will be used in treating first-passage properties, such as Fourier and Laplace transforms, the generating function, and basic asymptotic analysis.

1.3.1. Discrete Space and Time

Consider a particle that hops at discrete times between neighboring sites on a one-dimensional chain with unit spacing. Let $P(x, N)$ be the probability that the particle is at site x at the Nth time step. The evolution of this occupation probability is described by the master equation

$$P(x, N + 1) = p P(x - 1, N) + q P(x + 1, N). \qquad (1.3.1)$$

This states that the probability for the particle to be at x at time $N + 1$ is simply p times the probability of being at $x - 1$ at time N (contribution of a step to the right) plus q time the probability of being at $x + 1$ at time N (step to the left). The case $p = q = \frac{1}{2}$ is the symmetric random walk, whereas for $p > q$ there is a uniform bias to the right and for $p < q$ a bias to the left (see Fig. 1.3).

Because of translational invariance in both space and time, it is natural to solve this equation by transform techniques. For this example, we therefore

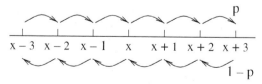

Fig. 1.3. The hopping processes for the one-dimensional nearest-neighbor random walk defined by Eq. (1.3.1).

define the combined generating function and Fourier transform:

$$P(k, z) = \sum_{N=0}^{\infty} z^N \sum_{x=-\infty}^{\infty} e^{ikx} P(x, N).$$

In all the discussions that follow, either the arguments of a function or the context (when obvious) will be used to distinguish the transform from the function itself.

Applying this combined transform to the master equation, and multiplying by one additional factor of z, gives

$$\sum_{N=0}^{\infty} \sum_{x=-\infty}^{\infty} z^{N+1} e^{ikx} [P(x, N + 1) = pP(x - 1, N) + qP(x + 1, N)].$$

(1.3.2)

The left-hand side is just the generating function $P(k, z)$, except that the term $P(x, N = 0)$ is missing. Similarly, on the right-hand side the two factors are just the generating function at $x - 1$ and at $x + 1$ times an extra factor of z. The Fourier transform then converts these shifts of ± 1 in the spatial argument to the phase factors $e^{\pm ik}$, respectively. These steps give

$$P(k, z) - \sum_{x=-\infty}^{\infty} P(x, N = 0)e^{ikx} = z(pe^{ik} + qe^{-ik})P(k, z).$$

$$\equiv zu(k)P(k, z), \qquad (1.3.3)$$

where $u(k)$ is the Fourier transform of the single-step hopping probability. For our biased nearest-neighbor random walk, $u(k)$ is $u(k) = pe^{ik} + qe^{-ik}$. The subtracted term on the left-hand side of Eq. (1.3.3) compensates for the absence of a term of the order of z^0 in Eq. (1.3.2). Thus for the initial condition of a particle initially at the origin, $P(x, N = 0) = \delta_{x,0}$, the joint Fourier transform and generating function of the probability distribution becomes

$$P(k, z) = \frac{1}{1 - zu(k)}. \qquad (1.3.4)$$

We can now reconstruct the original probability distribution by inverting these transforms. When $P(k, z)$ is expanded in a Taylor series in z, the inverse of the generating function is simply $P(k, N) = u(k)^N$. Then the inverse Fourier transform is

$$P(x, N) = \frac{1}{2\pi} \oint e^{-ikx} u(k)^N \, dk, \qquad (1.3.5)$$

where the integration is around the unit circle in the complex plane. To evaluate the integral, we write $u(k)^N = (pe^{ik} + qe^{-ik})^N$ in a binomial series. This gives

$$P(x, N) = \frac{1}{2\pi} \oint e^{-ikx} \sum_{m=0}^{N} \binom{N}{m} p^m e^{ikm} q^{N-m} e^{-ik(N-m)} dk. \quad (1.3.6)$$

The only term that survives the integration is the one in which all the phase factors cancel. This is the term with $m = (N + x)/2$ and this ultimately leads to gives the classical binomial probability distribution of a discrete random walk:

$$P(x, N) = \frac{N!}{\left(\frac{N+x}{2}\right)!\left(\frac{N-x}{2}\right)!} p^{\frac{N+x}{2}} q^{\frac{N-x}{2}}. \quad (1.3.7)$$

Finally, using Stirling's approximation, this binomial approaches the Gaussian probability distribution in the long-time limit,

$$P(x, N) \to \frac{1}{\sqrt{2\pi Npq}} e^{-(x-Np)^2/2Npq}. \quad (1.3.8)$$

In fact, a Gaussian distribution arises for *any* hopping process in which the mean displacement $\langle x \rangle$ and the mean-square displacement $\langle x^2 \rangle$ in a single step of the walk is finite. This is the statement of the *central-limit theorem* (Gnedenko and Kolmogorov (1954)). When $\langle x \rangle$ and $\langle x^2 \rangle$ are both finite, $u(k)$ has the small-k series representation

$$u(k) = 1 + ik\langle x \rangle - \frac{1}{2}k^2\langle x^2 \rangle + \dots,$$

$$\sim e^{ik\langle x \rangle - \frac{1}{2}k^2\langle x^2 \rangle}, \quad k \to 0.$$

When this result for $u(k)$ is substituted into Eq. (1.3.5), the main contribution to the integral comes from the region $k \to 0$. We can perform the resulting Gaussian integral by completing the square in the exponent; this leads to a Gaussian form for the probability distribution of a general nonsingular hopping process:

$$P(x, N) \to \frac{1}{\sqrt{2\pi N \langle x^2 \rangle}} e^{-(x-\langle x \rangle)^2/2N\langle x^2 \rangle}. \quad (1.3.9)$$

This independence of the probability distribution on details of the single-step hopping is a manifestation of the central-limit theorem. It is akin to the universality hypothesis of critical phenomena in which short-range details of a system do not affect large-scale properties [Stanley (1971)]. This same type of universality typifies random walks with short-range memory and/or correlations. The only property that is relevant is that $\langle x \rangle$ and $\langle x^2 \rangle$ are both finite. All such processes can be universally described by diffusion in the long-time limit.

As a complement to the above solution, which is based on $P(k, z) \rightarrow P(k, N) \rightarrow P(x, N)$, we first invert the Fourier transform in Eq. (1.3.4) to obtain the generating function for the occupation probability and then invert the latter to obtain the probability distribution; that is, we follow $P(k, z) \rightarrow P(x, z) \rightarrow P(x, N)$. This is conceptually identical to, although technically more complicated than, the inversion in the reverse order. However, there are applications in which the generating function $P(x, z)$ is indispensable (see Chap. 3) and it is also useful to understand how to move easily between generating functions, Laplace transforms, and real-time quantities. More on this will be presented in Section 1.4.

Starting with $P(k, z)$ in Eq. (1.3.4), the inverse Fourier transform is

$$P(x, z) = \frac{1}{2\pi} \int_{-\pi}^{\pi} \frac{e^{-ikx}}{1 - z(pe^{ik} + qe^{-ik})} \, dk. \qquad (1.3.10)$$

We may perform this integral by introducing the variable $w = e^{-ik}$ to recast Eq. (1.3.10) as a contour around the unit circle in the complex plane. By applying basic methods from residue calculus, the value of the integral is

$$P(x, z) = \frac{1}{\sqrt{1 - 4pqz^2}} \left[\frac{1 - \sqrt{1 - 4pqz^2}}{2zq} \right]^{|x|}. \qquad (1.3.11)$$

Finally, we may invert this generating function formally by again applying residue calculus to give the basic inversion formula

$$P(x, N) = \frac{1}{2\pi i} \oint \frac{P(x, z)}{z^{N+1}} \, dz, \qquad (1.3.12)$$

where the clockwise contour encloses only the pole at the origin. We could also attempt to compute the power-series representation of $P(x, z)$ to arbitrary order to extract the original probability distribution. Neither approach appears to be elementary, however. This shows that the order of inverting the double transform $P(k, z)$ can be important: Here $P(k, z) \rightarrow P(k, N) \rightarrow P(x, N)$ is easy, whereas $P(k, z) \rightarrow P(x, z) \rightarrow P(x, N)$ is not. This is a useful lesson to keep in mind whenever one is dealing with combined Fourier–Laplace transforms.

1.3.2. Discrete Space and Continuous Time

For this representation, we replace the discrete time N and time increment $N \rightarrow N + 1$ with continuous time t and infinitesimal increment $t \rightarrow t + \delta t$

and Taylor expand the resulting master equation (1.3.1) to first order in δt. We also rewrite the spatial argument as the integer n to emphasize its discrete nature. These steps give

$$\frac{\partial P(n,t)}{\delta t} = w_+ P(n-1,t) + w_- P(n+1,t) - w_0 P(n,t), \quad (1.3.13)$$

where $w_+ = p/\delta t$ and $w_- = q/\delta t$ are the hopping rates to the right and to the left, respectively, and $w_0 = 1/\delta t$ is the total hopping rate from each site. This hopping process satisfies detailed balance, as the total hopping rates *to* a site equal the hopping rate *from* the same site.

By Fourier transforming, this master equation becomes an ordinary differential equation for the Fourier-transformed probability distribution

$$\frac{dP(k,t)}{dt} = (w_+ e^{-ik} + w_- e^{ik} - w_0)\, P(k,t),$$

$$\equiv w(k)\, P(k,t). \quad (1.3.14)$$

For the initial condition $P(n, t = 0) = \delta_{n,0}$, the corresponding Fourier transform is $P(k, t = 0) = 1$, and the solution to Eq. (1.3.14) is simply $P(k,t) = e^{w(k)t}$. To invert this Fourier transform, it is useful to separate $w(k)$ into symmetric and antisymmetric components by defining $w_\pm \equiv \frac{1}{2}w_0 \pm \delta w$ to give $w(k) = w_0(\cos k - 1) - 2i\,\delta w \sin k$, and then use the generating function representations for the Bessel function [Abramowitz & Stegun (1972)],

$$e^{z\cos k} = \sum_{n=-\infty}^{\infty} e^{ikn} I_n(z),$$

$$e^{z\sin k} = \sum_{n=-\infty}^{\infty} (-i)^n e^{ikn} I_n(z),$$

where $I_n(z)$ is the modified Bessel function of the first kind of order n. After some simple steps, we find

$$P(k,t) = e^{-w_0 t} \sum_{\ell,m=-\infty}^{\infty} (-i)^m\, e^{ik(\ell+m)} I_\ell(w_0 t)\, I_m(2i\,\delta w t). \quad (1.3.15)$$

By extracting the coefficient of e^{ikn} in this double series, we obtain

$$P(n,t) = e^{-w_0 t} \sum_{m=-\infty}^{\infty} (-i)^{n-m} I_m(w_0 t)\, I_{n-m}(2i\,\delta w t). \quad (1.3.16)$$

In particular, for the symmetric nearest-neighbor random walk, where $w_+ = w_- = w_0/2 = 1/2$, this reduces to the much simpler form

$$P(n,t) = e^{-t} I_n(t). \quad (1.3.17)$$

In this discrete-space continuous-time representation, the asymptotic behavior may be conveniently obtained in the limit of $t \to \infty$, but with n *fixed*. For example, the probability that site n is occupied, as $t \to \infty$, is

$$P(n,t) = \frac{1}{\sqrt{2\pi t}}\left[1 - \frac{4n^2 - 1}{8t} + \cdots\right]. \qquad (1.3.18)$$

This gives the well-known result that the probability for a random walk to remain at the origin vanishes as $t^{-1/2}$. However, this discrete-space and continuous-time representation is awkward for determining the Gaussian scaling behavior of occupation probability, in which n and t both diverge but n^2/t remains constant. For this limit, the continuum formulation is much more appropriate.

To determine the probability distribution in this interesting scaling limit, we introduce the Laplace transform of the probability distribution $P(n,s) = \int_0^\infty e^{-st} P(n,t)\,dt$ and for simplicity consider the symmetric random walk. Applying this Laplace transform to master equation (1.3.13) then gives

$$\int_0^\infty dt\, e^{-st}\left[\frac{\partial P(n,t)}{\partial t} = \frac{1}{2}P(n+1,t) + \frac{1}{2}P(n-1,t) - P(n,t)\right].$$

This Laplace transform reduces the master equation to the discrete recursion formula

$$sP(n,s) - P(n,t=0) = \frac{1}{2}P(n+1,s) + \frac{1}{2}P(n-1,s) - P(n,s). \qquad (1.3.19)$$

For $n \neq 0$, we have $P(n,s) = a[P(n+1,s) + P(n-1,s)]$, with $a = 1/2(s+1)$. We solve this difference equation by assuming the exponential solution $P(n,s) = A\lambda^n$ for $n > 0$. For the given initial condition, the solution is symmetric in n; hence $P(n,s) = A\lambda^{-n}$ for $n < 0$. Substituting this form into the recursion for $P(n,s)$ gives a quadratic characteristic equation for λ whose solution is $\lambda_\pm = (1 \pm \sqrt{1-4a^2})/2a$. For all $s > 0$, $\lambda_\pm$ are both real and positive, with $\lambda_+ > 1$ and $\lambda_- < 1$. For regularity at $n = \infty$, we must reject the solution that grows exponentially with n, thus giving $P_n = A\lambda_-^n$. Finally, we obtain the constant A from the $n = 0$ boundary master equation:

$$sP(0,s) - 1 = \frac{1}{2}P(1,s) + \frac{1}{2}P(-1,s) - P(0,s)$$

$$= P(1,s) - P(0,s). \qquad (1.3.20)$$

The -1 on the left-hand side arises from the initial condition, and the second equality follows by spatial symmetry. Substituting $P(n,s) = A\lambda_-^n$ into

Eq. (1.3.20) gives A, from which we finally obtain

$$P(n, s) = \frac{1}{s + 1 - \lambda_-} \lambda_-^n. \tag{1.3.21}$$

This Laplace transform diverges at $s = 0$; consequently, we may easily obtain the interesting asymptotic behavior by considering the limiting form of $P(n, s)$ as $s \to 0$ limit. Because $\lambda_- \approx 1 - \sqrt{2s}$ as $s \to 0$, we find

$$P(n, s) \approx \frac{(1 - \sqrt{2s})^n}{\sqrt{2s} + s}$$
$$\sim \frac{e^{-n\sqrt{2s}}}{\sqrt{2s}}. \tag{1.3.22}$$

We can now compute the inverse Laplace transform $P(n, t) = \int_{s_0-i\infty}^{s_0+i\infty} P(n, s)e^{st} \, ds$ by elementary means by using the integration variable $u = \sqrt{s}$. This immediately leads to the Gaussian probability distribution quoted in Eq. (1.3.9) for the case $\langle x \rangle = 0$ and $\langle x^2 \rangle = 1$.

1.3.3. Continuous Space and Time

Finally, consider master equation (1.3.1) in both continuous time and space. By expanding this equation in a Taylor series to lowest nonvanishing order – second order in space x and first order in time t – we obtain the fundamental *convection–diffusion equation*,

$$\frac{\partial c(x, t)}{\partial t} + v\frac{\partial c(x, t)}{\partial x} = D \frac{\partial^2 c(x, t)}{\partial x}, \tag{1.3.23}$$

for the particle concentration $c(x, t)$. This should be viewed as the continuum analog of the occupation probability of the random walk; we will therefore use $c(x, t)$ and $P(x, t)$ interchangeably. Here $v = (p - q)\delta x/\delta t$ is the bias velocity and $D = \delta x^2/2\delta t$ is the diffusion coefficient. For the symmetric random walk, the probability distribution obeys the simpler *diffusion equation*

$$\frac{\partial c(x, t)}{\partial t} = D \frac{\partial^2 c(x, t)}{\partial x}. \tag{1.3.24}$$

Note that in the convection–diffusion equation, the factor v/D diverges as $1/\delta x$ in the continuum limit. Therefore the convective term $\partial c/\partial x$ invariably dominates over the diffusion term $\partial^2 c/\partial x^2$. To construct a nonpathological continuum limit, the bias $p - q$ must be proportional to δx as $\delta x \to 0$ so that both the first- and the second-order spatial derivative terms are simultaneously finite. For the diffusion equation, we obtain a nonsingular continuum limit merely by ensuring that the ratio $\delta x^2/\delta t$ remains finite as both δx and δt approach zero. Roughly speaking, any stochastic hopping process in which the distribution of step lengths is well behaved has a continuum description in terms of the convection–diffusion or the diffusion equation. Much more about this relation between discrete hopping processes and the continuum can be found, for example, in Gardiner (1985), Weiss (1994), or van Kampen (1997).

Several basic approaches are now given for solving the diffusion and the convection–diffusion equations.

1.3.3.1. Scaling Solution

Scaling provides a relatively cheap but general approach for solving wide classes of partial differential equations that involve diverging characteristic scales, such as coarsening, aggregation, fragmentation, and many other nonequilibrium processes [see, e.g., Lifshitz & Slyozov (1961), Ernst (1985), and Cheng & Redner (1990) for examples of each]. For the diffusion equation, the scaling solution is based on the observation that there is a *single* length scale that characterizes the particle displacement. Consequently, the probability density is not a function of x and t separately, but rather, is a function of the scaling variable $u \equiv x/X(t)$, where $X(t)$ is the characteristic length scale of the spatial spread of the probability distribution. We may then separate the dependences on u and t to give two single-variable equations – one for the time dependence and another for the functional form of the probability distribution. For the convection–diffusion equation, two length scales are needed to characterize the probability distribution, and, although a scaling approach is still tractable, it no longer has the same degree of simplicity.

The scaling ansatz for the concentration in the diffusion equation is

$$c(x, t) = \frac{1}{X(t)} f[x/X(t)]. \qquad (1.3.25)$$

The prefactor $1/X(t)$ ensures that the spatial integral of $c(x, t)$ is normalized, $\int c(x, t)\,dx = 1$, as is evident by dimensional analysis, and the function f encodes the dependence on the scaled distance $u = x/X(t)$. Substituting this ansatz into the diffusion equation gives, after some elementary algebra,

$$X(t)\dot{X}(t) = -D\frac{f''(u)}{f(u) + uf'(u)}, \qquad (1.3.26)$$

where the prime denotes differentiation with respect to u and the overdot denotes the time derivative. Because the left-hand side is a function of time only whereas the right-hand side is a function of u only, both sides must equal a constant. The scaling ansatz thus leads to variable separation. Strikingly, the value of the separation constant drops out of the solution to the diffusion equation. However, in more complex situations, such as aggregation and coarsening, the separation constant plays an essential role in characterizing the solution, and additional physical considerations must be imposed to determine the solution completely. These are beyond the scope of the present discussion.

Solving for the time dependence gives $X\dot{X} = A$, or $X(t)^2 = 2At$. When this is used in Eq. (1.3.26), the scaling function satisfies

$$f'' = -\frac{A}{D}(f + uf') = -\frac{A}{D}(uf)'.$$

Integrating once gives $f' = -Auf/D + \text{const}$, but the symmetry condition $f' = 0$ at $u = 0$, which corresponds to the random walk starting at the origin, means that the constant is zero. Integrating once again gives the scaling function

$$f(u) = f(0)e^{-Au^2/2D},$$

where the prefactor $f(0)$ is most easily determined by invoking normalization. With $u = x/X(t)$, the final result for the concentration is

$$c(x, t) = \frac{1}{\sqrt{4\pi Dt}} e^{-x^2/4Dt}, \tag{1.3.27}$$

where, as advertised, the separation constant A drops out of this solution. Note that for $D = 1/2$ this solution agrees with expression (1.3.9), with $\langle x^2 \rangle = 1$ and $N \to t$.

1.3.3.2. Fourier Transform Solution

Here we solve the convection–diffusion equation by first introducing the Fourier transform

$$c(k, t) = \int c(x, t)e^{ikx}\,dx$$

to simplify the convection–diffusion equation to

$$\dot{c}(k, t) = (ikv - Dk^2)c(k, t). \tag{1.3.28}$$

The solution is

$$c(k, t) = c(k, 0)e^{(ikv-Dk^2)t}$$
$$= e^{(ikv-Dk^2)t} \tag{1.3.29}$$

for the initial condition $c(x, t = 0) = \delta(x)$. We then obtain the probability distribution by the inverse Fourier transform

$$c(x, t) = \frac{1}{2\pi} \int_{-\infty}^{\infty} e^{(ikv - Dk^2)t - ikx} \, dk.$$

We may perform this integral by completing the square in the exponential, and the final result is the Gaussian probability distribution

$$c(x, t) = \frac{1}{\sqrt{4\pi Dt}} e^{-(x - vt)^2/4Dt}. \tag{1.3.30}$$

1.3.3.3. Laplace Transform Solution

An alternative approach is to first perform a Laplace transform in the time domain. For the convection–diffusion equation, this yields the ordinary differential equation

$$sc(x, s) - \delta(x) + vc(x, s) = Dc''(x, s), \tag{1.3.31}$$

where the delta function again reflects the initial condition of a particle at the origin at $t = 0$. This equation may be solved separately in the half-spaces $x > 0$ and $x < 0$. In each subdomain Eq. (1.3.31) reduces to a homogeneous constant-coefficient differential equation that has exponential solutions in x. The corresponding solution for the entire line has the form $c_+(x, s) = A_+ e^{-\alpha_- x}$ for $x > 0$ and $c_-(x, s) = A_- e^{\alpha_+ x}$ for $x < 0$, where $\alpha_\pm = (v \pm \sqrt{v^2 + 4Ds})/2D$ are the roots of the characteristic polynomial. The complementary divergent term in each of these half-space solutions has been discarded. When these two solutions are joined at the origin, they yield the global solution. The appropriate joining conditions are continuity of $c(x, s)$ at $x = 0$, and a discontinuity in $(\partial c/\partial x)$ at $x = 0$ whose magnitude is determined by integrating Eq. (1.3.31) over an infinitesimal domain that includes the origin.

The continuity condition trivially gives $A_+ = A_- \equiv A$, and the condition for the discontinuity in $c(x, s)$ is

$$D(c'_+|_{x=0} - c'_-|_{x=0}) = -1.$$

This gives $A = 1/\sqrt{v^2 + 4Ds}$. Thus the Laplace transform of the probability distribution is

$$c_\pm(x, s) = \frac{1}{\sqrt{v^2 + 4Ds}} e^{-\alpha_\mp |x|}. \tag{1.3.32}$$

For zero bias, this coincides with Eq. (1.3.22) and thus recovers the Gaussian probability distribution.

1.3.3.4. Fourier–Laplace Transform Solution

Perhaps the simplest approach for solving the convection–diffusion equation is to apply the combined Fourier–Laplace transform

$$c(k, s) = \int_{-\infty}^{\infty} dx\, e^{ikx} \int_{0}^{\infty} dt\, c(x, t)e^{-st}$$

to recast the differential equation into the purely algebraic equation $sc(k, s) - 1 = -(ivk + Dk^2)c(k, s)$. The solution to the latter is

$$c(k, s) = \frac{1}{s + ivk + Dk^2}. \tag{1.3.33}$$

Performing the requisite inverse transforms, we again recover the Gaussian probability distribution as a function of space and time.

In summary, all of the representations of the probability distribution for the convection–diffusion equation are interrelated and equivalent, as represented graphically below. The nature of the problem usually dictates which representation is best to use in solving a specific problem.

$$P(x, t) = \frac{e^{-(x - vt)^2/4Dt}}{\sqrt{4\pi Dt}} \qquad\qquad P(k, t) = e^{-(ikv + Dk^2 t)}$$

$$(x, t) \overset{\text{Fourier}}{\Longleftrightarrow} (k, t)$$

$$\updownarrow \qquad\qquad \updownarrow \text{Laplace}$$

$$(x, s) \Longleftrightarrow (k, s)$$

$$P(x, s) = \frac{e^{-(v \mp \sqrt{v^2 + 4Ds})\,|x|/2D}}{\sqrt{v^2 + 4Ds}} \qquad\qquad P(k, s) = \frac{1}{s + ikv + Dk^2}$$

1.4. Relation between Laplace Transforms and Real-Time Quantities

In first-passage phenomena, we typically seek the asymptotic behavior of a time-dependent quantity when only its generating function or its Laplace transform is readily available. A typical example is a function $F(t)$ whose generating function has the form $F(z) \sim (1 - z)^{\mu-1}$, with $\mu < 1$ as $z \to 1$ from below. We will show that the corresponding time dependence is $F(t) \sim t^{-\mu}$ as $t \to \infty$. Although there are well-established and rigorous methods available for inverting such transforms [see, e.g., Titschmarsh (1945), Hardy (1947), and the discussion in Weiss (1994) for specific applications to random walks], if we are interested in long-time properties only, then basic asymptotic features can be gained through simple and intuitively appealing means. Although lacking in rigor, they provide the correct behavior for all cases of physical interest. The most useful of these methods are outlined in this section.

Let us first determine the long-time behavior of a function $F(t)$ when only the associated generating function $F(z) = \sum_{t=0}^{\infty} F(t) z^t$ is available. There are two fundamentally different cases: (a) $\sum_{t=0}^{\infty} F(t)$ diverges or (b) $\sum_{t=0}^{\infty} F(t)$ converges.

We relate the generating function $F(z)$ to $F(t)$ by the following steps:

$$
\begin{aligned}
F(z) &= \sum_{t=0}^{\infty} F(t) z^t \\
&\approx \int^{\infty} F(t) e^{-t \ln(1/z)} \, dt \\
&\sim \int^{t^*} F(t) \, dt,
\end{aligned}
\tag{1.4.1}
$$

with $t^* = [\ln(1/z)]^{-1}$. These simple algebraic steps involve several important simplifications and approximations that we can justify as follows:

- Converting the sum to an integral. In many cases of relevance to first-passage phenomena, $F(t) \sim t^{-\mu}$ for $t \to \infty$ but rapidly goes to zero for $t \to 0$. Further, it is usually only the long-time behavior of $F(t)$ that is readily obtainable in the continuum limit. If we were to use the continuum expression for $F(t)$ in the integral for all times, a spurious singularity could arise from the contribution to the integral at short times. We eliminate this unphysical singularity by replacing the lower limit with a value of the order of 1. Because asymptotic behavior is unaffected by this detail, we often leave the lower limit as indefinite with the understanding that it is of the order of 1.
- From the second line of Eq. (1.4.1), we see that $F(z)$ is essentially the Laplace transform of $F(t)$ with $s = \ln(1/z)$. Because the $s \to 0$ limit of the Laplace transform corresponds to long-time limit of $F(t)$, we immediately infer that long-time behavior may also be obtained from the limit $z \to 1$ from below in the generating function.
- Sharp cutoff. In the last step, we replace the exponential cutoff in the integral, with characteristic lifetime $t^* = [\ln(1/z)]^{-1}$, with a step function at t^*. Although this crude approximation introduces numerical errors of the order of 1, we shall see that the essential asymptotic behavior is preserved when $\sum_{t=0}^{\infty} F(t)$ diverges.

We now use the approach of Eq. (1.4.1) to determine the asymptotic relation between $F(t)$ and $F(z)$, considering separately the two fundamental cases in which $\sum_{t=0}^{\infty} F(t)$ diverges and in which it converges. In the former case, it is natural to use $s \equiv (1 - z)$ as the basic variable, since the generating function diverges as $z \to 1$. Then $t^* = [\ln(1/z)]^{-1} \sim 1/s$ for small s. Thus the basic relation between $F(t)$ and $F(z)$ is simply

$$F(z) \sim \int^{1/s} F(t)\,dt. \qquad (1.4.2)$$

Now if $F(t) \to t^{-\mu}$ with $\mu < 1$ as $t \to \infty$, then $F(z)$ diverges as

$$F(z) \sim \int^{1/s} t^{-\mu}\,dt \sim s^{\mu-1} = (1 - z)^{\mu-1} \qquad (1.4.3)$$

as $z \to 1$ from below. In summary, the fundamental connection is

$$F(t) \sim t^{-\mu} \longleftrightarrow F(z) \sim (1 - z)^{\mu-1}. \qquad (1.4.4)$$

As a useful illustration of the equivalence between the generating function as $z \to 1$ from below and the Laplace transform, let us reconsider the formidable-looking generating function of the one-dimensional random walk

$$P(x, z) = \frac{1}{\sqrt{1 - z^2}} \left[\frac{1 - \sqrt{1 - z^2}}{z} \right]^{|x|}. \qquad (1.4.5)$$

By taking the limit $z \to 1$ from below, we obtain

$$P(x, z) \approx \frac{[1 - \sqrt{2(1 - z)}]^{|x|}}{\sqrt{2(1 - z)}}$$
$$\sim \frac{e^{-x\sqrt{2s}}}{\sqrt{2s}}, \quad \text{letting } s = 1 - z, \qquad (1.4.6)$$

which reproduces the Laplace transform in approximation (1.3.22).

Parenthetically, another useful feature of the generating function is that it can simplify integral relations between two functions of the generic form $P(t) = \sum_{t'=0}^{t} F(t')$. By introducing the generating functions for P and F, we easily obtain the corresponding relation between the generating functions $P(z) = F(z)/(1 - z)$. In continuous time, the corresponding relation $P(t) = \int_0^t F(t')\,dt'$ translates to $P(s) = F(s)/s$ for the Laplace transforms.

Finally, let us discuss the general connection between the time integral of a function, $\mathcal{F}(t) \equiv \int_0^t F(t)\,dt$, and the generating function $F(z)$. For $z = 1 - 1/t^*$ with $t^* \to \infty$, approximation (1.4.2) becomes

$$F(z = 1 - 1/t^*) \sim \int_0^{t^*} F(t)\,dt = \mathcal{F}(t^*). \qquad (1.4.7)$$

Thus a mere variable substitution provides an approximate, but asymptotically correct, algebraic relation between the generating function (or Laplace transform) and the time integral of the function itself. For this class of examples, there is no need to perform an integral to relate a function and its Laplace transform.

Conversely, when $\sum_t^\infty F(t)$ converges, the simplifications outlined in Eq. (1.4.1) no longer apply and we must be more careful in constructing the relation between $F(t)$ and $F(z)$. We again suppose that $F(t) \sim t^{-\mu}$ as $t \to \infty$, but now with $\mu > 1$ so that $F(z)$ is finite as $z = 1$. Then $F(z)$ typically has the form $F(z) = F(1) + p_1(1 - z)^{\alpha_1} + \cdots$ as $z \to 1$, where p_1 is a constant. To determine the exponent α_1 in the first correction term, consider the difference $F(1) - F(z)$. We again use $z = 1 - s$ when $z \to 1$, so that

$$
\begin{aligned}
F(1) - F(z) &\propto \sum_t^\infty t^{-\mu}(1 - z^t), \\
&\sim \int^\infty t^{-\mu}(1 - e^{-st})\,dt, \\
&\sim \int_{1/s}^\infty t^{-\mu}\,dt, \\
&\sim (1 - z)^{\mu-1}.
\end{aligned}
\tag{1.4.8}
$$

The exponential cutoff in the integrand has again been replaced with a sharp cutoff at $t = 1/s$ in the third line. We therefore infer the asymptotic behavior

$$
F(z) \sim F(1) + a_1(1 - z)^{\mu-1} + \ldots,
\tag{1.4.9}
$$

where the a_1 is a detail-dependent constant.

Parallel results exist for the Laplace transform. For a generic power-law form, $F(t) \sim t^{-\mu}$, the Laplace transform $F(s)$ has the small-s expansion:

$$
F(s) \sim F(s = 0) + (A_1 s^\alpha + A_2 s^{\alpha+1} + \cdots) + (B_1 s^1 + B_2 s^2 + \cdots).
\tag{1.4.10}
$$

There are again several possibilities. When $\mu \neq 1$, asymptotic behavior is governed by the nonanalytic term. In this case, the exponent $\alpha = \mu - 1$. Note also that when α is less than one, then the first moment and indeed all positive integer moments of $F(t)$ are divergent. This is reflected by the fact that $F(t)$ has the power-law tail $t^{-\mu}$ with $\mu = \alpha + 1 < 2$. Additionally, note that the zeroth-order term in Eq. (1.4.9) is $\int_0^\infty F(t)\,dt$. Thus if $F(t)$ is the first-passage probability to a given point, $F(s = 0)$ is the probability of *eventually* hitting this point.

On the other hand, if all positive moments of $F(t)$,

$$\langle t^n \rangle = \frac{\int_0^\infty t^n \, F(t) \, dt}{\int_0^\infty F(t) \, dt}, \tag{1.4.11}$$

exist, then $F(s)$ in Eq. (1.4.10) contains only the Taylor series terms. These generate the positive integer moments of $F(t)$ by means of

$$
\begin{aligned}
F(s) &= \int_0^\infty F(t) \, e^{-st} \, dt \\
&= \int_0^\infty F(t) \left(1 - st + \frac{s^2 t^2}{2!} - \frac{s^3 t^3}{3!} + \cdots \right) \\
&= \mathcal{F}(\infty) \left(1 - s\langle t \rangle + \frac{s^2}{2!} \langle t^2 \rangle - \frac{s^3}{3!} \langle t^3 \rangle + \cdots \right).
\end{aligned} \tag{1.4.12}
$$

Thus the Laplace transform is a *moment generating function*, as it contains *all* the positive integer moments of the probability distribution $F(t)$. This is one of the reasons why the Laplace transform is such a useful tool for first-passage processes.

In summary, the small-s behavior of the Laplace transform, or, equivalently, the $z \rightarrow 1$ behavior of the generating function, are sufficient to determine the long-time behavior of the function itself. Because the transformed quantities are usually easy to obtain by the solution of an appropriate boundary-value problem, the asymptotic methods outlined here provide a simple route to obtain long-time behavior.

1.5. Asymptotics of the First-Passage Probability

We now use the techniques of Section 1.4 to determine the time dependence of the first-passage probability in terms of the generating function for the occupation probability. For simplicity, consider an isotropic random walk that starts at the origin. From the Gaussian probability distribution given in Eq. (1.3.27), $P(\vec{r} = 0, t) = (4\pi Dt)^{-d/2}$ in d dimensions. Then Eq. (1.4.1) gives the corresponding generating function:

$$P(0, z) \approx \int^\infty P(0, t) \, z^t \, dt \sim \int^\infty (4\pi Dt)^{-d/2} z^t \, dt. \tag{1.5.1}$$

As discussed in Section 1.4, this integral has two fundamentally different behaviors, depending on whether $\int^\infty P(0, t) \, dt$ diverges or converges. In the former case, we apply the last step in Eq. (1.4.1) to obtain

$$P(0, z) \propto \int^{t^*} (4\pi Dt)^{-d/2} \, dt \sim \begin{cases} \mathcal{A}_d (t^*)^{1-d/2} = \mathcal{A}_d (1-z)^{d/2-1}, & d < 2 \\ \mathcal{A}_2 \ln t^* = -\mathcal{A}_2 \ln(1-z), & d = 2 \end{cases},$$

$$(1.5.2)$$

where the dimension-dependent prefactor $\mathcal{A}_d$ is of the order of 1 and does not play any role in the asymptotic behavior.

For $d > 2$, the integral $\int^\infty P(0, t) \, dt$ converges and we apply Eq. (1.4.8) to compute the asymptotic behavior of $P(0, 1) - P(0, z)$. By definition, $F(0, 1) = \sum_t F(0, t) = 1 - [P(0, 1)]^{-1}$. Further, $\sum_t F(0, t)$ is just the eventual probability $\mathcal{R}$ that a random walk reaches the origin, so that $P(0, 1) = (1 - \mathcal{R})^{-1}$. In the asymptotic limit $z = 1 - s$, with $s \to 0$, $P(0, 1) - P(0, z)$ may therefore be written as

$$P(0, 1) - P(0, z) \propto \sum_{t=1}^{\infty} t^{-d/2} (1 - z^t),$$

$$\sim (1 - z)^{d/2-1}. \qquad (1.5.3)$$

Thus $P(0, z)$ has the asymptotic behavior

$$P(0, z) \sim (1 - \mathcal{R})^{-1} + B_d (1 - z)^{d/2-1} + \ldots, \quad d > 2, \qquad (1.5.4)$$

where B_d is another dimension-dependent constant of the order of 1. Using these results in Eq. (1.2.3), we infer that the generating function for the first-passage probability has the asymptotic behaviors

$$F(0, z) = 1 - \sqrt{1 - z^2}, \quad d = 1, \qquad (1.5.5)$$

whereas, for $d > 1$,

$$F(0, z) \sim \begin{cases} 1 - \dfrac{1}{A_d (1 - z)^{d/2-1}}, & d < 2 \\[2mm] 1 + \dfrac{1}{A_2 \ln(1 - z)}, & d = 2 \\[2mm] \mathcal{R} + B_d (1 - \mathcal{R})^2 (1 - z)^{d/2-1}, & d > 2 \end{cases} \qquad (1.5.6)$$

From this generating function, we determine the time dependence of the survival probability by approximation (1.4.7); that is,

$$F(0, z = 1 - 1/t^*) \sim \int_0^{t^*} F(0, t) \, dt$$

$$= \text{first-passage probability up to time } t^*$$

$$\equiv T(t^*) = 1 - S(t^*), \qquad (1.5.7)$$

where $S(t)$ is the survival probability and $T(t)$ is the complementary probability that the particle is trapped, that is, reaches the origin by time t. From Eqs. (1.5.6) and (1.5.7) we thus find

$$S(t) \sim \begin{cases} \dfrac{1}{A_d t^{1-d/2}}, & d < 2 \\[2ex] \dfrac{1}{A_2 \ln t}, & d = 2, \\[2ex] (1 - \mathcal{R}) + C_d (1 - \mathcal{R})^2 t^{1-d/2}, & d > 2 \end{cases} \qquad (1.5.8)$$

where C_d is another d-dependent constant of the order of 1. Finally, the time dependence of the first-passage probability may be obtained from the basic relation $1 - S(t) \sim \int^t F(0, t)\, dt$ to give

$$F(0, t) = -\frac{\partial S(t)}{\partial t} \propto \begin{cases} t^{d/2-2}, & d < 2 \\[2ex] \dfrac{1}{t \ln^2 t}, & d = 2. \\[2ex] t^{-d/2}, & d > 2 \end{cases} \qquad (1.5.9)$$

Equations (1.5.8) and (1.5.9) are the fundamental results of this section – the asymptotic time dependences of the survival and the first-passage probabilities. There are several important features that deserve emphasis. First, asymptotic time dependence is determined by the spatial dimension only, and not on any other properties of diffusive motion. Note, in particular, the change in behavior as a function of spatial dimension. For $d \leq 2$, the survival probability $S(t)$ ultimately decays to zero. This means that a random walk is *recurrent*, that is, certain to eventually return to its starting point, and indeed visit *any* site of an infinite lattice. Finally, because a random walk has no memory, it is "renewed" every time a specific lattice site is reached. Thus recurrence also implies that every site is visited infinitely often.

There is a simple physical basis for this efficient visitation of lattice sites. After a time t, a random walk explores a roughly spherical domain of radius $\sqrt{Dt}$. The total number of sites visited during this exploration is also proportional to t. Consequently, in d dimensions, the density of visited sites within this exploration sphere is $\rho \propto t/t^{d/2} \propto t^{1-d/2}$. Because this diverges as $t \to \infty$ for $d < 2$, a random walk visits each site within the sphere infinitely often; this is termed *compact exploration* [de Gennes (1983)]. Paradoxically, although every site is visited with certainty, these visitations take forever because the mean time to return to the origin, $\langle t \rangle = \int t F(0, t)\, dt$, diverges for all $d \leq 2$.

On the other hand, for $d > 2$, Eq. (1.5.8) predicts that there is a nonzero probability for a diffusing particle to not return to its starting point. More generally, there is a nonzero probability for a random walk to miss most of the lattice sites. This incomplete visitation follows from the density of visited sites ρ within the exploration sphere tending to zero as $t \to \infty$. Thus the random-walk trajectory within this sphere is relatively "transparent." This behavior is more commonly known as *transient* [Pólya (1921) and Feller (1968)].

The distinction between recurrence and transience has important physical implications. As a basic example, consider a diffusion-controlled reaction in three dimensions. Because the trajectory of each diffusing reactant is transparent, the probability for two molecules to meet is essentially independent of their initial separation. This means that a molecule reacts with any other reactant molecule in the system with almost equal probability. Such an egalitarian process corresponds to the mean-field limit. Conversely, for dimension $d \leq 2$, nearby reactants are most likely to meet and this naturally induces slow kinetics and spatial correlations (see Chap. 8).

For a biased random walk, first-passage characteristics are largely dominated by the bias. For $d > 1$, a biased random walk visits only those lattice sites within a narrow cone along the bias whose length is proportional to t and whose width is proportional to $t^{1/2}$. The density of visited sites within this cone is therefore $t/t^{1+(d-1)/2} \sim t^{(1-d)/2}$. Thus a biased random walk is transient for all $d > 1$. For $d = 1$ there is a trivial version of recurrence in which each downstream site is visited with certainty, but only a finite number of times.

1.6. Connection between First-Passage and Electrostatics

1.6.1. Background

The fundamental connection between the first-passage properties of diffusion and electrostatics is now outlined. Basic questions of first passage include *where* a diffusing particle is absorbed on a boundary and *when* does this absorption event occur. These are *time-integrated* attributes, obtained by integration of a time-dependent observable over all time. For example, to determine when a particle is absorbed, we should compute the first-passage probability to the boundary and then integrate over all time to obtain the *eventual hitting probability*. However, it is more elegant to reverse the order of calculation and first integrate the equation of motion over time and then compute the outgoing flux at the boundary. This first step transforms the diffusion equation to the simpler Laplace equation. Then, in computing the flux, the exit probability is just the electric field at the boundary point. Thus there is a complete correspondence between a first-passage problem and an electrostatic problem in the same geometry. This mapping is simple yet powerful, and can be adapted to compute related time-integrated properties, such as the splitting probabilities and the moments of the exit time. Some of these simplifications are also discussed in Gardiner (1985) and Risken (1988).

1.6.2. The Green's Function Formalism

Let us recall the conventional method to compute the first-passage probability of diffusion. Consider a diffusing particle that starts at $\vec{r}_0$ within a domain with boundary B. We compute the hitting probability to a point $\vec{r}_B$ on this boundary by first solving for the Green's function for the diffusion equation

$$\frac{\partial c(\vec{r}, t; \vec{r}_0)}{\partial t} = D\nabla^2 c(\vec{r}, t; \vec{r}_0), \tag{1.6.1}$$

with the initial condition $c(\vec{r}, 0; \vec{r}_0) = \delta(\vec{r} - \vec{r}_0)$ and the absorbing boundary condition $c(\vec{r}, t; \vec{r}_0)|_{\vec{r} \in B} = 0$. This condition accounts for the fact that once a particle reaches the boundary it leaves the system. Because the initial condition is invariably a single particle at $\vec{r}_0$, we will typically not write this argument in the Green's function. The outgoing flux at $\vec{r}_B$, $j(\vec{r}_B, t)$, is simply

$$j(\vec{r}_B, t) = -D\frac{\partial c(\vec{r}, t)}{\partial \hat{n}}\bigg|_{\vec{r} = \vec{r}_B}, \tag{1.6.2}$$

where $\hat{n}$ is a unit outward normal at $\vec{r}_B$. Finally, the eventual hitting probability at $\vec{r}_B$ is

$$\mathcal{E}(\vec{r}_B) = \int_0^\infty j(\vec{r}_B, t)\, dt. \tag{1.6.3}$$

In this direct approach, we first solve the diffusion equation, which has a first derivative in time, and then effectively "undo" the differentiation by integrating over all time to find the eventual hitting probability.

1.6.2.1. Hitting Probability

We now rederive this hitting probability by mapping the time-integrated diffusive system to electrostatics. First we integrate Eq. (1.6.1) over all time to give

$$c(\vec{r}, t = \infty) - c(\vec{r}, t = 0) = D\nabla^2 \mathcal{C}_0(\vec{r}), \tag{1.6.4}$$

where $C_0(\vec{r}) = \int_0^\infty c(\vec{r}, t)\, dt$ is the time integral of the Green's function. (The reason for including the subscript 0 will become apparent shortly.) We now consider spatial domains for which eventual absorption is certain, that is, $c(\vec{r}, t = \infty) = 0$. Additionally, at $t = 0$ the Green's function just reduces to the initial condition $c(\vec{r}, t = 0) = \delta(\vec{r} - \vec{r}_0)$. Thus $C_0(\vec{r})$ obeys the Laplace equation

$$D\nabla^2 C_0(\vec{r}) = -\delta(\vec{r} - \vec{r}_0), \tag{1.6.5}$$

with homogeneous Dirichlet boundary conditions. This defines an electrostatic system with a point charge of magnitude $1/(D\Omega_d)$ at $\vec{r}_0$ (where Ω_d is the surface area of a d-dimensional unit sphere). The absorbing boundaries in the diffusive system are equivalent to grounded conducting surfaces in the corresponding electrostatic problem. More general initial conditions can easily be considered by the linearity of the basic equations.

In terms of C_0, the eventual hitting probability $\mathcal{E}(\vec{r}_B)$ is given by

$$\begin{aligned}
\mathcal{E}(\vec{r}_B) &= \int_0^\infty j(\vec{r}_B, t)\, dt \\
&= -D \int_0^\infty \frac{\partial c(\vec{r}, t)}{\partial \hat{n}}\, dt \\
&= -D \frac{\partial C_0}{\partial \hat{n}}.
\end{aligned} \tag{1.6.6}$$

On the other hand, from Eq. (1.6.5) this normal derivative is just the electric field associated with the initial charge distribution. This leads to the following fundamental conclusion:

- For a diffusing particle that is initially at $\vec{r}_0$ inside a domain with absorbing boundary conditions, the eventual hitting probability to a boundary point $\vec{r}_B$ equals the electric field at this same location when a point charge of magnitude $1/(D\Omega_d)$ is placed at $\vec{r}_0$ and the domain boundary is grounded.

This simplifies the computation of the exit probability significantly, as it is much easier to solve the time-independent Laplace equation rather than the corresponding diffusion equation.

1.6.2.2. Hitting Time

This electrostatic formalism can be extended to integer moments of the mean time to exit or hit the boundary. By definition, the nth moment of the exit time is

$$\langle t^n \rangle = \int_0^\infty t^n F(t) \, dt, \tag{1.6.7}$$

where $F(t)$ is the first-passage probability to the boundary at time t. Here we are again tacitly considering situations for which the particle is certain to eventually reach the exit boundary. If $\langle t^0 \rangle = \int_0^\infty F(t) \, dt < 1$, as would occur if we were considering the first passage to a subset of a composite boundary, then the formulas given below need to be normalized in an obvious way by dividing the moments by $\langle t^0 \rangle$.

Using the fact that $F(t) = -[\partial S(t)/\partial t]$, we integrate by parts to obtain

$$\begin{aligned}
\langle t^n \rangle &= -\int_0^\infty t^n \frac{\partial S(t)}{\partial t} \, dt, \\
&= -t^n S(t) \Big|_0^\infty + n \int_0^\infty t^{n-1} S(t) \, dt \\
&= n \int_0^\infty t^{n-1} \, dt \int_V c(\vec{r}, t) \, d\vec{r}.
\end{aligned} \tag{1.6.8}$$

In the second line, the integrated term is trivially equal to zero at the lower limit and equals zero at the upper limit if the system is finite and $S(t) \to 0$ faster than a power law as $t \to \infty$. In the last line, the integral is over the domain volume. Note the important special case of $n = 1$ for which the mean exit time is simply $\langle t \rangle = \int_0^\infty S(t) \, dt$.

We now recast this derivation for $\langle t^n \rangle$ as a time-independent problem by reversing the order of the temporal and the spatial integrations at the outset. First, we define the time-integrated moments of the probability distribution

$$\mathcal{C}_n(\vec{r}) = \int_0^\infty c(\vec{r}, t) t^n \, dt. \tag{1.6.9}$$

Then, from the last line of Eq. (1.6.8), the nth moment of the mean exit time is just the time-independent expression

$$\langle t^n \rangle = n \int_V \mathcal{C}_{n-1}(\vec{r}) \, d\vec{r}. \tag{1.6.10}$$

We now show that each of these integrated moments satisfies the Poisson equation with an n-dependent source term. Consider

$$\int_0^\infty t^n \left[\frac{\partial c(\vec{r}, t)}{\partial t} = D\nabla^2 c(\vec{r}, t) \right] dt. \tag{1.6.11}$$

Integrating the left-hand side by parts, we find that the integrated term,

$t^n c(\vec{r}, t)|_0^\infty$, vanishes, except for $n = 0$, where the lower limit coincides with the initial condition. The remaining term, $-\int_0^\infty n\, t^{n-1} c(\vec{r}, t)\, dt$, is proportional to the time-integrated moment of order $n - 1$. The right-hand side is just $D\nabla^2 \mathcal{C}_n(\vec{r})$. Therefore $\mathcal{C}_n(\vec{r})$ obeys the equation hierarchy

$$DV^2\mathcal{C}_0(\vec{r}) = -\delta(\vec{r} - \vec{r}_0),$$
$$DV^2\mathcal{C}_1(\vec{r}) = -\mathcal{C}_0(\vec{r}),$$
$$DV^2\mathcal{C}_2(\vec{r}) = -2\mathcal{C}_1(\vec{r})$$
$$\vdots \qquad\qquad\qquad (1.6.12)$$

Thus each $\mathcal{C}_n$ is the electrostatic "potential" generated by the "charge" distribution $\mathcal{C}_{n-1}$, with $\langle t^n \rangle = n \int_V \mathcal{C}_{n-1}(\vec{r})\, d\vec{r}$. This provides all moments of the first-passage time in terms of an associated hierarchy of electrostatic potentials.

1.6.3. Laplacian Formalism

Another useful version of the correspondence between diffusion and electrostatics is based on encoding the initial condition as the spatial argument of an electrostatic potential rather than as an initial condition. This Laplacian approach gives the splitting probability for composite domain boundaries in a natural fashion and also provides *conditional* hitting times, namely, the times to eventually hit a specific subset of the boundary.

1.6.3.1. Splitting Probabilities

For simplicity, we start with a symmetric nearest-neighbor random walk in the finite interval $[x_-, x_+]$ and then take the continuum limit after developing the formalism. We define $\mathcal{E}_-(x)$ as the probability for a particle, which starts at x, to eventually hit x_- *without* hitting x_+. The generalization to higher dimensions and to more general boundaries follows by similar reasoning. Pictorially, we obtain the eventual hitting probability $\mathcal{E}_-(x)$ by summing the probabilities for all paths that start at x and reach x_- without touching x_+ (Fig. 1.4). A parallel statement holds for $\mathcal{E}_+(x)$. Thus

$$\mathcal{E}_\pm(x) = \sum_{p_\pm} \mathcal{P}_{p_\pm}(x), \qquad (1.6.13)$$

where $\mathcal{P}_{p_\pm}(x)$ denotes the probability of a path from x to $x_\pm$ that avoids $x_\mp$. As illustrated in Fig. 1.4, the sum over all such restricted paths can be decomposed into the outcome after one step and the sum over all path remainders from the

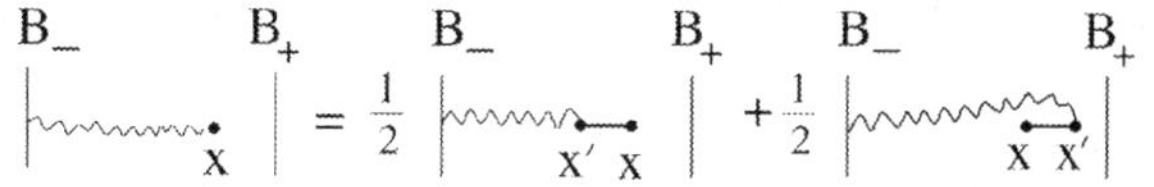

$$B_- \qquad B_+ \qquad B_- \qquad B_+ \qquad B_- \qquad B_+$$

Fig. 1.4. Decomposition of random-walk paths from x to the absorbing subset B_- (the point x_-) into the outcome after one step and the remainder from x' to B_-. The factors $1/2$ account for the probabilities associated with the first step of the decomposed paths.

intermediate point x' to $x_\pm$. This gives

$$\mathcal{E}_\pm(x) = \sum_{p_\perp} \left[\frac{1}{2} \mathcal{P}_{p_\perp}(x + \delta x) + \frac{1}{2} \mathcal{P}_{p_\perp}(x - \delta x) \right]$$

$$= \frac{1}{2} [\mathcal{E}_\pm(x + \delta x) + \mathcal{E}_\pm(x - \delta x)]. \tag{1.6.14}$$

By a simple rearrangement, this is equivalent to

$$\Delta^{(2)} \mathcal{E}_\pm(x) = 0, \tag{1.6.15}$$

where $\Delta^{(2)}$ is the discrete second-difference operator, which is defined by $\Delta^{(2)} f(x) = f(x - \delta x) - 2f(x) + f(x + \delta x)$. Note the opposite sense of this recursion formula compared with master equation Eq. (1.3.1) for the probability distribution. Here $\mathcal{E} \pm (x)$ is expressed in terms of *output from* x, whereas in the master equation, the occupation probability at x is expressed in terms of *input to* x.

The basic Laplace equation (1.6.15) is subject to the boundary conditions $\mathcal{E}_-(x_-) = 1$, $\mathcal{E}_-(x_+) = 0$, and correspondingly $\mathcal{E}_+(x_-) = 0$, $\mathcal{E}_+(x_+) = 1$. In the continuum limit, Eq. (1.6.15) reduces to the one-dimensional Laplace equation $\mathcal{E}''_\pm(x) = 0$. For isotropic diffusion in d spatial dimensions, the corresponding equation is

$$\nabla^2 \mathcal{E}_\pm(\vec{r}) = 0, \tag{1.6.16}$$

subject to the boundary conditions $\mathcal{E}_-(\vec{r}_-) = 1$ and $\mathcal{E}_-(\vec{r}_+) = 0$; that is, $\mathcal{E}_- = 1$ on the exit subset of the absorbing boundary and $\mathcal{E}_- = 0$ on the complement of this boundary. These conditions are interchanged for $\mathcal{E}_+$.

The functions $\mathcal{E}_\pm$ are *harmonic* because $\mathcal{E}_\pm(x)$ equals the average of $\mathcal{E}_\pm$ at neighboring points [Eq. (1.6.14)]; that is, $\mathcal{E}_\pm$ is in "harmony" with its local environment. This is a basic feature of solutions to the Laplace equation. Because $\mathcal{E}_\pm$ satisfies the Laplace equation, we can transcribe well-known results from electrostatics to less familiar, but corresponding, first-passage properties. This approach will be applied repeatedly in later chapters.

We can easily extend the Laplacian formalism to a general random-walk process in which the probability of hopping from $\vec{r}$ to $\vec{r}'$ in a single step is

$p_{\vec{r} \to \vec{r}'}$. The appropriate generalization of Eq. (1.6.14) is

$$\mathcal{E}_{\pm}(\vec{r}) = \sum_{\vec{r}'} p_{\vec{r} \to \vec{r}'} \mathcal{E}_{\pm}(\vec{r}'), \qquad (1.6.17)$$

which, in the continuum limit, reduces to

$$D(\vec{r}) \nabla^2 \mathcal{E}_{\pm}(\vec{r}) + \vec{v}(\vec{r}) \cdot \vec{\nabla} \mathcal{E}_{\pm}(\vec{r}) = 0, \qquad (1.6.18)$$

where the local diffusion coefficient $D(\vec{r})$ is just the mean-square displacement and the local velocity $\vec{v}(\vec{r})$ is the mean displacement after a single step when starting from $\vec{r}$ in this hopping process. The existence of the continuum limit requires that the range of the hopping is finite. This equation should be solved subject again to the boundary condition of $\mathcal{E}_{\pm} = 1$ on the exit boundary and $\mathcal{E}_{\pm} = 0$ on the complement of the exit.

In summary, the hitting, or exit, probability coincides with the electrostatic potential when the boundary conditions of the diffusive and the electrostatic systems are the same. This statement applies for both continuum diffusion and the discrete random walk. This approach can be extended in an obvious way to more general hopping processes with both a spatially varying bias and diffusion coefficient. The consequences of this simple equivalence are powerful. As an example, consider a diffusing particle that is initially at radius r_0 exterior to a sphere of radius a centered at the origin in d dimensions. By the electrostatic formalism, the probability that this particle eventually hits the sphere is simply the electrostatic potential at r_0, $\mathcal{E}_-(r_0) = (a/r_0)^{d-2}$! Conversely, if a diffusing particle starts in the interior of a bounded domain, it is physically obvious that the boundary is eventually reached. This means that $\mathcal{E}(\vec{r}) = 1$ for any interior point. This also follows from the fact that $\nabla^2 \mathcal{E}(\vec{r}) = 0$ within the domain, subject to the boundary condition $\mathcal{E}(\vec{r}) = 1$ on the boundary. The solution is clearly $\mathcal{E}(\vec{r}) = 1$ for any interior point. This corresponds to the well-known fact that the electrostatic potential in the interior of a conductor is constant, or equivalently, the electric field is zero.

1.6.3.2. *Unconditional and Conditional Mean First-Passage Times*

We now extend the Laplacian approach to determine the mean exit time from a domain with composite boundaries. Here we distinguish between the *unconditional* mean exit time, namely, the time for a particle to reach any point on an absorbing boundary B, and the *conditional* mean exit time, namely, the time for a particle to reach a specified subset of the absorbing boundary B_- *without* touching the complement boundary $B_+ = B - B_-$. This conditional exit time is closely related to the splitting probability of Subsection 1.6.3.1.

We again treat a symmetric random walk on the finite interval $[x_-, x_+]$; the generalization to higher dimensions and to general hopping processes is straightforward. Let the time increment between successive steps be δt, and let $t(x)$ denote the mean time to exit at *either* boundary component when a particle starts at x. This quantity is simply the time for each exit path times the probability of the path, averaged over all particle trajectories. This leads to the analog of Eq. (1.6.13):

$$t(x) = \sum_p \mathcal{P}_p(x) t_p(x), \qquad (1.6.19)$$

where $t_p(x)$ is the exit time of a specific path to the boundary that starts at x.

In analogy with Eq. (1.6.14), this mean exit time obeys the recursion formula

$$t(x) = \frac{1}{2} \{[t(x + \delta x) + \delta t] + [t(x - \delta x) + \delta t]\}, \qquad (1.6.20)$$

with the boundary conditions $t(x_-) = t(x_+) = 0$, which correspond to the exit time being equal to zero if the particle starts at the boundary. This recursion relation expresses the mean exit time starting at x in terms of the outcome one step in the future, for which the initial walk can be viewed as restarting at either $x + \delta x$ or at $x - \delta x$, each with probability $1/2$, but also with the time incremented by δt. In the continuum limit, the Taylor expansion of this recursion formula to lowest nonvanishing order yields the Poisson equation $Dt''(x) = -1$. For diffusion in a d-dimensional domain with absorption on the boundary B, the corresponding Poisson equation for the exit time is

$$D\nabla^2 t(\vec{r}) = -1, \qquad (1.6.21)$$

subject to the boundary condition $t(\vec{r}) = 0$ for $\vec{r} \in B$. This Poisson is often termed the *adjoint equation* for the mean exit, or mean first-passage, time.

These results can be extended to a general short-range hopping process with single-step hopping probability $p_{\vec{r} \to \vec{r}'}$. In this case, the analog of Eq. (1.6.20) is

$$t(\vec{r}') = \sum_{\vec{r}'} p_{\vec{r} \to \vec{r}'} [t(\vec{r}') + \delta t], \qquad (1.6.22)$$

which in the continuum limit, becomes the Poisson-like equation

$$D\nabla^2 t(\vec{r}) + \vec{v}(\vec{r}) \cdot \vec{\nabla} t(\vec{r}) = -1. \qquad (1.6.23)$$

Note, in particular, that the determination of the mean exit time has been recast as a time-independent electrostatic problem. As we shall see in the next chapters, this device greatly simplifies the computation of the mean exit times and also provides useful physical insights.

Finally, we may extend this electrostatic formalism to the conditional exit times in which we discriminate the exit time by which part of the boundary is reached. Thus let $t_-(x)$ be the conditional mean exit time for a random walk that starts at x and exits at B_- without hitting B_+. Similarly let $t_+(x)$ be the conditional mean exit time for a random walk that starts at x and exits at B_+ without hitting B_-. By its definition, $t_-(x)$ can be written as

$$t_-(x) = \frac{\sum_{p_-} \mathcal{P}_p(x)\, t_p(x)}{\sum_{p_-} \mathcal{P}_{p_-}(x)} = \frac{\sum_{p_-} \mathcal{P}_p(x)\, t_p(x)}{\mathcal{E}_-(x)}, \qquad (1.6.24)$$

where the sum over paths p_- denotes only the allowed paths that start at x and reach B_- without touching B_+. By decomposing each path as the outcome after one step and its remainder, and then applying Eq. (1.6.14), we may write

$$
\begin{aligned}
\mathcal{E}_-(x)t_-(x) = \sum_{p_-} \frac{1}{2} & [\mathcal{P}_{p_-}(x - \delta x)(t_{p_-}(x - \delta x) + \delta t) \\
& + \mathcal{P}_{p_-}(x + \delta x)(t_{p_-}(x + \delta x) + \delta t)] \\
= \delta t\, \mathcal{E}_-(x) + \frac{1}{2} & [\mathcal{E}_-(x - \delta x)t_-(x - \delta x) \\
& + \mathcal{E}_-(x + \delta x)t_-(x + \delta x)] \\
\approx \delta t\, \mathcal{E}_-(x) + & \left[\mathcal{E}_-(x)t_-(x) + \frac{\delta x^2}{2}\frac{\partial^2 \mathcal{E}_-(x)t_-(x)}{\partial x^2}\right]. \quad (1.6.25)
\end{aligned}
$$

In the continuum limit, this leads to a Poisson equation for the conditional mean first-passage time (now written for general spatial dimension)

$$D\nabla^2[\mathcal{E}_-(\vec{r})t_-(\vec{r})] = -\mathcal{E}_-(\vec{r}). \qquad (1.6.26)$$

with $D = 2(\delta r)^2/2\delta t$ and subject to the boundary conditions $\mathcal{E}_-(\vec{r})t_-(\vec{r}) = 0$ both on B_- (where t_- vanishes) and on B_+ (where $\mathcal{E}_-$ vanishes). The governing equations and boundary conditions for $t_+(\vec{r})$ are entirely analogous. Finally, if there is a bias in the hopping process, then Eq. (1.6.26) is generalized to

$$D\nabla^2[\mathcal{E}_\pm(\vec{r})t_\pm(\vec{r})] + \vec{v}(\vec{r}) \cdot \vec{\nabla}[\mathcal{E}_\pm(\vec{r})t_\pm(\vec{r})] = -\mathcal{E}_\pm(\vec{r}). \qquad (1.6.27)$$

With this formalism, we can obtain eventual hitting probabilities and mean hitting times (both unconditional and conditional) by solving time-independent electrostatic boundary-value problems. Thus this electrostatic connection will allow us to obtain subtle conditional first-passage properties in a relatively simple manner.

1.7. Random Walks and Resistor Networks

1.7.1. Introduction

In parallel with the connection between diffusive first-passage properties and the Laplace and the Poisson equations of electrostatics, there is a deep relation between first-passage properties of random walks and the discrete analog of electrostatics. The latter is naturally expressed as the current-carrying properties of a suitably defined resistor network. This is the basis of an appealing mapping between the first-passage characteristics of a random walk on a particular graph and the current-carrying properties of a resistor network whose elements consist of the same graph. We shall show that the voltages at each site and the overall network resistance are directly and simply related to the exit probabilities of a corresponding random walk. Our discussion closely follows that given in Doyle and Snell (1984).

1.7.2. The Basic Relation

To develop the resistor network connection, consider a discrete random walk on a finite lattice graph with hopping between nearest-neighbor sites (see Fig. 1.5). We divide the boundary points into two classes, B_+ and B_-. As usual, we are interested in the exit probabilities to B_+ and to B_- as functions of the initial position x of the random walk. As shown in Section 1.6, these exit probabilities, $\mathcal{E}_+(x)$, and $\mathcal{E}_-(x)$, respectively, are governed by the discrete Laplace equation $\Delta^{(2)}\mathcal{E}_\pm(x) = 0$, subject to the boundary conditions $\mathcal{E}_+ = 1$ on B_+ and $\mathcal{E}_+ = 0$ on B_- and vice versa for $\mathcal{E}_-$.

This Laplace equation has a simple resistor network interpretation. The basic connection is that if all the lattice bonds are viewed as resistors (not necessarily identical), then Kirchhoff's laws for steady current flow in the

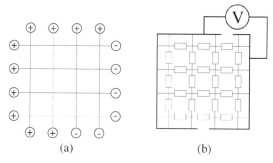

(a) (b)

Fig. 1.5. (a) A lattice graph with boundary sites B_+ or B_-. (b) Corresponding resistor network in which each bond (with rectangle) is a 1-Ω resistor. The sites in B_+ are all fixed at potential $V = 1$, and sites in B_- are all grounded.

network are identical to the discrete Laplace equation for $\mathcal{E}$. Suppose that the boundary sites in B_+ are fixed at unit potential while the sites in B_- are grounded. The net current at each interior site i of the network must be zero, as there is current input and output only at the boundary. This current conservation condition is

$$\sum_j g_{ij}(V_i - V_j) = 0, \qquad (1.7.1)$$

where g_{ij} is the conductance of the bond between sites i and j, V_i is the voltage at site i, and the sum runs over the nearest neighbors j of site i. Solving for V_i gives

$$V_i = \frac{\sum_j g_{ij} V_j}{\sum_j g_{ij}} \rightarrow \frac{1}{4} \sum_j V_j, \qquad (1.7.2)$$

where the last step applies for a homogeneous network. Thus, in the steady state, the voltage at each interior site equals the weighted average of the voltages at the neighboring sites; that is, V_i is a harmonic function with respect to the weight function g_{ij}. The voltage must also satisfy the boundary conditions $V = 1$ for sites in B_+ and $V = 0$ for sites in B_-.

We can also give a random-walk interpretation for the process defined by Eq. (1.7.2). Consider a lattice random walk in which the probability of hopping from site i to site j is $P_{ij} = g_{ij}/\sum_j g_{ij}$. Then the probability that this walk eventually reaches B_+ without first reaching B_- is just given by $\mathcal{E}_+(i) = \sum_{p_+} \mathcal{P}_{p_+}(i)$ [see Eqs. (1.6.13)–(1.6.15)]. Because both V_i and $\mathcal{E}_+(i)$ are harmonic functions that satisfy the same boundary conditions, these two functions are identical. Thus we find the basic relation between random walks and resistor networks:

- Let the boundary sets B_+ and B_- in a resistor network be fixed at voltages 1 and 0 respectively, with g_{ij} the conductance of the bond between sites i and j. Then the voltage at any interior site i is the same as the probability for a random walk that starts at i to reach B_+ before reaching B_- when the hopping probability from i to j is $P_{ij} = g_{ij}/\sum_j g_{ij}$.

1.7.3. Escape Probability, Resistance, and Pólya's Theorem

An important extension of this relation between escape probability and site voltages is to infinite networks. This provides a simple connection between the recurrence/transience transition of random walks on a given network and the electrical resistance of this same network. Suppose that the voltage V at the boundary sites B_+ is set to one. Then the total current entering the network is given by

$$I = \sum_j (V - V_j) g_{+j}$$
$$= \sum_j (1 - V_j) P_{+j} \sum_j g_{+j}. \tag{1.7.3}$$

Here g_{+j} is the conductance between the positive terminal of the voltage source and the neighboring sites j and $P_{+j} = g_{+j}/\sum_j g_{+j}$. Because the voltage V_j also equals the probability for the corresponding random walk to reach B_+ without reaching B_-, the term $V_j P_{+j}$ is just the probability that a random walk starts at B_+, makes a single step to the sites j (with hopping probabilities P_{ij}), and then returns to B_+ without reaching B_-. We therefore deduce that

$$I = \sum_j g_{+j} \sum_j (1 - V_j) P_{+j}$$
$$= \sum_j g_{+j} \times (1 - \text{return probability})$$
$$= \sum_j g_{+j} \times \text{escape probability}. \tag{1.7.4}$$

Here "escape" means reaching the opposite terminal of the voltage source without returning to the starting point.

On the other hand, the input current and the voltage drop across the network are simply related to the conductance G between the two boundary sets by $I = GV = G$. From this and Eq. (1.7.4) we obtain the fundamental result:

$$\text{escape probability} \equiv P_{\text{escape}} = \frac{G}{\sum_j g_{+j}}. \tag{1.7.5}$$

Perhaps the most interesting situation is when a current I is injected at a single point of an infinite network with outflow at infinity (see Fig. 1.6). Then the probability for a random walk that starts at this input point to escape, that is, never return to its starting point, is simply proportional to the network conductance G. It is amazing that a subtle feature of random walks is directly related to currents and voltages in a resistor network!

One appeal of this connection is that network conductances can be computed easily. In one dimension, the conductance of an infinitely long chain of identical resistors is clearly zero. Thus $P_{\text{escape}} = 0$ or, equivalently, $P_{\text{return}} = 1$; that is, a random walk in one dimension is recurrent. In higher dimensions,

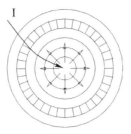

Fig. 1.6. Decomposition of a conducting medium into concentric shells, each of which consists of fixed-conductance blocks. A current I is injected at the origin and flows radially outward through the medium.

the conductance between one point and infinity in an infinite resistor lattice is a beautiful and more demanding problem that has inspired many physicists [see e.g., van der Pol & Bremmer (1955), see also Atkinson & van Steenwijk (1999) and Cserti (2000) for more recent pedagogical accounts]. However, for merely determining the recurrence or transience of a random walk, a crude physical estimate suffices. The basic idea of this estimate is to replace the discrete lattice with a continuum medium of constant conductivity. We then compute the conductance of the infinite system by further decomposing the medium into a series of concentric shells of fixed thickness dr. A shell at radius r can be regarded as a parallel array of r^{d-1} volume elements, each of which has a fixed conductance. The conductance of one such a shell is then simply proportional to its surface area, and the overall resistance is the sum of the shell resistance. This gives

$$R \sim \int^{\infty} R_{\text{shell}}(r)\, dr \sim \int^{\infty} \frac{dr}{r^{d-1}}$$

$$\tag{1.7.6}$$

$$= \begin{cases} \infty & \text{for } d \leq 2 \\ \left(P_{\text{escape}} \sum_j g_{+j} \right)^{-1} & \text{for } d > 2 \end{cases} .$$

$$\tag{1.7.7}$$

This provides a remarkably easy solution to the recurrence/transience transition of random walks. For $d > 2$, the conductance between a single point and infinity in an infinite homogeneous resistor network is nonzero and is simply related to the random-walk escape probability. For $d \leq 2$, the conductance to infinity is zero, essentially because there are an insufficient number of independent paths from the origin to infinity. Correspondingly, the escape probability is zero and the random walk is recurrent. The case $d = 2$ is more delicate because the integral in expression (1.7.6) diverges only logarithmically at the upper limit. Nevertheless, the conductance to infinity slowly goes to zero as the radius of the network diverges and the corresponding escape probability is still zero.

1.8. Epilogue

The approaches outlined in this chapter provide the tools to determine where a diffusing particle gets trapped on, or exits from, an absorbing boundary and how long it takes for this event to occur. Often we can determine these properties most elegantly by mapping the diffusion problem onto a corresponding electrostatic problem or onto a resistor network problem (in the case of random walks). These mappings have powerful implications yet are generally easy to apply. This use of the analogy to electrostatics underlies much of the mathematical literature on first passage [see, e.g., Dynkin & Yushkevich (1969), Spitzer (1976), and Salminen & Borodin (1996)].

Another related message that will become even more apparent in the next chapter is that it is invariably much easier to deal with continuum diffusion and its corresponding electrostatics, rather than with discrete random walks. Although the generating function formalism is an elegant and powerful way to treat the dynamics of random walks and their associated first-passage properties [Montroll (1965), Montroll & Weiss (1965), and Weiss & Rubin (1983)], the corresponding properties are much easier to treat in the continuum-diffusion approximation. Thus we will concentrate primarily on diffusion and return to discrete random walks only for pedagogical completeness or when random walks provide the most appropriate description of a particular system. Our main goal in the following chapters is to elucidate the first-passage properties of diffusion and its physical implications for a variety of physically relevant systems, both by direct methods and by exploiting the connections with electrostatics.

2

First Passage in an Interval

2.1. Introduction

We now develop the ideas of the previous chapter to determine basic first-passage properties for both continuum diffusion and the discrete random walk in a finite one-dimensional interval. This is a simple system with which we can illustrate the physical implications of first-passage processes and the basic techniques for their solution. Essentially all of the results of this chapter are well known, but they are scattered throughout the literature. Much information about the finite-interval system is contained in texts such as Cox and Miller (1965), Feller (1968), Gardiner (1985), Risken (1988), and Gillespie (1992). An important early contribution for the finite-interval system is given by Darling and Siegert (1953). Finally, some of the approaches discussed in this chapter are similar in spirit to those of Fisher (1988).

For continuum diffusion, we start with the direct approach of first solving the diffusion equation and then computing first-passage properties from the time dependence of the flux leaving the system. Much of this is classical and well-known material. These same results will then be rederived more elegantly by the electrostatic analogies introduced in Chap. 1. This provides a striking illustration of the power of these analogies and sets the stage for their use in higher dimensions and in more complex geometries (Chaps. 5–7).

We also derive parallel results for the discrete random walk. One reason for this redundancy is that random walks are often more familiar than diffusion because the former often arise in elementary courses. It will therefore be satisfying to see the essential unity of their first-passage properties. It is also instructive to introduce various methods for analyzing the recursion relations for the discrete random walk. This will also illustrate an important and general lesson that was alluded to in the previous chapter – it is generally much easier to solve a continuum-diffusion process rather than the corresponding random walk.

2.1.1. Basic Questions

Consider a diffusing particle or a random walk in the interval $[0, L]$ with at least one absorbing end point. Eventually the particle is absorbed, and our basic goal is to characterize the time dependence of this trapping phenomenon. Depending on the nature of the boundary conditions, there are three generic cases, which we term (i) *absorption mode*, (ii) *transmission mode*, and (iii) *reflection mode*.

(i) *Absorption Mode:* Both boundaries are absorbing. The following are the basic first-passage questions:
  - What is the time dependence of the survival probability $S(t)$? This is the probability that a diffusing particle does not touch either absorbing boundary before time t.
  - What is the time dependence of the first-passage probabilities, or the exit probabilities, to either 0 or to L as a function of the starting position of the particle? Integrating these probabilities, over all time gives the eventual hitting probability to a specified boundary, or equivalently, the *splitting probability*. What is the dependence of the splitting probability to either 0 or to L as a function of the starting position of the particle?
  - What is the mean exit time, that is, the mean time until the particle hits either of the absorbing boundaries, again as a function of starting position? What are the conditional exit times, that is, the mean time for the particle to hit a specified boundary as a function of the starting position?

(ii) *Transmission Mode:* In this case, there is one absorbing and one reflecting boundary. The particle typically starts at the reflecting boundary and leaves the system when it reaches the absorbing end. Because the time to traverse the system is the same as that required for the particle to detect the finiteness of the system, a single time scale characterizes the first-passage properties of the system.

(iii) *Reflection Mode:* Here the particle starts near an absorbing boundary, while the other end is reflecting. At relatively short times, the first-passage properties are identical to those of a semi-infinite system and thus exhibit power-law behaviors (see Chap. 3). After a sufficiently long time, however, a diffusing particle can reach the reflecting boundary before absorption. The subsequent first-passage properties must also reflect this finite-size cutoff.

The first-passage questions for these latter two modes are similar to the those of absorption mode, except that there is only one absorbing boundary:

- What is the survival probability $S(t)$?
- What is the first-passage probability $F(t)$ to the output?
- How long does it take the particle to reach the output?
- How do each of these quantities depend on the initial particle position?

The goal of the remainder of this chapter is to answer all of these fundamental questions.

2.2. Time-Dependent Formulation

2.2.1. Survival Probability: Absorption Mode

For a normalized initial concentration, that is, $\int_0^L c(x, t = 0)\,dx = 1$, the survival probability equals the spatial integral of the concentration $c(x, t)$ over the interval,

$$S(t) \equiv \int_0^L c(x, t)\,dx. \tag{2.2.1}$$

The concentration is determined from the solution to the diffusion equation. Once the diffusive spread of the concentration becomes comparable to the interval size, the flux of probability through the boundaries becomes significant and a rapid decay of the density inside the interval ensues. We determine this long-time decay for both isotropic and biased diffusion.

2.2.1.1. Isotropic Diffusion

The diffusion equation for the concentration is

$$\frac{\partial c(x, t)}{\partial t} = D \frac{\partial^2 c(x, t)}{\partial x^2}. \tag{2.2.2}$$

This equation is defined on $0 \leq x \leq L$, subject to the absorbing boundary conditions $c(0, t) = c(L, t) = 0$. For simplicity, we consider the initial condition $c(x, t = 0) = \delta(x - x_0)$ with $0 < x_0 < L$, corresponding to a particle starting at x_0. Because of the absorbing boundary condition, only those trajectories that do not touch the boundaries contribute to the survival probability. Note also that the diffusion equation is identical in form to the time-dependent Schrödinger equation for a quantum particle in the infinite square-well potential, $V(x) = 0$ for $0 < x < L$ and $V(x) = \infty$ otherwise, with the diffusion coefficient D playing the role of $-\hbar^2/2m$ in the Schrödinger equation.

Therefore the long-time concentration for diffusion in the absorbing interval is closely related to the low-energy eigenstates of the quantum-well system.

We solve Eq. (2.2.2) by making the variable separation $c(x, t) = f(t)g(x)$ and substituting to give

$$\frac{\dot{f}}{f} = D\frac{g''}{g},$$

where the prime and the overdot denote spatial and temporal differentiation, respectively. For this equation to hold for all x and t, each side of the equation must equal a constant. This separation constant is taken to be negative to conform with the fact that the probability density decays with time. Then the single-variable functions satisfy

$$\dot{f} = -kf,$$
$$Dg'' = -kg.$$

The solutions to the spatial equation that satisfy the absorbing boundary condition are of the form $\sin(n\pi x/L)$, with n an integer. The solution to the temporal equation is exponential decay, with the decay rate equal to $k_n = D(n\pi/L)^2$ for the nth spatial eigenmode.

In general, the solution to the diffusion equation may be written as the eigenfunction expansion

$$c(x, t) = \sum_{n=1}^{\infty} A_n \sin\left(\frac{n\pi x}{L}\right) e^{-(\frac{n\pi}{L})^2 Dt}, \qquad (2.2.3)$$

where the constants A_n are determined by the initial condition. Because the functions $\sin[(n\pi x)/L]$ are a complete set of states for the given boundary conditions, any initial condition can be represented as an expansion of this form.

Each eigenmode decays exponentially in time, with a numerically different decay time $\tau_n = k_n^{-1} = L^2/n^2\pi^2 D$. Because the large-$n$ eigenmodes decay more rapidly with time, only the most slowly decaying eigenmode remains in the long-time limit. As a result, the asymptotic survival probability decays as

$$S(t) \propto e^{-D\pi^2 t/L^2} \equiv e^{-t/\tau_1}. \qquad (2.2.4)$$

As we shall see, the longest decay time $\tau_1 = L^2/D\pi^2$ characterizes diffusive dynamics within the absorbing interval.

2.2.1.2. Biased Diffusion

For biased diffusion the concentration is governed by the convection-diffusion equation

$$\frac{\partial c(x,t)}{\partial t} + v\frac{\partial c(x,t)}{\partial x} = D\frac{\partial^2 c(x,t)}{\partial x^2}. \tag{2.2.5}$$

The presence of the convection term fundamentally alters the asymptotic behavior of diffusion, with the competition between convection and diffusion leading to a subtle crossover in the behavior of the survival probability.

When we use the same variable separation as that in isotropic diffusion, the separated functions satisfy

$$\dot{f} = -kf,$$
$$Dg'' - vg' = -kg.$$

Assuming the exponential solution $g(x) \propto e^{\alpha x}$ and substituting into the equation for g leads to a characteristic equation for α with the solution

$$\alpha_\pm = \frac{v}{2D} \pm \frac{1}{2D}\sqrt{v^2 - 4kD}$$
$$\equiv \frac{v}{2D} \pm iw.$$

Hence $g(x)$ has the form

$$e^{vx/2D}\sin wx,$$

with w again quantized as $w_n = n\pi/L$ to satisfy the absorbing boundary condition. This then gives $k_n = v^2/4D + D(n\pi/L)^2$ for the spectrum of decay rates.

As in isotropic diffusion, the survival probability can be represented as an eigenfunction expansion of decaying exponentials in time. Only the lowest eigenmode remains in the long-time limit, from which the asymptotic decay time is $\tau_1 = k_1^{-1} = \{(L^2/D\pi^2)/[1 + (vL/2\pi D)^2]\}$. Thus for weak bias, the lifetime approaches that of isotropic diffusion, whereas for strong bias the decay time is proportional to D/v^2. The crossover between these two limits is determined by the dimensionless *Péclet number*

$$Pe \equiv vL/2D.$$

This fundamental parameter characterizes the relative importance of diffusion and convection in biased diffusion.

At first sight, it seems paradoxical that the decay time for strong bias is simply D/v^2, independent of the interval length. Naively, we might anticipate that the decay time should be of the order of the mean exit time. In Subsection 2.2.2, we will show that this exit time equals $(L - x_0)/v$, as one might naturally

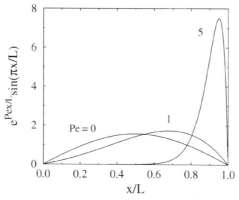

Fig. 2.1. Lowest eigenmode of the probability distribution, $e^{Pex/L} \sin(\pi x/L)$, for a diffusing particle in the absorbing interval $[0, L]$ for Péclet numbers 0, 1, and 5. Each eigenmode has been normalized to unit integral.

expect. The apparently too-small decay time arises because the eigenfunctions are skewed and sharply peaked at a small distance of the order of D/v from the right boundary of the interval (Fig. 2.1). Because the "skin depth" of each eigenmode is system-size independent, the decay rate of each eigenmode is similarly system-size independent. However, for a particle that is initially at x_0, a much larger mean exit time is generated by the contributions of eigenmodes that are peaked about the initial position rather than those within the skin depth.

In summary, for a finite-size interval the survival probability decays exponentially in time. The characteristic time of this decay is proportional to $\tau_D = L^2/D$ for isotropic diffusion, independent of the initial particle position. As we shall see, the mean time for a diffusing particle to exit the interval is also of the order of τ_D, *unless* the particle starts at a distance of less than $1/L$ from a boundary. For biased diffusion, the situation is more subtle. The spectrum of decay times depends on the Péclet number, and the smallest decay time does not depend on the interval length. Nevertheless, we will see that the mean exit time for strongly biased diffusion is proportional to $(L - x_0)/v$, in agreement with naive expectations.

2.2.2. First-Passage Probability and Mean Exit Times

At a finer level of description, we investigate not only *when* a particle is trapped, but also *where* it is trapped, as embodied by the first-passage probability and moments of the exit time to a specified boundary.

2.2.2.1. Isotropic Diffusion

To compute the first-passage probability on the finite interval, it is simpler to use the Laplace transform rather than the Fourier transform because the concentration may be expressed as a single function rather than as an infinite Fourier series. We now derive this first-passage probability by first solving for the Green's function of the diffusion equation.

Applying a Laplace transform recasts diffusion equation (2.2.2) as

$$sc(x, s) - c(x, t = 0) = Dc''(x, s), \qquad (2.2.6)$$

where the prime denotes differentiation with respect to x. Once again, the argument s in $c(x, s)$ indicates that this function is the Laplace transform. In each subdomain $(0, x_0)$ and (x_0, L), Eq. (2.2.6) is homogeneous, and the solution is the sum of exponentials of the form $c(x, s) = A \exp(+x\sqrt{s/D}) + B \exp(-x\sqrt{s/D})$, with the constants A and B dependent on the boundary conditions.

Absorption Mode. We can simplify the algebra considerably by choosing the appropriate linear combination of exponential solutions that manifestly satisfies the boundary conditions at the outset. The absorbing boundary condition at $x = 0$ suggests an antisymmetric combination of exponentials. Further, the form of the Green's function as $x \to L$ must be identical to that as $x \to 0$. These considerations lead to

$$c_<(x, s) = A \sinh\left(\sqrt{\frac{s}{D}}x\right), \quad x < x_0,$$

$$c_>(x, s) = B \sinh\left[\sqrt{\frac{s}{D}}(L - x)\right], \quad x > x_0 \qquad (2.2.7)$$

for the subdomain Green's functions $c_<$ and $c_>$, where A and B are constants.

When the continuity condition $c_<(x_0, s) = c_>(x_0, s)$ is imposed, the Green's function reduces to the symmetrical form

$$c(x, s) = A \sinh\left(\sqrt{\frac{s}{D}}x_<\right) \sinh\left[\sqrt{\frac{s}{D}}(L - x_>)\right], \qquad (2.2.8)$$

with $x_> = \max(x, x_0)$ and $x_< = \min(x, x_0)$. This notation allows us to write the Green's function in the entire domain as a single expression. With practice, one should be able to immediately write down this nearly complete form of the Green's function for these simple boundary-value problems.

We determine the remaining constant by integrating Eq. (2.2.6) over an infinitesimal interval that includes x_0. This fixes the magnitude of the discontinuity in the first derivative of the Green's function at $x = x_0$ to be

$$c'(x, s)\big|_{x=x_0^+} - c'(x, s)\big|_{x=x_0} = -\frac{1}{D}.$$

Applying this condition determines A, and we finally obtain

$$c(x, s) = \frac{\sinh\left(\sqrt{\frac{s}{D}}x_<\right)\sinh\left[\sqrt{\frac{s}{D}}(L - x_>)\right]}{\sqrt{sD}\sinh\left(\sqrt{\frac{s}{D}}L\right)}. \qquad (2.2.9)$$

From this Green's function, first-passage properties follow directly. For example, the Laplace transform of the flux at $x = 0$ and at $x = L$ are

$$j_-(s; x_0) \equiv +D\frac{\partial c(x, s)}{\partial x}\bigg|_{x=0} = \frac{\sinh\left[\sqrt{\frac{s}{D}}(L - x_0)\right]}{\sinh\left(\sqrt{\frac{s}{D}}L\right)}, \qquad (2.2.10a)$$

$$j_+(s; x_0) \equiv -D\frac{\partial c(x, s)}{\partial x}\bigg|_{x=L} = \frac{\sinh\left(\sqrt{\frac{s}{D}}x_0\right)}{\sinh\left(\sqrt{\frac{s}{D}}L\right)}. \qquad (2.2.10a)$$

The subsidiary argument x_0 is written to emphasize that the flux depends on the initial position of the particle. Because the initial condition is normalized, the flux is identical to the first-passage probability to the boundaries.

For $s = 0$, these Laplace transforms reduce to the time-integrated first-passage probabilities at 0 and at L. These quantities therefore coincide with the respective splitting probabilities, $\mathcal{E}_-(x_0)$ and $\mathcal{E}_+(x_0)$, as a function of the initial position x_0. Thus

$$\mathcal{E}_-(x_0) = j_-(s = 0; x_0) = 1 - \frac{x_0}{L},$$

$$\mathcal{E}_+(x_0) = j_+(s = 0; x_0) = \frac{x_0}{L}. \qquad (2.2.11)$$

Thus the splitting probability is given by an amazingly simple formula – the probability of reaching one end point is just the fractional distance to the other end point!

To illustrate this result, consider a two-person betting game in which players A and B flip an unbiased coin repeatedly, and A takes \$1 from B if a coin flip is heads and vice versa for tails. The game ends when one player loses all his initial capital. This is also popularly known as the *gambler's ruin* problem. From Eqs. (2.2.11), if the initial capitals of each player are X_A and X_B, then the probability that player A goes broke is $X_B/(X_A + X_B)$ and the probability that player B goes broke is $X_A/(X_A + X_B)$. An important feature of this result is that these ruin probabilities depend on only the ratio of the initial capitals and *not* on the total amount of capital.

The astute reader may notice that this betting game is discrete whereas the splitting probabilities were derived for a continuum-diffusion problem. As we will derive in Section 2.4, this distinction turns out to be immaterial on the finite interval; there is complete parallelism between diffusion and the random walk for this system.

Let us now determine the lifetime of a diffusing particle in the interval, that is, the mean time until one of the boundaries is hit. This lifetime may be obtained from the power-series expansion of the flux in the Laplace domain. By the definition of the Laplace transform,

$$
\begin{aligned}
j_{\pm}(s; x_0) &= \int_0^\infty j_{\pm}(t; x_0) e^{-st}\, dt \\
&= \int_0^\infty j_{\pm}(t; x_0) \left(1 - st + \frac{s^2 t^2}{2!} - \frac{s^3 t^3}{3!} + \cdots \right) dt \\
&= \mathcal{E}_{\pm}(x_0) \left[1 - s\langle t(x_0)\rangle_{\pm} + \frac{s^2}{2!}\langle t(x_0)^2\rangle_{\pm} - \frac{s^3}{3!}\langle t(x_0)^3\rangle_{\pm} + \cdots \right],
\end{aligned}
$$

(2.2.12)

where

$$
\mathcal{E}_{\pm}(x_0) = \int_0^\infty j_{\pm}(t; x_0)\, dt \tag{2.2.13}
$$

is the exit probability to $x = 0$ or $x = L$, respectively, when starting at x_0, and

$$
\langle t^n(x_0)\rangle_{\pm} = \frac{\int_0^\infty t^n\, j_{\pm}(t; x_0)\, dt}{\int_0^\infty j_{\pm}(t; x_0)\, dt} \tag{2.2.14}
$$

are the corresponding moments of the mean first-passage time to these boundaries. Note that these moments are conditioned on the diffusing particle hitting the specified boundary without touching the opposite boundary.

By expanding the first-passage probabilities in Eqs. (2.2.10) in power series in s, we obtain

$$
\begin{aligned}
j_{-}(s; x_0) &= \left(1 - \frac{x_0}{L} \right) \left[1 - s\frac{\left(2Lx_0 - x_0^2\right)}{6D} + \cdots \right], \\
j_{+}(s; x_0) &= \frac{x_0}{L} \left[1 - s\frac{\left(L^2 - x_0^2\right)}{6D} + \cdots \right].
\end{aligned}
$$

(2.2.15)

Comparing the coefficients of like powers of s in Eqs. (2.2.12) and (2.2.15), we find that the conditional mean exit times to the left and right boundaries are, respectively,

$$\langle t(x_0)\rangle_- = \frac{2Lx_0 - x_0^2}{6D} = \frac{L^2}{6D}u_0(2 - u_0),$$

$$\langle t(x_0)\rangle_+ = \frac{L^2 - x_0^2}{6D} = \frac{L^2}{6D}(1 - u_0)^2, \qquad (2.2.16)$$

where $u_0 = x_0/L$. Finally, the unconditional mean exit time (independent of which side is exited) is just the appropriately weighted average of the conditional mean exit times to each boundary:

$$\langle t(x_0)\rangle = \mathcal{E}_-(x_0)\langle t(x_0)\rangle_- + \mathcal{E}_+(x_0)\langle t(x_0)\rangle_+$$

$$= \frac{1}{2D}x_0(L - x_0)$$

$$= \frac{L^2}{2D}u_0(1 - u_0). \qquad (2.2.17)$$

A basic feature of this exit time is that it is of the order of L^2/D, unless that initial position is close to one of the boundaries, in which case the exit time is of the order of L (Fig. 2.2). This behavior has a simple physical explanation. If the particle starts at a small distance ℓ from an absorbing boundary, then from Eqs. (2.2.11) and (2.2.16), the nearer boundary is hit with probability $1 - \ell/L$ in a time of the order of L. On the other hand, the particle hits the opposite boundary with probability ℓ/L in a time of the order

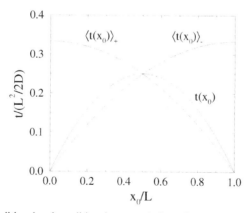

Fig. 2.2. Unconditional and conditional mean exit times from the finite interval $[0, L]$, normalized by $L^2/2D$, as a function of the dimensionless initial position x_0/L. The exit times are of the order of L for a particle that starts a distance of the order of 1 from an absorbing boundary and of the order of L^2 for a particle which starts in the interior of the interval.

of L^2. The average over these two possibilities gives the unconditional exit time $\langle t(\ell) \rangle = \langle t(L - \ell) \rangle \propto L$. Note, in particular, that as $\ell \to 0$, this exit time remains linear in L!

By computing more terms in the series expansion of the fluxes, we can obtain higher moments of the exit times. As an example, the second moment of the unconditional exit time is

$$\langle t(x_0)^2 \rangle = \frac{L^4}{12D^2} u_0 (1 - u_0)(1 + u_0 - u_0^2). \qquad (2.2.18)$$

The behavior of arbitrary-order moments is similar. If the particle starts near the center of the interval, then $\langle t^k \rangle \propto (L^2/D)^k$. On the other hand, if the particle starts near an end point (either u_0 or $1 - u_0$ of the order of ℓ/L), then the nearer boundary is hit with probability $1 - \ell/L$ after a time of the order of L and the more distant boundary is hit with probability ℓ/L after a time of the order of L^2. The average of these two possibilities gives $\langle t^k \rangle \approx (1 - \ell/L) \times L^k + (\ell/L) \times L^{2k} \sim \ell \, L^{2k-1}$.

As a final note, these results for the splitting probabilities and exit times can be extended by *Wald's identity* [Wald (1947)] to more general hopping processes, in which individual steps are drawn from a distribution. As long as the mean step length is small compared with the interval length, then splitting probabilities and exit times are essentially the same as those in diffusion. This result is treated in various books on stochastic processes [see, in particular, Weiss (1994)] and also lies outside our main emphasis on diffusion processes. Nevertheless, the results are worth quoting for completeness. Specifically, consider a discrete-time isotropic hopping process that starts at a point x_0 within the interval $[0, L]$. The individual step lengths are drawn from a distribution with a mean-square step length equal to $\sigma^2 \ll L^2$. This last inequality implies that, to an excellent approximation, one can view the particle as landing exactly at the boundary when it first exits the interval. With this assumption, Wald's identity then gives the fundamental results that are parallel to those of pure diffusion:

$$\mathcal{E}_-(x_0) = 1 - \frac{x_0}{L},$$
$$\mathcal{E}_+(x_0) = \frac{x_0}{L},$$
$$\langle t(x_0) \rangle = x_0(L - x_0)/\sigma^2. \qquad (2.2.19)$$

Reflection Mode. It is also very instructive to consider the mixed boundary conditions of reflection at $x = L$ and absorption at $x = 0$. If the particle starts close to $x = 0$, the effect of the distant reflecting boundary will not be apparent until a time of the order of L^2/D. The first-passage probability should therefore vary as $t^{-3/2}$ for short times, as in the infinite system; this fundamental result is a central theme of Chap. 3. Subsequently, however, the first-passage probability should decay exponentially with time. This crossover provides a nice illustration of how to extract asymptotic behavior from the Laplace transform without performing an explicit inversion.

When the techniques outlined in Subsection 2.2.2 are applied, the Green's function for this mixed boundary condition system, when the particle starts at x_0, is

$$c(x, s) = \frac{\sinh\left(\sqrt{\frac{s}{D}}x_<\right)\cosh\left[\sqrt{\frac{s}{D}}(L - x_>)\right]}{\sqrt{Ds}\cosh\left(\sqrt{\frac{s}{D}}L\right)}. \qquad (2.2.20)$$

From this, the Laplace transform of the flux at $x = 0$ is

$$j_-(s; x_0) = Dc'(x, s)|_{x=0} = \frac{\cosh\sqrt{\frac{s}{D}}(x_0 - L)}{\cosh\sqrt{\frac{s}{D}}L}. \qquad (2.2.21)$$

To appreciate the meaning of Eq. (2.2.21), consider first the limit $L \to \infty$, for which $j_-(s; x_0) \to e^{-x_0\sqrt{s/D}}$. The corresponding inverse Laplace transform is

$$j_-(t; x_0) = \frac{x_0}{\sqrt{4\pi Dt^3}}e^{-x_0^2/4Dt}, \qquad (2.2.22)$$

which is the well-known first-passage probability to the origin for a diffusing particle that begins at x_0 in a semi-infinite interval. The situation for finite L is more delicate, with different asymptotic behaviors emerging for $t \ll L^2/D$ and $t \gg L^2/D$. To understand this behavior, we expand $j_-(s; x_0)$ for small s, but for large but finite L. From Eq. (2.2.21),

$$j_-(s; x_0) = \cosh\sqrt{\frac{s}{D}}x_0 - \sinh\sqrt{\frac{s}{D}}x_0\tanh\sqrt{\frac{s}{D}}L,$$

$$\approx \begin{cases} 1 - \sqrt{\frac{sx_0}{D}} + \frac{sx_0^2}{2D} + \dots, & \text{for } \frac{sL^2}{D} \gg 1 \text{ and } \frac{sx_0^2}{D} \ll 1 \\ 1 + \frac{s}{D}\left(\frac{x_0^2}{2} - x_0L\right) + \dots, & \text{for } \frac{sL^2}{D} \ll 1 \end{cases}.$$

$$(2.2.23)$$

The former limit corresponds to $x_0^2/D \ll t \ll L^2/D$. This ensures both that the particle has had sufficient time to diffuse to the absorbing boundary and that the reflecting boundary plays no role. Thus the first-passage behavior should coincide with that of the infinite system in the long-time limit. To check this, note that the small-s expansion of the flux for $sL^2/D \gg 1$ is $j(s; x_0)_- \approx 1 - \sqrt{sx_0/D}$. From the discussion of Section 1.4, the corresponding inverse

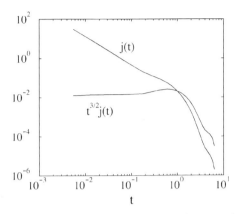

Fig. 2.3. First-passage probability to $x = 0$, $j_-(t; x_0)$, for a particle that starts at $x_0 = 0.05$ in a unit length interval with absorption at $x = 0$ and reflection at $x = 1$. The diffusion coefficient $D = 1$. Shown are $j(t; x_0)$ and $t^{3/2} j(t; x_0)$. The latter shows the enhancement that is due to the reflected contribution before the asymptotic decay sets in.

Laplace transform is $j_-(t; x_0) \propto x_0/t^{3/2}$, which coincides with Eq. (2.2.22) in the long-time limit.

On the other hand, for $t \gg L^2/D$, the reflecting boundary strongly influences the asymptotics. Here, the limiting behavior of $j_-(s; x_0)$ is simply $1 - s x_0 (L - x_0)/D$. This corresponds to an exponential decay of $j_-(t; x_0)$ in time, with a decay time $x_0(L - x_0)/D$. Finally, when $t \approx L^2/D$, an enhancement in the first-passage probability should arise that stems from the contribution of trajectories that arrive after reflection from the far boundary. These features are illustrated in Fig. 2.3, where $j(t; x_0)$ is computed from a numerical inversion of Eq. (2.2.21) by the Stehfest algorithm [Stehfest (1970)].

This crossover between the intermediate-time power law and the long-time exponential decay can be formulated more generally by an approach similar to that given in Chap. 1 for determining the asymptotics of generating functions. Consider the situation in which the first-passage probability has the generic form

$$j(t) = t^{-\alpha} e^{-t/\tau},$$

with $\tau \gg 1$, and where $j(t)$ is vanishingly small for $t \ll 1$. The power law represents the asymptotic behavior of an infinite system, and the exponential factor represents a finite-size cutoff. The short-time cutoff arises because a particle has not yet had time to diffuse to the boundary. The corresponding

Laplace transform is

$$j(s) \approx \int_{\epsilon}^{\infty} t^{-\alpha} e^{-t(s+1/\tau)} \, dt,$$

where the lower limit ϵ accounts crudely for the short-time cutoff in $j(t)$. This cutoff allows us to use the long-time form for $j(t)$ in the integral with negligible errors in determining the small-s behavior of $j(s)$.

To compute the asymptotic behavior of $j(s)$ in this example, we rewrite the above integral in the dimensionless form,

$$j(s) = (s + 1/\tau)^{\alpha-1} \int_{\epsilon(s+1/\tau)}^{\infty} u^{-\alpha} e^{-u} \, du,$$

which has two limiting behaviors as $s \to 0$. If $s \gg 1/\tau$, then $j(s) \sim s^{\alpha-1}$, whereas if $s \ll 1/\tau$, then $j(s)$ varies linearly with s. This corresponds, in real time, to power-law behavior $j(t) \sim t^{-\alpha}$ in the intermediate-time range $\epsilon \ll t \ll \tau$ and exponential decay of $j(t)$ for $t \gg \tau$. Thus our rough approximations capture the crossover between power-law and exponential behavior in a relatively simple manner.

The mixed boundary system also provides a nice illustration of the relation between initial conditions for the concentration and for the flux. Thus far, the initial condition has invariably been a single particle located at a point x_0 in the system interior. This necessitates the use of joining conditions at x_0 to solve for the Green's function. However, for x_0 close to a reflecting boundary, it is simpler to consider the system as initially empty and inject a delta function pulse of flux at the reflecting end. This represents a considerable simplification as we need to solve a homogeneous, rather than an inhomogeneous, differential equation.

We now show how this approach reproduces the corresponding Green's function results. For an initially empty system with a unit current pulse injected at $x = L$, the initial condition is $c(x, t = 0) = 0$ and the boundary conditions are $c(0, t) = 0$ and $j(L, t) = \delta(t)$. In the Laplace domain, we therefore solve $sc = Dc''$, subject to $c(0, s) = 0$ and $j(L, s) = 1$. The general solution is again $c(x, s) = A \exp(x\sqrt{s/D}) + B \exp(-x\sqrt{s/D})$. The boundary condition at $x = 0$ implies that this linear combination should be antisymmetric, and the condition $j(x = L, s) = 1$ fixes the overall amplitude. We thereby find

$$c(x, s) = \frac{\sinh \sqrt{\frac{s}{D}} x}{\sqrt{sD} \cosh \sqrt{\frac{s}{D}} L}. \tag{2.2.24}$$

This coincides with the Green's function of Eq. (2.2.20) with $x_>$ set to L, but the present approach requires much less computation.

The flux at $x = 0$ is just

$$j_-(s) = \operatorname{sech}\sqrt{\frac{s}{D}}L. \qquad (2.2.25)$$

From this fundamental result, all first-passage properties follow easily. Because $j_-(s)$ is a function of sL^2/D, we immediately determine that the flux in the time domain is a function of $Dt/L^2 \equiv t/\tau$. Thus the nth moment of the first-passage time scales as τ^n.

We may also invert the Laplace transform explicitly. Because $\cosh x = \cos ix$, $j(s)$ has simple poles when $i\sqrt{(s/D)}L = (n + \frac{1}{2})\pi$, or $s \equiv s_n = -(n + \frac{1}{2})^2\pi^2 D/L^2$, and the corresponding residue is $(-1)^{n+1}(2n + 1)\pi D/L^2$. To compute the inverse Laplace transform, we close the contour with an infinite semicircle in the left-half-plane to reduce the integral to the sum of the residues at these singularities. This gives

$$
\begin{aligned}
j_-(t) &= \frac{1}{2\pi i}\int_{s_0-i\infty}^{s_0+i\infty} j_-(s)\,e^{st}\,ds \\
&= \sum_{n=0}^{\infty} \text{residues of } \sec\left(iL\sqrt{\frac{s}{D}}\right) \times e^{st}\Big|_{s=-(n+\frac{1}{2})^2\pi^2 D/L^2} \\
&= \sum_{n=0}^{\infty}(-1)^{n+1}\frac{L^2}{(2n+1)\pi D}e^{-(n+\frac{1}{2})^2\pi^2 Dt/L^2}.
\end{aligned} \qquad (2.2.26)
$$

In the long-time limit, the asymptotics of the first-passage probability are determined by the singularity in $j_-(s)$ which is closest to the origin. This leads to $j_-(t)$ having an exponentially decaying controlling factor $j_-(t) \to e^{-\pi^2 Dt/4L^2}$.

2.2.2.2. Biased Diffusion

We now consider the role of bias on first-passage characteristics. As a rule, the bias dominates first-passage properties for sufficiently long systems or for long times. This crossover from isotropic to biased behavior is naturally quantified by the Péclet number. Although qualitative features of this crossover can be obtained by intuitive physical arguments, the complete solution reveals many subtle and interesting properties.

Absorption Mode. Parallel to our discussion of isotropic diffusion, consider first a biased diffusing particle that starts at x_0 and is absorbed when it hits either end of the interval. To determine first-passage properties for this system, we solve convection–diffusion equation (2.2.5), subject to the initial condition $c(x, t = 0) = \delta(x - x_0)$ and the boundary conditions $c(0, t) = c(L, t) = 0$. We compute the Green's function by following the same steps as in isotropic diffusion. Introducing the Laplace transform reduces the convection–diffusion equation to

$$sc(x, s) + vc'(x, s) = Dc''(x, s). \qquad (2.2.27)$$

The elemental solutions to this equation are $c(x, s) = A_+ e^{\alpha_+ x} + A_- e^{\alpha_- x}$, with

$$\alpha_\pm = (v \pm \sqrt{v^2 + 4Ds})/2D \equiv v/2D \pm w$$

and the constants $A_\pm$ determined from the boundary conditions. Following the same calculational steps as those given for isotropic diffusion, we find the Green's function

$$c(x, s) = \frac{e^{v(x-x_0)/2D}}{Dw \sinh wL} \sinh(wx_<) \sinh[w(L - x_>)]. \qquad (2.2.28)$$

Given the complexity of this expression, we defer discussing its first-passage properties until we derive the corresponding results from the much simpler time-integrated formulation of Section 2.3.

Transmission Mode. This mixed system is an example of a *hydrodynamic-dispersion* process, in which passive Brownian particles are driven by a steady flow and develop a distribution of exit times under the combined influence of convection and diffusion. To determine this distribution of exit times, we solve the convection–diffusion equation subject to the initial condition $c(x, t = 0) = 0$ and the boundary conditions $j(x = 0, t) = \delta(t)$ and $c(x = L, t) = 0$, corresponding to a unit input of flux into an initially empty system. We then determine the outlet flux $j = vc - D\frac{\partial c}{\partial x}|_{x=L}$. The sense of the boundary conditions is opposite to that used in our discussion of pure diffusion, but is natural in the present context. Note also that there are ostensibly two contributions to the flux – one from convection (vc) and another from diffusion ($-D\frac{dc}{dx}$). However, the absorbing boundary condition means that the convective component of the flux vanishes at $x = L$.

The solution for $c(x, s)$ has the general form as in absorption mode. The absorbing boundary condition at $x = L$ gives

$$c(x, s) = A\left[e^{\alpha_+(x-L)} - e^{\alpha_-(x-L)}\right],$$

and the reflecting boundary condition at $x = 0$ fixes the constant to be

$$A = \frac{1}{D(\alpha_- e^{-\alpha_+ L} - \alpha_+ e^{-\alpha_- L})}.$$

The outlet flux, or equivalently, the Laplace transform for the first-passage probability, is

$$j(x = L, s) = -D\frac{\partial c(x, s)}{\partial x}\Big|_{x=L}$$

$$= \frac{\alpha_+ - \alpha_-}{D(\alpha_+ e^{-\alpha_- L} - \alpha_- e^{-\alpha_+ L})}, \quad (2.2.29)$$

where we have used the simplification $v - D\alpha_\pm = D\alpha_\mp$.

We can recast this expression into a convenient dimensionless form by writing $\alpha_\pm L = (v \pm \sqrt{v^2 + 4Ds})L/2D \equiv Pe \pm P_s$, where $Pe = vL/2D$ is again the Péclet number. With these definitions, the outlet flux is

$$j(L, s) = \frac{P_s e^{Pe}}{Pe \sinh P_s + P_s \cosh P_s}, \quad (2.2.30)$$

which reduces to Eq. (2.2.25) in the limit of zero Péclet number.

The Taylor series expansion of Eq. (2.2.30) in powers of s then gives the moments of the mean exit time. The first two moments are

$$\langle t \rangle = \frac{L^2}{D}\left[\frac{1}{2Pe} - \frac{1}{4Pe^2}(1 - e^{-2Pe})\right],$$

$$\langle t^2 \rangle = \left(\frac{L^2}{D}\right)^2\left\{\frac{1}{4Pe^2} - \frac{1}{8Pe^4}[2 - (6Pe + 1)e^{-2Pe} - e^{-4Pe}]\right\}. \quad (2.2.31)$$

The limiting forms of these moments are instructive. The first moment is

$$\langle t \rangle \to \begin{cases} \dfrac{L^2}{2D}, & Pe \to 0 \\[2ex] \dfrac{L}{v} - \dfrac{D}{v^2}, & Pe \to +\infty \\[2ex] \dfrac{D}{v^2}e^{|v|L/D}, & Pe \to -\infty \end{cases} \quad (2.2.32)$$

In fact, the arbitrary-order reduced moments $\langle t^k \rangle^{1/k}$ also behave as $\langle t \rangle$ for positive integer k. This arises because the first-passage probability is a scaling function for arbitrary bias.

For no bias, the mean exit time in an interval of length L with the particle starting at the reflecting edge is the same as that for a particle that starts in the middle of an absorbing interval of length $2L$ [see Eq. (2.2.17)]. For large positive bias, the mean exit time reduces to the obvious result – the length divided by the bias velocity. Conversely, for large negative bias, the mean exit time scales as $e^{|v|L/D}$. This arises because the density decays exponentially as $e^{-|v|x/D}$ along the interval.

This last result is important because it corresponds to a simple barrier-crossing problem. A particle trying to exit "into the wind" is equivalent to moving up a potential gradient. From equilibrium statistical mechanics, the probability that the particle can penetrate a distance L scales as $e^{-E/kT} = e^{-|v|L/D}$, where D is playing the role of the thermal energy. Thus we can interpret the last of expressions (2.2.32) as the inverse of the equilibrium probability of penetrating a distance L up a potential barrier. More general barrier-crossing problems are discussed in Chap. 4.

For $Pe \to -\infty$, we can also obtain the asymptotic time dependence of the first-passage probability. For this purpose, we need to consider only the nearest singularity to the origin along the negative s axis in the expression of $j(L, s)$ in Eq. (2.2.30). The condition for the denominator of $j(L, s)$ to vanish can be written as

$$\tanh \sqrt{Q^2 + s} = \sqrt{1 + s/Q^2},$$

which gives the criterion $s = -4Pe^2 e^{-2|Pe|} \equiv 1/\tau$. Consequently, in the long-time limit, the time dependence of the outlet flux is simply $j(t) \sim e^{-t/\tau}$.

2.3. Time-Integrated Formulation

Although the above direct approaches are conceptually straightforward and comprehensive, they lead to unpleasant and unnecessary computations if we are interested in only time-integrated first-passage characteristics, such as eventual hitting probabilities or first-passage times. As discussed in Chap. 1, it is much simpler to first integrate the equation of motion over time to obtain the Laplace equation as the governing equation for the time-integrated concentration. Then exit probabilities, mean exit times, and related properties can be found from suitable moments of this integrated concentration. We now illustrate this general approach by revisiting our previous examples.

2.3.1. Splitting Probabilities and Unconditional Exit Time: Absorption Mode

We start by determining the exit probabilities to $x = 0$ or at $x = L$, as well as the corresponding exit times, as functions of starting position, by exploiting the electrostatic formulation of Section 1.6.

2.3.1.1. Isotropic Diffusion

To apply the electrostatic formalism, we first compute the electrostatic potential in the interval with grounded end points due to a point charge of magnitude

$1/2D$ at $x = x_0$. Thus we solve [see also Eq. (1.6.5)]

$$DC_0''(x) = -\delta(x - x_0), \tag{2.3.1}$$

subject to the boundary conditions $C_0(0) = C_0(L) = 0$. Here $C_0(x)$ is the time-integrated concentration $C_0(x) = \int_0^\infty c(x, t)\,dt$. The Green's function is

$$C_0(x) = \frac{1}{D} x_< \left(1 - \frac{x_>}{L}\right). \tag{2.3.2}$$

From this, the electric fields at each end are

$$\mathcal{E}_-(x_0) = +D \frac{\partial C_0}{\partial x}\bigg|_{x=0} = 1 - \frac{x_0}{L},$$
$$\mathcal{E}_+(x_0) = -D \frac{\partial C_0}{\partial x}\bigg|_{x=L} = \frac{x_0}{L}. \tag{2.3.3}$$

According to the electrostatic formalism, these are also the exit probabilities to each end, thus reproducing Eqs. (2.2.11).

Next we determine the unrestricted mean exit time by integrating the electrostatic potential over the interval [Eq. (1.6.10)]. Thus

$$
\begin{aligned}
t(x_0) &= \int_0^L C_0(x)\,dx \\
&= \frac{1}{D} \int_0^{x_0} x \left(1 - \frac{x_0}{L}\right) dx + \int_{x_0}^L x_0 \left(1 - \frac{x}{L}\right) dx \\
&= \frac{1}{2D} x_0(L - x_0) = \frac{L^2}{2D} u_0(1 - u_0),
\end{aligned} \tag{2.3.4}
$$

with $u_0 = x_0/L$. This coincides with Eq. (2.2.17).

We can compute higher moments of the exit time similarly. Given C_0, we use Eqs. (1.6.12) to compute C_1. By straightforward integration of the differential equation, this function is

$$C_1(x) = \begin{cases} \dfrac{L^3}{6}[uu_0(1 - u_0)(2 - u_0) - u^3(1 - u_0)] & u < u_0 \\[2mm] \dfrac{L^3}{6}\left[u_0(1 - u_0^2)(1 - u_0) - u_0(1 - u)^3\right] & u > u_0 \end{cases}, \tag{2.3.5}$$

with $u = x/L$. Then, after some elementary calculation, $\langle t(x_0)^2 \rangle = \int_0^L C_1(x)\,dx$ reproduces Eq. (2.2.18)

2.3.1.2. Biased diffusion

We merely sketch calculational steps for first-passage properties because of the conceptual similarity with isotropic diffusion. Starting from the

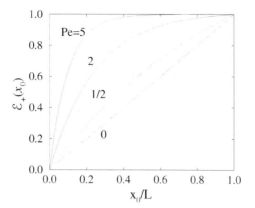

Fig. 2.4. Dependence on $u_0 = x_0/L$ for the exit probability to the right boundary of a finite absorbing interval for various Péclet numbers Pe.

convection–diffusion equation and integrating over all time, $\mathcal{C}_0(x)$ satisfies

$$D\mathcal{C}_0'' - v\mathcal{C}_0' = -\delta(x - x_0). \tag{2.3.6}$$

Note again the opposite signs of the diffusive and the convective terms compared with the convection–diffusion equation itself. By standard methods, the Green's function is

$$\mathcal{C}_0(x) = -\frac{1}{v} \frac{\left(1 - e^{v x_< /D}\right)\left[1 - e^{v(x_> - L)/D}\right]}{e^{v x_0/D}(1 - e^{-vL/D})}. \tag{2.3.7}$$

Then the exit probabilities at $x = 0$ and $x = L$ are given by

$$\mathcal{E}_-(x_0) = +D\mathcal{C}_0'(x)|_{x=0} = \frac{e^{-v x_0/D} - e^{-vL/D}}{1 - e^{-vL/D}},$$

$$\mathcal{E}_+(x_0) = -D\mathcal{C}_0'(x)|_{x=L} = \frac{1 - e^{-v x_0/D}}{1 - e^{-vL/D}} = 1 - \mathcal{E}_-(x_0). \tag{2.3.8}$$

In terms of the Péclet number and dimensionless initial position $u_0 = x_0/L$, these are

$$\mathcal{E}_-(u_0) = \frac{e^{-2u_0 Pe} - e^{-2Pe}}{1 - e^{-2Pe}}, \quad \mathcal{E}_+(u_0) = \frac{1 - e^{-2u_0 Pe}}{1 - e^{-2Pe}}. \tag{2.3.9}$$

A plot of $\mathcal{E}_+(u_0)$ as a function of the initial position is shown in Fig. 2.4. For small Pe, $\mathcal{E}_+ \approx u_0$, which is the result for pure diffusion. Conversely, for large Pe, $\mathcal{E}_+$ rapidly increases to one over a small length scale of the order of D/v^2 and is nearly constant thereafter. This reflects the obvious fact that in the presence of a strong bias, exit at $x = 0$ will occur only if the particle is initially within a skin depth D/v^2 of $x = 0$.

To appreciate the practical meaning of this result, consider again our two-person gambling game in which each person starts with the same capital, but in which A wins \$1 with probability 51% in each trial. The ultimate outcome of the game now depends both on the relative initial condition *and* on the amount of capital. For this game, the Péclet number is

$$Pe = \frac{vL}{2D}$$
$$= \frac{(p-q)L}{2} \frac{\delta x/\delta t}{\delta x^2/\delta t}$$
$$= (p-q)L.$$

Here $\delta x = 1$ and $\delta t = 1$ are the increments in capital and in time, respectively, after each trial, and L is the sum of the capitals of both players. From Eqs. (2.3.9) the probability that B is eventually ruined is 0.622 for $Pe = 1$, corresponding to each player starting with \$25; 0.731 for $Pe = 2$, each player starts with \$50; and 0.924 for $Pe = 5$, each player starts with \$125. Therefore a slightly biased game leads to an overwhelming global bias if the scale of the game is sufficiently large, that is, $Pe \gg 1$. The moral of this example is that you should not play in a casino (where the odds are slightly unfavorable to you and their capital is much greater than yours) for very long.

Continuing with the electrostatic formulation, the unrestricted mean exit time when starting at x_0 is just the integral of $C_0(x)$ over the interval. After some algebra, we find

$$t(x_0) = \frac{L\left(1 - e^{-vx_0/D}\right) - x_0\left(1 - e^{-vL/D}\right)}{v\left(1 - e^{-vL/D}\right)},$$

$$= \frac{L^2}{2D} \frac{1}{Pe} \frac{(1 - e^{-2u_0 Pe}) - u_0(1 - e^{-2Pe})}{(1 - e^{-2Pe})}. \qquad (2.3.10)$$

This reduces to the "obvious" limits $t(x_0) \to (L - x_0)/v$ as $v \to \infty$ and to $t(x_0) \to x_0(L - x_0)/2D$ as $v \to 0$ [Eq. (2.2.17)].

To help understand these results pictorially, consider the Péclet number dependences of x_{equal}, the initial location that leads to equal exit probabilities through either end of the interval, and x_{max}, the initial location that maximizes the exit time (Fig. 2.5). Roughly speaking, at these points the "effective" distance to each boundary is the same. Setting $\mathcal{E}_{\pm}(x_0) = 1/2$ in Eqs. (2.3.9), we find

$$\frac{x_{equal}}{L} = -\frac{1}{2Pe} \ln \frac{1}{2}(1 + e^{-2Pe})$$

$$\to \begin{cases} \frac{1}{2}(1 - Pe), & Pe \to 0 \\ \frac{1}{2Pe} \ln 2, & Pe \to \infty \end{cases}. \qquad (2.3.11)$$

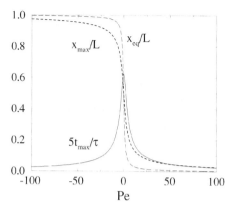

Fig. 2.5. Péclet number dependence of the starting position that gives equal exit probabilities to either end, x_{eq}/L (long-dashed curve), and that maximizes the exit time, x_{max}/L (short-dashed curve). Also shown is the normalized maximal exit time t_{max}/τ (multiplied by 5 for visibility), where $\tau = L^2/2D$.

Clearly, equal splitting occurs at the midpoint for $v = 0$ and quickly approaches the left boundary of the interval for large bias.

Similarly, maximizing $t(x_0)$ with respect to x_0, the maximal exit time occurs when the particle is initially located at

$$\frac{x_{max}}{L} = -\frac{1}{2Pe} \ln \left[\frac{1}{2Pe}(1 - e^{-2Pe}) \right]$$

$$\rightarrow \begin{cases} \frac{1}{2}\left(1 - \frac{Pe}{12}\right), & Pe \to 0 \\ \frac{1}{2Pe} \ln 2Pe, & Pe \to \infty \end{cases} . \qquad (2.3.12)$$

The corresponding maximal first-passage time is

$$t(x_{max}) = \frac{L^2}{D} \frac{1}{2Pe} \left(\frac{1}{2Pe} \left\{ \ln \left[\frac{1}{2Pe}(1 - e^{-2zPe}) \right] - 1 \right\} + \frac{1}{1 - e^{-2Pe}} \right)$$

$$\rightarrow \begin{cases} \frac{L^2}{2D}\left(\frac{1}{4} - \frac{Pe^2}{72} + \cdots\right), & Pe \to 0 \\ \frac{L}{v}\left(1 - \frac{\ln 2Pe}{2Pe} + \cdots\right), & Pe \to \infty \end{cases} . \qquad (2.3.13)$$

Note that the maximal time is largest at $Pe = 0$ and quickly decreases as the bias increases, even though the optimal starting location moves to $x = 0$ to compensate for the effect of the bias.

2.3.1.3. The Freely Accelerated Particle

A more general version of escape from an interval can be posed for a Brownian particle with inertia (i.e., nonzero mass) that is also subject to an external force field. This general question appears to have been originally considered by Wang and Uhlenbeck (1945). It is a striking fact that this natural generalization leads to an extraordinarily difficult mathematical physics problem. Nevertheless, some of the most basic results are simple and nicely complement the discussion of escape from an interval for a Brownian particle without inertia.

The average position of a forced, massive Brownian particle is most conveniently described by a Langevin equation of the form [see, e.g., Gardiner (1985) and van Kampen (1997) for a general discussion of this equation]

$$\frac{d^2x}{dt^2} + \beta \frac{dx}{dt} + F(x) = \xi(t), \qquad (2.3.14)$$

where β is the damping constant, $F(x)$ is the external force, and $\xi(t)$ is the random component of the force that arises from thermal fluctuations. As is standard, we take these fluctuations to have a Gaussian distribution with zero mean, $\langle \xi(t) \rangle = 0$, and with their temporal correlations obeying $\langle \xi(t)\xi(t') \rangle = D\delta(t - t')$. Note that for the case of a random walk or freely diffusing particle, that is, no force and no inertia, the Langevin equation reduces to $\dot{x} = \xi(t)$.

Here we consider the simple but still very intriguing situation of a *freely accelerated particle*, in which there is no external force and no damping. Thus the equation of motion is simply

$$\frac{d^2x}{dt^2} = \xi(t). \qquad (2.3.15)$$

If such a particle is placed in an interval $[0, L]$ with absorbing boundaries, we are now interested in the following basic first-passage questions:

- What is the mean time to exit the interval as a function of the starting position of the particle?
- What is the probability to exit from either end of the interval, again as a function of the initial particle position?

Although exact answers to these questions have been obtained [Franklin & Rodemich (1968), Marshall & Watson (1985, 1987), Hagan, Doering, & Levermore (1989a, 1989b), and Masoliver & Porrà (1995, 1996)] the computations represent a *tour de force*. However, the main results are intriguing and relatively easy to appreciate.

By following standard methods [Gardiner (1985) and van Kampen (1997)], we find that the Langevin equation for the freely accelerated particle corresponds to the following Fokker–Planck equation for $P(x, v, t)$, the probability that a particle is at position x with velocity v at time t:

$$\frac{\partial P}{\partial t} + v\frac{\partial P}{\partial x} = D\frac{\partial^2 P}{\partial v^2}. \tag{2.3.16}$$

The operator on the left-hand side represents the usual time derivative with convection at velocity v, and the right-hand side represents diffusion in velocity space [from $\dot{v} = \xi$ in Eq. (2.3.15)]. Corresponding to this process, the mean exit time obeys the Poisson-like equation [compare with Eq. (1.6.23) for the signs of the terms in Eq. (2.3.16)]

$$D\frac{\partial^2 T}{\partial v^2} + v\frac{\partial T}{\partial x} = -1. \tag{2.3.17}$$

Here $T \equiv \langle t(x, v)\rangle$ is the mean exit time through either end of the interval for a particle with initial position and velocity x and v, respectively. The subtle aspect of this equation is that it is subject to the "incomplete" boundary conditions

$$\langle t(0, v)\rangle = 0 \quad \text{for } v \leq 0, \quad \langle t(L, v)\rangle = 0 \quad \text{for } v \geq 0, \tag{2.3.18}$$

but with $\langle t(0, v > 0)\rangle$ and $\langle t(L, v < 0)\rangle$ undetermined. These latter conditions just reflect that a particle will survive if it is at the boundary but moving toward the center of the interval. This incompleteness is part of the reason why this problem and general problems of this class are so difficult to solve. However, for the special case in which the initial particle velocity is zero, the final result is compact and easy to visualize (Fig. 2.6). The mean exit time is [Franklin & Rodemich (1968), and Masoliver & Porrà (1995, 1996)]

$$\langle t(x, 0)\rangle = \frac{(16/3)^{1/6}}{\Gamma(7/3)}\left(\frac{L^2}{D}\right)^{1/3}\left(\frac{x}{L}\right)^{1/6}\left(1 - \frac{x}{L}\right)^{1/6}$$
$$\times \left[{}_2F_1\left(1, -\frac{1}{3}, \frac{7}{6}, \frac{x}{L}\right) + {}_2F_1\left(1, -\frac{1}{3}, \frac{7}{6}, 1 - \frac{x}{L}\right)\right], \tag{2.3.19}$$

where ${}_2F_1$ is the Gauss hypergeometric function [Abramowitz & Stegun (1972)]. The expression for the exit time for general velocity is given in Franklin and Rodemich (1968) and Masoliver and Porrà (1995, 1996).

There are two salient features of this result that can be understood on simple grounds. First, from a dimensional analysis of the Langevin equation

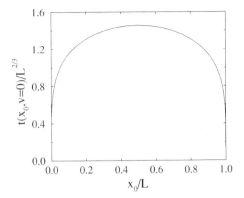

Fig. 2.6. Dependence of the mean exit time on starting position for a freely accelerated particle with zero initial velocity.

$\ddot{x} = \xi$, $x(t)$ must scale as $t^{3/2}$. This follows because the dimension of the noise is proportional $t^{-1/2}$, as required by the condition $\langle \xi(t)\xi(t') \rangle \propto \delta(t - t')$. Because of the scaling $x \propto t^{3/2}$, the exit time should vary as $L^{2/3}$, as quoted in Eq. (2.3.19). Second, note that the exit time is strongly singular near the boundaries. Roughly speaking, this arises because a particle in the interval quickly develops a large velocity (since x scales as $t^{3/2}$). For a very rapidly moving particle, the exit time should then depend weakly on its starting position.

This last feature may be better understood by consideration of the exit probability. This exit probability obeys the Laplace-like equation [compare with Eq. (1.6.18)]

$$v\frac{\partial \mathcal{E}_+}{\partial x} + \frac{\partial^2 \mathcal{E}_+}{\partial v^2} = 0, \qquad (2.3.20)$$

where $\mathcal{E}_+(x, v)$ now denotes the probability for the particle to exit at $x = L$ when the particle starts at x with initial velocity v. This equation is subject to the boundary condition $\mathcal{E}_+(L, v) = 1$. The corresponding exit probability is [Bicout & Burkhardt (2000)],

$$\mathcal{E}(x, 0) = \frac{6\Gamma(1/3)}{[\Gamma(1/6)]^2} \left(\frac{x_0}{L}\right)^{1/6} {}_2F_1\left(\frac{1}{6}, \frac{5}{6}, \frac{7}{6}, \frac{x_0}{L}\right). \qquad (2.3.21)$$

As shown in Fig. 2.7, this exit probability is roughly independent of the particle position, except for narrow boundary layers near $x = 0$ and $x = L$. This is again a manifestation of the rapid particle motion in the interval. Unless a particle is absorbed almost immediately, it is likely to develop a large velocity so that its exit probability will be relatively insensitive to the starting location.

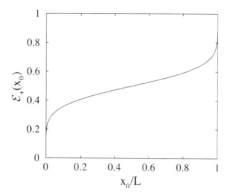

Fig. 2.7. Dependence of the eventual hitting probability at $x = L$ on starting position for a freely accelerated particle with zero initial velocity.

2.3.2. Conditional Mean Exit Times: Absorption Mode

If we are interested in *only* the mean time until the particle reaches a boundary, the Laplacian formalism of Chap. 1 is the most appropriate calculational approach. There are again two basic situations to consider: (a) the unconditional mean first-passage time to *either* boundary, and (b) the conditional mean first-passage time to a *specific* boundary, in which visits to the other boundary not allowed.

As a preliminary, we reconsider the unconditional exit time $t(x)$ for a particle that starts at x with bias velocity v and diffusion coefficient D. For notational simplicity, we drop the subscript 0 of the initial particle location. As shown in Subsection 1.6.3.2, $t(x)$ satisfies

$$Dt''(x) + vt'(x) = -1. \qquad (2.3.22)$$

The solution to this equation, subject to the boundary conditions $t(0) = t(L) = 0$, is just Eq. (2.3.10).

The conditional mean exit times are more subtle. To determine $\mathcal{E}_{\pm}(x)$ and $t_{\pm}(x)$, we need to solve the basic Laplace and Poisson equations for the exit probabilities and conditional exit times, Eqs. (1.6.18) and (1.6.29), respectively,

$$D\mathcal{E}_{\pm}(x)'' + v\mathcal{E}_{\pm}(x)' = 0,$$
$$D\left[\mathcal{E}_{\pm}(x)t_{\pm}(x)\right]'' + v\left[\mathcal{E}_{\pm}(x)t_{\pm}(x)\right]' = -\mathcal{E}_{\pm}(x), \qquad (2.3.23)$$

subject to the boundary conditions $\mathcal{E}_{+}(0) = 0$ and $\mathcal{E}_{+}(L) = 1$ and vice versa for $\mathcal{E}_{-}(x)$, as well as $\mathcal{E}_{\pm}(x)t_{\pm}(x) = 0$ for both $x = 0$ and $x = L$. When the scaled coordinate $u = x/L$ and the Péclet number $Pe = vL/2D$ are

introduced and when all times are measured in units of the diffusion time
$\tau_D = L^2/D$, Eqs. (2.3.23) can be written in the dimensionless form,

$$\mathcal{E}_\pm(u)'' + 2Pe\mathcal{E}_\pm(u)' = 0,$$

$$[\mathcal{E}_\pm(u)t_\pm(u)]'' + 2Pe\,[\mathcal{E}_\pm(u)t_\pm(u)]' = -\mathcal{E}_\pm(u),$$

where the prime now denotes differentiation with respect to scaled coordinate u.

The solution to the first of these equations was given in Eqs. (2.3.9). We
now outline the steps to compute $t_+(u)$; the steps for $t_-(u)$ are similar. First
we define

$$g_+(z) = 4Pe^2(1 - e^{-2Pe})[\mathcal{E}_+(u)t_+(u)],$$

with $z = 2uPe$ to simplify the second of Eqs. (2.3.23) to $g_+'' + g_+' = -(1 - e^{-z})$, where the prime now denotes differentiation with respect to z.
Then $h_+ = g_+ + z$ satisfies $(h_+'e^z)' = 1$, and this may be solved by elementary methods. Finally, imposing the boundary conditions $g_+(0) = g_-(L) = 0$
gives $g_+(z)$, from which the result for $t_+(x)$ is (see Fig. 2.8)

$$t_+(x) = \frac{L}{v}\left(\frac{1 + e^{-vL/D}}{1 - e^{-vL/D}}\right) - \frac{x}{v}\left(\frac{1 + e^{-vx/D}}{1 - e^{-vx/D}}\right). \qquad (2.3.24)$$

Similarly,

$$t_-(x) = \frac{x}{v}\left[\frac{1 + e^{v(x-L)/D}}{1 - e^{v(x-L)/D}}\right] - \frac{2L/v}{1 - e^{v(x-L)/D}}\left(\frac{e^{vx/D} - 1}{e^{vL/D} - 1}\right). \qquad (2.3.25)$$

In the limit $v \to 0$, these conditional exit times reduce to the forms previously derived in Eqs. (2.2.16):

$$t_+(x) = \frac{L^2 - x^2}{6D}, \quad t_-(x) = \frac{xL}{6D}(2 - x/L). \qquad (2.3.26)$$

Note also that $t_-(x) = t_+(L - x)$. For $v \to \infty$, the time to exit "downstream"
is $t_+(x) \to (L - x)/v$, as expected naively. On the other hand, the mean time
to exit "upstream" is just the distance to the exit divided by the "wrong-way"
velocity; $t_-(x) \to x/v$! We can give only a suggestion to justify this strange
result. Clearly if a particle exits against the bias, this event must happen
quickly, if it is to occur at all. The only time scale with this behavior is the
wrong-way time x/v.

As a final note, we may verify that the weighted sum of conditional exit
times, $\mathcal{E}_+(x)t_+(x) + \mathcal{E}_-(x)t_-(x)$, reproduces the unconditional exit time of
Eq. (2.3.10).

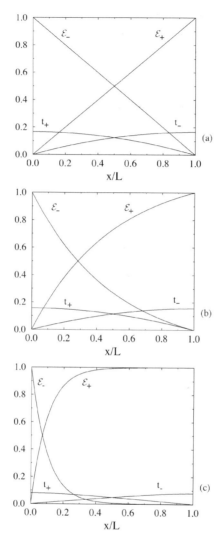

Fig. 2.8. Dependence on x of the conditional exit probabilities $\mathcal{E}_\pm(x)$ and exit times $t_\pm(x)$ on the initial position x for Péclet numbers (a) $Pe = 0$, (b) $Pe = 2$, and (c) $Pe = 10$.

2.3.3. Transmission Mode

Let us now derive the mean time to exit an interval in transmission mode when there is a net bias. In this case, the equation of motion for the time-integrated concentration is

$$v\frac{\partial \mathcal{C}_0(x)}{\partial x} = D\frac{\partial^2 \mathcal{C}_0(x)}{\partial x^2}, \tag{2.3.27}$$

subject to the boundary conditions $C_0(x) = 0$ at $x = L$ and $\mathcal{J}(x) = 1$ at $x = 0$. Here $\mathcal{J}(x) = \int_0^\infty j(x, t)\, dt$ is the time-integrated flux. Solving this equation and imposing the boundary conditions immediately gives

$$C_0(x) = \frac{1}{v}\left[1 - e^{v(x-L)/D}\right], \tag{2.3.28}$$

and the mean exit time is

$$\langle t \rangle = \int_0^L C_0(x)\, dx = \frac{L}{v} - \frac{D}{v^2}\left(1 - e^{-vL/D}\right), \tag{2.3.29}$$

which reproduces the first of Eqs. (2.2.31).

2.3.4. Biased Diffusion as a Singular Perturbation

The sharp transition in the behavior of the exit probabilities $\mathcal{E}_\pm(L)$ for large Pe is an example of a *boundary layer* phenomenon in which the character of the solution changes rapidly over an infinitesimal range of the domain as $v \to \infty$. This arises because as $v \to \infty$ the diffusion term represents a *singular perturbation* in the convection–diffusion equation. We now briefly discuss the consequences of this singular perturbation for biased diffusion. A nice general discussion of singular perturbation methods in given in Hinch (1991).

To keep the discussion as simple as possible, consider biased diffusion in a one-dimensional interval of linear size L, where the concentration $c(x, t)$ obeys the convection–diffusion equation

$$\frac{\partial c}{\partial t} + v\frac{\partial c}{\partial x} = D\frac{\partial^2 c}{\partial x^2}.$$

It is useful to express all lengths in units of L and all times in units of the convection time L/v. Thus we introduce the dimensionless length $\tilde{x} = x/L$ and time $\tilde{t} = t/(L/v)$. It is then natural to measure the concentration in the dimensionless units $\tilde{c} = cL$. In these units, the convection–diffusion equation is

$$\frac{\partial \tilde{c}}{\partial \tilde{t}} + \frac{\partial \tilde{c}}{\partial \tilde{x}} = Pe^{-1}\frac{\partial^2 \tilde{c}}{\partial \tilde{x}^2}, \tag{2.3.30}$$

where, for this discussion only, we define the Péclet number as $Pe = vL/D$ (without the factor of 2).

For $Pe^{-1} \to 0$, it seems natural to treat the second derivative as a small perturbation in the convection–diffusion equation. Unfortunately, this natural expectation is erroneous. In fact, the nature of the solution as $Pe^{-1} \to 0$ is completely different from that we would obtain by setting $Pe^{-1} = 0$ at the outset.

This discontinuity is a hallmark of a singular perturbation. If we proceed naively and ignore the diffusive term in Eq. (2.3.30), then the remaining *unperturbed* equation is first order. However, the original second-order equation requires two boundary conditions to yield a well-posed problem. Typically, the unperturbed equation satisfies one of the boundary conditions but not the other.

On the other hand, the second boundary condition *is* satisfied by the full solution, in which the concentration varies rapidly within a thin layer about this boundary. In this boundary layer, the second derivative is sufficiently large that the product $Pe^{-1}(\partial^2 c/\partial x^2)$ is no longer negligible in the convection–diffusion equation. This subtle feature explains why a naive perturbation approach is inapplicable for this class of problems.

Before we turn to a specific example, it is instructive to discuss the singular nature of viscous fluid flow, which is formally analogous to biased diffusion. For an incompressible fluid of density ρ and viscosity coefficient η, the velocity field $\vec{v}(\vec{r})$ is governed by the Navier–Stokes equation [see, e.g., Faber (1995) for a general discussion]

$$\rho\left(\frac{\partial \vec{v}}{\partial t} + (\vec{v} \cdot \vec{\nabla})\vec{v}\right) = -\vec{\nabla}P + \eta\nabla^2\vec{v},$$

where $P(\vec{r})$ is the pressure at $\vec{r}$. We introduce the vorticity $\vec{\Omega} = \vec{\nabla} \times \vec{v}$ to eliminate the pressure-gradient term and obtain an equation for $\vec{\Omega}$ whose dimensional structure is similar to the convection–diffusion equation:

$$\frac{\partial \vec{\Omega}}{\partial t} + \vec{\nabla} \times (\vec{\Omega} \times \vec{v}) = \nu\nabla^2\vec{\Omega},$$

where $\nu = \eta/\rho$ is the kinematic viscosity.

Now we express lengths in units of the system size L, velocities in units of the external velocity V, and times in units of the convection time L/V to obtain the dimensionless vorticity equation (where tildes have been dropped for notational simplicity):

$$\frac{\partial \vec{\Omega}}{\partial t} + \vec{\nabla} \times (\vec{\Omega} \times \vec{v}) = \mathcal{R}^{-1}\nabla^2\vec{\Omega}.$$

A nice discussion of this approach and its ramifications appears in the Feynman lecture series book [Feynman, Leighton, & Sands (1963)]. In this equation $\mathcal{R} = vL/\nu$ is the *Reynolds number* and it is entirely analogous to the Péclet number in the visually similar convection–diffusion equation Eq. (2.3.30).

The basic fact is that the inverse Reynolds number multiplies the highest power of the derivative in the vorticity equation and therefore also represents a singular perturbation. Thus in hydrodynamics, the equation of motion with

$\mathcal{R}^{-1}$ set to zero exhibits different behavior from the case $\mathcal{R}^{-1} \to 0$. The former leads to so-called "dry" hydrodynamics [Feynman, Leighton, & Sands (1963)], in which flow patterns are laminar and there is no dissipation. On the other hand, the limit $\mathcal{R}^{-1} \to 0$ is typically turbulent, with the velocity field changing rapidly over small distance scales.

Returning to the convection–diffusion equation, a very simple situation that illustrates the singular nature of the bias is the steady-state concentration $c(x)$ on $[0, L]$, with $c = c_0$ at $x = 0$ and with $c = 0$ at $x = L$. This concentration satisfies $Pe^{-1}c'' - c' = 0$ (in dimensionless units), with the solution

$$c(x) = c_0 \frac{1 - e^{v(x-L)/D}}{1 - e^{-vL/D}}. \tag{2.3.31}$$

From the exact solution, the concentration is nearly constant over most of the interval; this corresponds to the unperturbed solution of the zeroth-order equation $c' = 0$. However, in a narrow region of width D/v about $x = L$, the concentration sharply decreases. In this boundary layer, the naive zeroth-order approximation is invalid because both terms in the convection–diffusion equation are comparable.

In the complementary situation in which Pe^{-1} is small but negative, the probability distribution rapidly decreases to nearly zero over a scale of D/v about $x = 0$ and remains vanishingly small thereafter (Fig. 2.9). This concentration profile can also be described by a boundary-layer analysis.

In short, the biased diffusive system provides a simple, pedagogical illustration of a singular perturbation and the role of boundary layers, in which the character of the probability distribution changes over a very short distance scale.

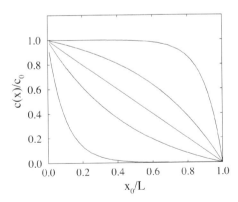

Fig. 2.9. Steady-state concentration for biased diffusion on $[0, L]$, with absorption at $x = L$ and $c = c_0$ at $x = 0$. Shown are the cases $Pe = 10, 2, 0, -2$, and -10 (top to bottom).

2.4. Discrete-Space Random Walk

We now turn to the first-passage characteristics of a discrete-space and -time random walk on the interval. It will be satisfying to see that the asymptotic first-passage properties of the discrete random walk and continuum diffusion are identical. Because of this equivalence and because the random walk is technically more complicated, it is generally advisable to study first-passage properties within continuum diffusion. However, the methods for solving the random-walk master equation are technically enlightening and therefore are presented in this section.

2.4.1. Illustration for a Short Chain

As a preliminary, the solution to random-walk master equations are illustrated by the very simple example of a five-site chain with both ends absorbing (Fig. 2.10). For a random walk that starts at the midpoint, let us compute the occupation probability at each lattice site and the first-passage probability to the end points.

2.4.1.1. Discrete Time

Let $P_i(n)$ be the probability that the random walk is at site i at the nth time step. At sites 0 and ± 1, the particles hops to the right or to the left with probability $1/2$. If the particle reaches either end point, it remains there forever. The occupation probabilities obey the master equations

$$P_{-2}(n + 1) = P_{-2}(n) + \frac{1}{2}P_{-1}(n),$$

$$P_{-1}(n + 1) = \frac{1}{2}P_0(n),$$

$$P_0(n + 1) = \frac{1}{2}P_{-1}(n) + \frac{1}{2}P_1(n),$$

$$P_1(n + 1) = \frac{1}{2}P_0(n),$$

$$P_2(n + 1) = P_2(n) + \frac{1}{2}P_1(n).$$

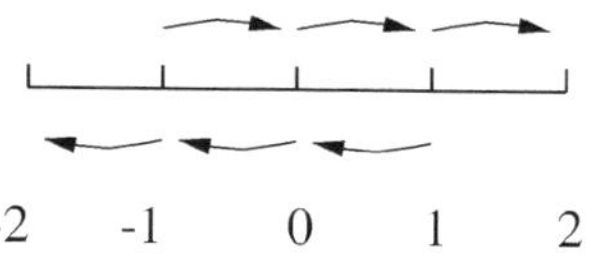

Fig. 2.10. A five-site chain, $-2, -1, 0, 1$, and 2. The arrows indicate the nonzero hopping probabilities (or nonzero hopping rates).

The first-passage probability to the end points, $F_{\pm 2}(n)$, obeys the master equation $F_{\pm 2}(n + 1) = \frac{1}{2} P_{\pm 1}(n)$. We solve these equations by the generating function approach. To apply this method, we multiply each equation by z^{n+1}, sum from $n = 0$ to ∞, and define $P_i(z) = \sum_{n=0}^{\infty} P_i(n) z^n$. On the left-hand side we have $P_i(z) - P_i(n = 0)$, and on the right-hand side each term is the corresponding generating function multiplied by z. Thus the master equations are transformed to the set of algebraic equations

$$P_{-2}(z) - P_{-2}(n = 0) = z P_{-2}(z) + \frac{z}{2} P_{-1}(z),$$

$$P_{-1}(z) - P_{-1}(n = 0) = \frac{z}{2} P_0(z),$$

$$P_0(z) - P_0(n = 0) = \frac{z}{2} P_{-1}(z) + \frac{z}{2} P_1(z),$$

$$P_1(z) - P_1(n = 0) = \frac{z}{2} P_0(z),$$

$$P_2(z) - P_2(n = 0) = z P_2(z) + \frac{z}{2} P_1(z),$$

and, equivalently,

$$F_{\pm 2}(z) - F_{\pm 2}(n = 0) = +\frac{z}{2} P_{\pm 1}(z).$$

Here again, the argument denotes either a function of discrete time (n) or the generating function (z).

Suppose the particle starts in the middle of the chain, so that $P_i(0) = \delta_{i,0}$. Then obviously $P_{-1} = P_1$ and $P_{-2} = P_2$, and the master equations reduce to

$$P_0(z) - 1 = z P_1(z),$$

$$P_1(z) = \frac{z}{2} P_0,$$

$$P_2(z) = z P_2(z) + \frac{z}{2} P_1(z), \quad F_2(z) = \frac{z}{2} P_1(z),$$

with solutions

$$P_0(z) = \frac{1}{1 - z^2/2},$$

$$P_1(z) = \frac{z/2}{1 - z^2/2},$$

$$P_2(z) = \frac{1}{1 - z} \frac{(z/2)^2}{1 - z^2/2}, \quad F_2(z) = \frac{(z/2)^2}{1 - z^2/2}. \tag{2.4.1}$$

To obtain $P_i(n)$, we expand these generating functions as power series in z and extract the coefficients of z^n. In general, we can do this by performing a contour integration of $P_i(z)/z^{n+1}$. For our example, however, these terms can easily be obtained by direct expansion. We thereby find

$$P_0(z) = 1 + \frac{z^2}{2} + \frac{z^4}{4} + \frac{z^6}{8} + \dots,$$

$$P_1(z) = \frac{z}{2}\left(1 + \frac{z^2}{2} + \frac{z^4}{4} + \frac{z^6}{8} + \cdots\right),$$

$$F_2(z) = \left(\frac{z}{2}\right)^2\left(1 + \frac{z^2}{2} + \frac{z^4}{4} + \frac{z^6}{8} + \cdots\right), \tag{2.4.2}$$

and

$$P_2(z) = \frac{1}{1-z}F_2(z). \tag{2.4.3}$$

This last relation has the important consequence that $P_2(n) = \sum_{k=1}^{n} F_2(k)$, a relation that is apparent physically. The corresponding results as a function of the number of steps are

$$P_0(n) = \begin{cases} 2^{-n/2}, & n \text{ even} \\ 0, & n \text{ odd} \end{cases};$$

$$P_1(n) = \begin{cases} 2^{-(n+1)/2}, & n \text{ odd} \\ 0, & n \text{ even} \end{cases};$$

$$P_2(n) = \begin{cases} \frac{1}{2}(1 - 2^{-n/2}), & n \text{ even} \\ \frac{1}{2}\left[1 - 2^{-(n-1)/2}\right], & n \text{ odd} \end{cases};$$

$$F_2(n) = \begin{cases} 2^{-(n+2)/2}, & n \text{ even} \\ 0, & n \text{ odd} \end{cases}. \tag{2.4.4}$$

As we might expect, the first-passage probability decays exponentially to zero as a function of the number of steps.

2.4.1.2. Continuous Time

Now consider continuous-time hopping on the five-site chain. At each interior site, define w as the hopping rate for a particle to move to either of its neighbors. The master equations for this system are

$$\dot{P}_0(t) = w[P_{-1}(t) + P_1(t) - 2P_0(t)],$$
$$\dot{P}_1(t) = w[P_0(t) - 2P_1(t)],$$
$$\dot{P}_2(t) = P_1(t),$$

with similar equations for P_{-2} and P_{-1}. We set the hopping rate to 1 by the time rescaling $t \rightarrow wt$ and introduce the Laplace transform to recast these differential equations as the algebraic relations

$$s P_0(s) - P_0(t = 0) = P_{-1}(s) + P_1(s) - 2 P_0(s),$$
$$s P_1(s) - P_1(t = 0) = P_0(s) - 2 P_1(s),$$
$$s P_2(s) - P_2(t = 0) = P_1(s).$$

For the initial condition $P_n(t = 0) = \delta_{n,0}$, the solution is

$$P_0(s) = \frac{s + 2}{(s + 2)^2 - 2},$$

$$P_1(s) = \frac{1}{(s + 2)^2 - 2},$$

$$P_2(s) = [1 - P_0(s) - 2 P_1(s)]/2. \tag{2.4.5}$$

Note that $P_0(s) + 2 P_1(s) + 2 P_2(s) = 1/s$, which is just the Laplace transform of the probability conservation statement $P_0(t) + 2 P_1(t) + 2 P_2(t) = 1$. By writing the probabilities in a partial fraction expansion, we immediately find the time dependences

$$P_0(t) = \frac{1}{2}(e^{-at} + e^{-bt}),$$

$$P_1(t) = \frac{1}{2\sqrt{2}}(e^{-bt} - e^{-at}),$$

$$P_2(t) = \frac{1}{2} + \left(\frac{1}{\sqrt{2}} - \frac{1}{4}\right)e^{-at} - \left(\frac{1}{\sqrt{2}} + \frac{1}{4}\right)e^{-bt}, \tag{2.4.6}$$

where $a = 2 - \sqrt{2}$ and $b = 2 + \sqrt{2}$.

As is obvious physically, for both the discrete- and the continuous-time systems the occupation probabilities at site 0 and ± 1 decay exponentially with time, whereas the occupation probabilities at the absorbing sites approach $1/2$ with an exponential time dependence. Correspondingly, the first-passage probability to the end points also decays exponentially with time. Although this short-chain example is elementary, it usefully illustrates the same methods that will also be used to solve first passage on more complex systems, such as the Cayley tree and other self-similar structures (Chap. 5).

2.4.2. First-Passage Probabilities

We now study the time dependence of the first-passage probability for a discrete-time symmetric random walk on a linear chain of sites $0, 1, 2, \ldots, N$. As in the continuum case, the two interesting situations are absorption mode – both ends absorbing, or transmission mode – one end reflecting and one end absorbing.

2.4.2.1. Absorption Mode

The random walk starts at x and is ultimately absorbed when either 0 or N is reached, with respective probabilities $\mathcal{E}_-(x)$ and $\mathcal{E}_+(x)$. By the Laplacian formalism of Subsection 1.6.3, these two quantities satisfy the discrete Laplace equation $\Delta^{(2)}\mathcal{E}_\pm(x) = 0$, subject to the boundary conditions $\mathcal{E}_-(0) = 1$, $\mathcal{E}_-(N) = 0$ and $\mathcal{E}_+(0) = 0$, $\mathcal{E}_+(N) = 1$. The solution to this problem is trivial and gives

$$\mathcal{E}_-(x) = 1 - \frac{x}{N}, \quad \mathcal{E}_+(x) = \frac{x}{N}, \tag{2.4.7}$$

as already derived in Eqs. (2.2.11).

For a biased random walk with p and $q = 1 - p$ the respective probabilities of hopping one step to the right and to the left, the eventual hitting probabilities now obey the recursion formula [see Eq. (1.6.17)]

$$\mathcal{E}_\pm(x) = p\mathcal{E}_\pm(x + 1) + q\mathcal{E}_\pm(x - 1). \tag{2.4.8}$$

For $p \neq q$, the solution is of the form $\mathcal{E}_\pm(x) = A_\pm\alpha_+^x + B_\pm\alpha_-^x$, with $\alpha_\pm = (1 \pm \sqrt{1 - 4pq})/2p$, so that $\alpha_+ = 1$ and $\alpha_- = q/p$. Now, applying the boundary conditions on $\mathcal{E}_\pm$, we obtain

$$\mathcal{E}_-(x) = \frac{\alpha_+^{x-N} - \alpha_-^{x-N}}{\alpha_+^{-N} - \alpha_-^{-N}} = \frac{\left(\frac{q}{p}\right)^x - \left(\frac{q}{p}\right)^N}{1 - \left(\frac{q}{p}\right)^N},$$

$$\mathcal{E}_+(x) = \frac{\alpha_+^x - \alpha_-^x}{\alpha_+^N - \alpha_-^N} = \frac{1 - \left(\frac{q}{p}\right)^x}{1 - \left(\frac{q}{p}\right)^N}. \tag{2.4.9}$$

These are the discrete analogs of Eqs. (2.3.9)

2.4.2.2. Transmission Mode

For this system, the eventual hitting probability is obviously one and the interesting feature is its time dependence. For simplicity, suppose that the random walk starts at site 0. If the walk returns to 0, we implement the reflecting boundary condition that the walk hops to site 1 one time step later.

We view the absorbing site N as "sticky," upon which the probability irreversibly accumulates. The hopping probabilities at all other sites are equal. The corresponding master equations are

$$P_0(n+1) = \frac{1}{2}P_1(n),$$

$$P_1(n+1) = P_0(n) + \frac{1}{2}P_2(n),$$

$$P_2(n+1) = \frac{1}{2}P_1(n) + \frac{1}{2}P_3(n),$$

$$\vdots$$

$$P_{N-2}(n+1) = \frac{1}{2}P_{N-3}(n) + \frac{1}{2}P_{N-1}(n),$$

$$P_{N-1}(n+1) = \frac{1}{2}P_{N-2}(n),$$

$$P_N(n+1) = \frac{1}{2}P_{N-1}(n) + P_N(n),$$

$$F(n+1) = \frac{1}{2}P_{N-1}(n).$$

Thus the occupation probability at site N obeys the master equation

$$P_N(n+1) = P_N(n) + F(n+1),$$

or, alternatively,

$$P_N(n) = \sum_{n'=1}^{n} F(n').$$

As usual, we introduce the generating functions to reduce the master equations to the set of algebraic equations

$$P_0 = 1 + a P_1,$$
$$P_1 = 2a P_0 + a P_2,$$
$$P_2 = a P_1 + a P_3,$$

$$\vdots$$

$$P_{N-2} = a P_{N-3} + a P_{N-1},$$
$$P_{N-1} = a P_{N-2},$$
$$P_N = a P_{N-1}/(1 - Z),$$
$$F = a P_{N-1}, \qquad\qquad\qquad (2.4.10)$$

where, for convenience, we define $a = z/2$. In terms of these generating functions, the relation between $P_N(n)$ and $F(n)$ is simply $P_N(z) = zF(z)/(1 - z)$.

Two complementary solutions are now presented for the first-passage probability. The first involves applying the same steps as those used for solving the first-passage probability in diffusion. The second is conceptually simpler but technically more involved and is based on successive variable elimination in the master equations.

2.4.2.3. Direct Solution

As a preliminary, note that by writing $P_{N-1} = a P_{N-2} + a P_N$ in Eqs. (2.4.10) and imposing the absorbing boundary condition $P_N = 0$, the recursion formulas for $2 \leq n \leq N - 1$ all have the same form $P_n = a(P_{n-1} + P_{n+1})$. This has the general solution $P_n = A_+\lambda_+^n + A_-\lambda_-^n$, where $\lambda_\pm = (1 \pm \sqrt{1 - z^2})/z$. This form is valid for $n = 2, 3, 4, \ldots, N$, whereas P_1 is distinct. We now impose the boundary conditions to fix $A_\pm$. The condition $P_N = 0$ allows us to write

$$P_n = A\left(\lambda_+^{n-N} - \lambda_-^{n-N}\right). \tag{2.4.11}$$

Next, we eliminate P_0 in the equation for P_1, giving $(1 - 2a^2)P_1 = 2a + a P_2$. Using Eq. (2.4.11) in this relation gives

$$A = \frac{2}{a} \frac{1}{\lambda_+^{-1} - \lambda_-^{-1}} \frac{1}{\lambda_+^{-N} + \lambda_-^{-N}}.$$

We use polar coordinates to simplify this expression. That is, we write $\lambda_\pm = (1 \pm i\sqrt{z^2 - 1})/z = e^{\pm i\phi}$, with $\phi = \tan^{-1}\sqrt{z^2 - 1}$. Consequently, the first-passage probability is

$$F = a P_{N-1},$$
$$= \frac{2}{\lambda_+^{-N} + \lambda_-^{-N}}$$
$$= \sec N\phi$$
$$= \sec(N \tan^{-1}\sqrt{z^2 - 1}). \tag{2.4.12}$$

Although this last expression appears formidable, it is completely equivalent to the first-passage probability for diffusion [Eq. (2.2.25)]. To see this equivalence, we take the continuum limit of the generating function by letting $z \to 1$. This gives $\phi = \tan^{-1}\sqrt{z^2 - 1} \to \sqrt{2is}$, with $s = z - 1$. Consequently, $\sec(N \tan^{-1}\sqrt{z^2 - 1})$ reduces to $\sec(N\sqrt{2s})$, thus coinciding with the Laplace transform of the continuum first-passage probability [Eq. (2.2.25)], with $D = 1/2$. Thus the first-passage probability for the discrete random walk and continuum diffusion in transmission mode are asymptotically identical.

2.4.2.4. Recursive Solution

For the recursive solution, our strategy is to first eliminate P_0 in the equation for P_1 to yield P_1 as a function of P_2 only. Continuing this procedure yields P_{n-1} as a function of P_n. This process then closes when the equation for P_N is reached. To simplify the continued fractions that arise in this successive elimination, we define $f_0 = 1$, $f_1 = 2a^2$ and $f_n = a^2/(1 - f_{n-1})$ for $n \geq 2$.

Eliminating P_0 in the equation for P_1 gives $P_1 = 2a(1 + aP_1) + aP_2$, or

$$
\begin{aligned}
P_1 &= \frac{2a}{1 - 2a^2} + \frac{a}{1 - 2a^2} P_2 \\
&= \frac{1}{a}(2f_2 + f_2 P_2).
\end{aligned}
\tag{2.4.13}
$$

Substituting this expression into the equation for P_2 yields $P_2 = (2f_2 + f_2 P_2) + aP_3$. This can be written as

$$
\begin{aligned}
P_2 &= \frac{2f_2}{1 - f_2} + \frac{a}{1 - f_2} P_3 \\
&= \frac{2}{a^2} f_2 f_3 + \frac{1}{a} f_3 P_3.
\end{aligned}
\tag{2.4.14}
$$

Continuing this procedure to site $N - 1$ gives

$$
P_{N-1} = \frac{2}{a^{N-2}} f_2 f_3 \cdots f_{N-1} + \frac{1}{a} f_{N-1} P_N.
$$

Then the equation for P_N gives

$$
\begin{aligned}
P_N &= aP_{N-1} \\
&= \frac{2}{a^{N-3}} f_2 f_3 \cdots f_{N-1} + f_{N-1} P_N,
\end{aligned}
$$

or

$$
P_N = \frac{2}{a^{N-1}} f_2 f_3 \cdots f_N,
$$

from which the formal solution for the first-passage probability is

$$
F = \frac{2}{a^{N-2}} f_2 f_3 \cdots f_N.
\tag{2.4.15}
$$

We now solve for f_n by converting the continued fraction for f_n to a linear second-order recursion, which may be solved by standard methods, as is nicely explained in Goldhirsch and Gefen (1986). In terms of $f_n = g_n/h_n$, the recursion for f_n becomes

$$\frac{g_n}{h_n} = \frac{a^2}{1 - \dfrac{g_{n-1}}{h_{n-1}}} = \frac{a^2 h_{n-1}}{h_{n-1} - g_{n-1}}.$$

This is equivalent to the two recursion formulas

$$g_n = (a^2 h_{n-1}) C_n,$$
$$h_n = (h_{n-1} - g_{n-1}) C_n.$$

Because we are ultimately interested in only the ratio of g_n / h_n, the coefficients C_n are irrelevant and we study the reduced recursion relations

$$g_n = a^2 h_{n-1},$$
$$h_n = h_{n-1} - g_{n-1} = h_{n-1} - a^2 h_{n-2}.$$

The general solution to the latter equation is, as usual, $h_n = A_+ \lambda_+^n + A_- \lambda_-^n$, with the coefficients A_+ and A_- determined by the boundary recursion relation. Because $f_1 = 2a^2$ and also $f_1 = a^2 h_0 / h_1$, we may choose $h_0 = 2$ and $h_1 = 1$, as it is only the ratio of h_0 to h_1 that is relevant. Matching these values of h_0 and h_1 gives $A_+ = A_- = 1$.

Finally, using $h_n = \lambda_+^n + \lambda_-^n$ and $f_n = a^2 h_{n-1} / h_n$ in Eq. (2.4.15) gives

$$F = \frac{2}{a^{N-2}} a^{2(N-1)} \frac{h_1}{h_2} \frac{h_2}{h_3} \cdots \frac{h_{N-1}}{h_N}$$

$$= \frac{2a^N}{\lambda_+^N + \lambda_-^N}. \qquad (2.4.16)$$

We can simplify this last result by rewriting $\lambda_\pm$ in polar coordinates as $\lambda_\pm = a e^{\pm i\phi}$, with $\phi = \tan^{-1} \sqrt{z^2 - 1}$, to give $\lambda_+^N + \lambda_-^N = 2a^N \cos N\phi$. Thus we find $F = \sec N\phi$, as derived above.

2.4.3. Mean First-Passage Time

Let us now determine the time for the random walk to exit the system for absorption mode (both end points absorbing) and transmission mode (one absorbing and one reflecting end point). For absorption mode, we compute the unrestricted mean time for a random walk to exit at either end of a finite chain $[0, 1, 2, \ldots, N]$. From Subsection 1.6.3.2, the mean exit time starting from site n, t_n, obeys the recursion relation

$$t_n = \frac{1}{2}(1 + t_{n-1}) + \frac{1}{2}(1 + t_{n+1}), \quad n = 1, 2, \ldots, N - 1, \qquad (2.4.17)$$

where the time for each step has been taken to be one and $t_0 = t_N = 0$. This can be rewritten as the discrete Laplace equation

$$\Delta^{(2)} t_n = -2, \tag{2.4.18}$$

whose general solution is the second-order polynomial $t_n = A + Bn - n^2$. Imposing the boundary conditions $t_0 = t_N = 0$ immediately gives $t_n = n(N - n)$. This is identical to the corresponding continuum result given in Eq. (2.2.17) when $D = 1/2$.

For transmission mode, we assume a reflecting point at $x = 0$ and an absorbing point at $x = N$. We may obtain the mean time to reach site N by solving Eq. (2.4.18) subject to the boundary conditions $t_0' = 0$ and $t_N = 0$. This gives $t_n = N^2 - n^2$. We can also find this first-passage time and indeed all integer moments of this time from the first-passage probability in Eq. (2.4.12). By the definition of the generating function,

$$
\begin{aligned}
\langle t \rangle &= \sum_{t=1}^{\infty} t\, F(t) \\
&= \sum_{t=1}^{\infty} t\, F(t) z^t \Big|_{z=1} \\
&= z \frac{d}{dz} F(z) \Big|_{z=1}.
\end{aligned}
\tag{2.4.19}
$$

To determine $F'(z)$, it is convenient to expand $F(z)$ to first order in $s = z - 1$. Thus

$$
\begin{aligned}
F(z) &\approx \sec(N \tan^{-1} \sqrt{2s}), \\
&\approx \sec N \sqrt{2s}, \\
&\approx 1 + N^2 s,
\end{aligned}
\tag{2.4.20}
$$

from which we immediately find $\langle t \rangle = N^2$. Higher moments of the first-passage time can be obtained similarly, but it is clear that $\langle t^k \rangle \propto N^{2k}$ for all positive integer k.

We can also determine the decay time in the asymptotic exponential decay of the first-passage probability, $F(t) \sim e^{-t/\tau}$, with $\tau \to \infty$ as the system size N diverges. We rewrite this time dependence as $F(t) \sim \mu^t$, with $\mu \to 1$ from below as $N \to \infty$. This leads to

$$F(z) \approx \sum_t \mu^t z^t \sim (1 - \mu z)^{-1} \equiv (1 - z/z_c)^{-1}. \tag{2.4.21}$$

Thus we seek the location of the singularity that is closest to $z = 1$ in $F(z) = \sec N\phi$. This is a simple pole whose location is determined by $\cos(N \tan^{-1} \sqrt{z^2 - 1}) = 0$. For large N, this gives $z_c \approx 1 + \pi^2/8N^2$ from which

$$F(t) \sim e^{-\pi^2 t/8N^2}, \qquad (2.4.22)$$

that is, the decay time of the first-passage probability is $\tau = 8N^2/\pi^2$. Thus the first-passage properties of the linear chain are accounted for by a single time scale that is of the order of N^2.

3

Semi-Infinite System

3.1. The Basic Dichotomy

A natural counterpart to the finite interval is the first-passage properties of the semi-infinite interval $[0, \infty]$ with absorption at $x = 0$. Once again, this is a classical geometry for studying first-passage processes, and many of the references mentioned at the outset of Chap. 2 are again pertinent. In particular, the text by Karlin and Taylor (1975) gives a particularly comprehensive discussion about first-passage times in the semi-infinite interval with arbitrary hopping rates between neighboring sites. Once again, however, our focus is on simple diffusion or the nearest-neighbor random walk. For these processes, the possibility of a diffusing particle making arbitrarily large excursions before certain trapping takes place leads to an infinite mean lifetime. On the other hand, the recurrence of diffusion in one dimension means that the particle *must* eventually return to its starting point. This dichotomy between infinite lifetime and certain trapping leads to a variety of extremely surprising first-passage-related properties both for the semi-infinite interval and the infinite system.

Perhaps the most amazing such property is the arcsine law for the probability of long leads in a symmetric nearest-neighbor random walk in an unbounded domain. Although this law applies to the unrestricted random walk, it is intimately based on the statistics of returns to the origin and thus fits naturally in our discussion of first-passage on the semi-infinite interval. Our natural expectation is that, for a random walk which starts at $x = 0$, approximately one-half of the total time would be spent on the positive axis and the remaining one-half of the time on the negative axis. Surprisingly, this is the *least* probable outcome. In fact, the arcsine law tells us that the most probable outcome is that the walk always remains entirely on the positive or on the negative axis.

We will begin by presenting the very appealing image method to solve the first-passage properties in the semi-infinite interval. This approach can be extended easily to treat biased diffusion. We then turn to the discrete random walk and explore some intriguing consequences of the fundamental dichotomy between certain-return and infinite return time, such as the arcsine law. We will also treat the role of partial trapping at the origin that leads naturally to a useful correspondence with diffusion in an attenuating medium and also the radiation boundary condition. Finally, we discuss the quasi-static approximation. This is a simple yet powerful approach for understanding diffusion near an absorbing boundary in terms of an effective, nearly time-independent system. This approach provides the simplest way to treat first-passage in higher dimensions, to be presented in Chap. 6.

3.2. Image Method

3.2.1. The Concentration Profile

We determine the first-passage probability to the origin for a diffusing particle that starts at $x_0 > 0$ subject to the absorbing boundary condition that the concentration $c(x, t)$ at the origin is zero. The most appealing method of solution is the familiar image method from electrostatics. In this method, a particle initially at $x = x_0$ and an image "antiparticle" initially at $x = -x_0$ that both diffuse freely on $[-\infty, \infty]$ give a resultant concentration $c(x, t)$ that obviously satisfies the absorbing boundary condition $c(0, t) = 0$. Thus in the half-space $x > 0$, the image solution coincides with the concentration of the initial absorbing boundary condition system.

Hence we may write the concentration (or probability distribution) for a diffusing particle on the positive half-line as the sum of a Gaussian and an anti-Gaussian:

$$c(x, t) = \frac{1}{\sqrt{4\pi Dt}} \left[e^{-(x-x_0)^2/4Dt} - e^{-(x+x_0)^2/4Dt} \right]. \tag{3.2.1}$$

In the long-time limit, it is instructive to rewrite this as

$$\begin{aligned} c(x, t) &= \frac{1}{\sqrt{4\pi Dt}} e^{-(x^2+x_0^2)/4Dt} \left(e^{xx_0/2Dt} - e^{-xx_0/2Dt} \right) \\ &\approx \frac{1}{\sqrt{4\pi Dt}} \frac{xx_0}{Dt} e^{-(x^2+x_0^2)/4Dt}, \end{aligned} \tag{3.2.2}$$

which clearly exhibits its short- and large-distance tails. We see that the concentration has a linear dependence on x near the origin and a Gaussian large-distance tail, as illustrated in Fig. 3.1.

A crude but useful way to visualize the concentration and to obtain first-passage characteristics quite simply is to replace the constituent Gaussian and anti-Gaussian in the image solution with simpler functions that have the

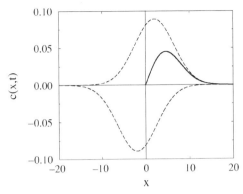

Fig. 3.1. Concentration profile of a diffusing particle in the absorbing semi-infinite interval $(0, \infty)$ at $Dt = 10$, with $x_0 = 2$. Shown are the component Gaussian and image anti-Gaussian (dashed curves) and their superposition in the physical region $x > 0$ (solid curve).

general scaling properties as these Gaussians. Thus we replace the initial Gaussian with a "blip" function centered at x_0, of width $\sqrt{Dt}$ and amplitude $1/\sqrt{Dt}$, and the anti-Gaussian with an "antiblip" at $-x_0$, of width $\sqrt{Dt}$ and amplitude $-1/\sqrt{Dt}$. Their superposition gives a constant distribution of magnitude $1/\sqrt{Dt}$ that extends from $\frac{1}{2}\sqrt{Dt} - x_0$ to $\frac{1}{2}\sqrt{Dt} + x_0$ (Fig. 3.2). From this picture, we immediately infer that the survival probability varies as $S(t) \approx 2x_0/\sqrt{Dt}$, whereas the mean displacement of the probability distribution that remains alive recedes from the trap as $\sqrt{Dt}$. Although this use of blip functions provides a modest simplification for the semi-infinite line, this cartoonlike approach is extremely useful for more complicated geometries, such as diffusion within an absorbing wedge (see Chap. 7).

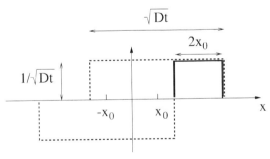

Fig. 3.2. Superposition (solid) of a blip and antiblip (dashed) to crudely represent the probability distribution for diffusion on $(0, \infty)$ with an absorber at $x = 0$.

As a concluding note to this section, it is worth pointing out an amusing extension of the image method. Let us work backward by introducing a suitably defined image and then determine the locus on which the concentration that is due to the particle and the image vanishes. This then immediately provides the solution to a corresponding boundary-value problem.

To provide a flavor for this method, consider the simple example of a particle that starts at x_0 and an image, with weight $-w \neq 1$, that starts at $-x_0$. The total concentration that is due to these two sources is

$$
\begin{aligned}
c(x, t) &= \frac{1}{\sqrt{4\pi D t}} \left[e^{-(x-x_0)^2/4Dt} - w e^{-(x+x_0)^2/4Dt} \right], \\
&= \frac{1}{\sqrt{4\pi D t}} e^{-(x-x_0)^2/4Dt} (1 - w e^{-xx_0/Dt}),
\end{aligned} \tag{3.2.3}
$$

which clearly vanishes on the locus $x(t) = (Dt/x_0) \ln w$. Therefore the solution to the diffusion equation in the presence of an absorbing boundary that moves at constant speed v is given by the sum of a Gaussian and an image anti-Gaussian of amplitude $w = e^{vx_0/D}$. By being judicious, this approach can be extended to treat a surprisingly wide variety of absorbing boundary geometries [Daniels (1969, 1982)].

3.2.2. First-Passage Properties

3.2.2.1. Isotropic Diffusion

Because the initial condition is normalized, the first-passage probability to the origin at time t is just the flux to this point. From the exact expression for $c(x, t)$ in Eq. (3.2.1), we find

$$
\begin{aligned}
F(0, t) &= +D \frac{\partial c(x, t)}{\partial t} \bigg|_{x=0} \\
&= \frac{x_0}{\sqrt{4\pi D t^3}} e^{-x_0^2/4Dt}, \quad t \to \infty.
\end{aligned} \tag{3.2.4}
$$

This distribution is sometimes termed the inverse Gaussian density [see Chikara & Folks (1989) for a general discussion about this distribution]. In the long-time limit $\sqrt{Dt} \gg x_0$, for which the diffusion length is much greater than the initial distance to the origin, the first-passage probability reduces to $F(0, t) \to x_0/t^{3/2}$, as we derived in Chap. 1 by extracting the first-passage probability directly from the underlying probability distribution. The existence of this long-time tail means that the mean time for the particle to reach the origin is infinite! That is

$$\langle t \rangle \equiv \int_0^\infty t \, F(0, t) \, dt \sim \int_0^\infty t \times t^{-3/2} \, dt = \infty. \qquad (3.2.5)$$

On the other hand, the particle is sure to return to the origin because $\int_0^\infty F(t) \, dt = 1$.

Although the mean time to return to the origin is infinite, suitably defined low-order moments *are* finite. In general, the kth moment is

$$\langle t^k \rangle = \frac{x_0}{\sqrt{4\pi D t}} \int_0^\infty t^{k-3/2} \, e^{-x_0^2/4Dt} \, dt. \qquad (3.2.6)$$

This integral converges for $k < 1/2$, so that moments of order less than $1/2$ are finite. By elementary steps, we explicitly find

$$\langle t^k \rangle = \frac{\Gamma\left(\frac{1}{2} - k\right)}{\sqrt{\pi}} \left(\frac{x_0^2}{4D}\right)^k. \qquad (3.2.7)$$

For $k < 1/2$, the units of $\langle t^k \rangle$ are determined by the time to diffuse a distance x_0, which is the only natural time scale in the problem. The finiteness of these low-order moments is yet another manifestation of the dichotomy between certain return and infinite return time for one-dimensional diffusion. An application of these results to neuron dynamics will be given in Section 4.2.

From the first-passage probability, the survival probability $S(t)$ may be found directly through $S(t) = 1 - \int_0^t F(0, t') \, dt'$, or it may be obtained by integration of the concentration over positive x. The former approach gives

$$S(t) = 1 - \int_0^t \frac{x_0}{\sqrt{4\pi D t'^3}} \, e^{-x_0^2/4Dt'} \, dt', \qquad (3.2.8)$$

and we may evaluate this integral easily by rewriting the integrand in terms of $u^2 = x_0^2/4Dt'$ to give

$$S(t) = 1 + \frac{2}{\sqrt{\pi}} \int_\infty^{x_0/\sqrt{4Dt}} e^{-u^2} \, du$$

$$= \frac{2}{\sqrt{\pi}} \int_0^{x_0/\sqrt{4Dt}} e^{-u^2} \, du$$

$$= \mathrm{erf}\left(\frac{x_0}{\sqrt{4Dt}}\right). \qquad (3.2.9a)$$

Alternatively, integrating of the probability distribution over the positive interval gives

$$S(t) = \int_0^\infty c(x, t)\, dx$$

$$= \frac{1}{\sqrt{4\pi Dt}} \left[\int_0^\infty dx\, e^{-(x-x_0)^2/4Dt} - \int_0^\infty dx\, e^{-(x+x_0)^2/4Dt} \right]$$

$$= \frac{1}{\sqrt{\pi}} \int_{-x_0/\sqrt{4Dt}}^\infty e^{-u^2}\, du - \frac{1}{\sqrt{\pi}} \int_{+x_0/\sqrt{4Dt}}^\infty e^{-u^2}\, du$$

$$= \frac{2}{\sqrt{\pi}} \int_0^{x_0/\sqrt{4Dt}} e^{-u^2}\, du$$

$$= \mathrm{erf}\left(\frac{x_0}{\sqrt{4Dt}} \right). \tag{3.2.9b}$$

From this result, the survival probability is nearly constant until the diffusion length $\sqrt{Dt}$ reaches the initial distance to the boundary, x_0. Subsequently the particle has the opportunity to reach the origin by diffusion, and trapping becomes significant. For $\sqrt{Dt} \gg x_0$, the asymptotic behavior of the error function [Abramowitz & Stegun (1972)] then gives

$$S(t) \sim \frac{x_0}{\sqrt{\pi Dt}}. \tag{3.2.10}$$

As expected by the recurrence of diffusion, the survival probability ultimately decays to zero. The $1/\sqrt{Dt}$ long-time behavior is clearly evident in Fig. 3.3.

Let us reinterpret these properties within the framework of the nervous-investor example of Chap. 1. An investor buys a share of stock at \$100 in

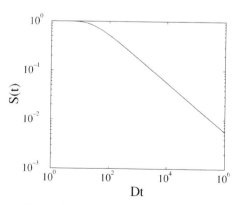

Fig. 3.3. Double logarithmic plot of survival probability $S(t)$ versus time for $x_0 = 10$. The survival probability remains close to one for $Dt < x_0^2$ and then decays as $t^{-1/2}$ thereafter.

a market in which the price changes equiprobably by a factor $f = 0.9$ or f^{-1} daily. After one day, the stock price drops to \$90 (one step away from the boundary), and the investor decides to sell out when the initial price is recovered. Unlike our example in Chap. 1, our investor now has infinite fortitude and will not sell until break even again occurs. That is, the process is now defined on the semi-infinite interval. Thus an average investor must be both patient and willing to accept an arbitrarily large intermediate loss to ensure a sale at the break-even point. After t trading days, a fraction $1 - 1/\sqrt{t}$ of the traders will break even, whereas the remaining unlucky fraction $1/\sqrt{t}$ of the traders will typically see their stock worth $\$100 \times f^{\sqrt{t}}$. For example, after 50 days, approximately 14% of the investors are still in the market and the stock price will typically sink to approximately \$47.

It is also useful to study the spatial dependence of the probability distribution. As seen most clearly in Fig. 3.2, the surviving fraction recedes from the origin with a mean displacement $\sqrt{Dt}$ – a particle must move away from the origin to be long lived. This is clearly evident in the mean displacement of the distribution of surviving particles. Following the same steps as those used to compute the survival probability, we find that this displacement is

$$
\begin{aligned}
\langle x \rangle &= \frac{1}{S(t)} \frac{1}{\sqrt{4\pi Dt}} \int_0^\infty x \left[e^{-(x-x_0)^2/4Dt} - e^{-(x+x_0)^2/4Dt} \right] dx \\
&= \frac{1}{S(t)} \frac{1}{\sqrt{\pi}} \left[\int_{-x_0/\sqrt{4Dt}}^\infty (u\sqrt{4Dt} + x_0) e^{-u^2} du \right. \\
&\quad \left. - \int_{x_0/\sqrt{4Dt}}^\infty (u\sqrt{4Dt} - x_0) e^{-u^2} du \right] \\
&= \frac{1}{S(t)} \left[\frac{x_0}{\sqrt{\pi}} \left(\int_{-x_0/\sqrt{4Dt}}^0 e^{-u^2} du + \int_{x_0/\sqrt{4Dt}}^0 e^{-u^2} du \right) \right. \\
&\quad \left. + \frac{\sqrt{16Dt}}{\sqrt{\pi}} \int_0^{x_0/\sqrt{4Dt}} u e^{-u^2} du \right] \\
&\approx \left(x_0 + \frac{1}{2} \frac{x_0^2}{\sqrt{\pi Dt}} \right) \times \frac{\sqrt{\pi Dt}}{x_0} \\
&= \sqrt{\pi Dt} + \frac{x_0}{2}, \tag{3.2.11}
\end{aligned}
$$

thus confirming our expectation from Fig. 3.2.

Note also that the asymptotic displacement is *independent* of the initial particle position. Once the diffusion length becomes much greater than x_0, this initial distance must disappear from the problem. Within a scaling approach, this fact alone is sufficient to give the $\sqrt{Dt}$ growth of the displacement. Because of the utility of this type of scaling argument, we use it to determine the behavior of $\langle x \rangle$. If the only lengths in the system are the initial particle position x_0 and the diffusion length $\sqrt{Dt}$, their ratio is the *only* intrinsic variable in the system. Thus we anticipate that the functional form of $\langle x \rangle$ may be expressed as

$$\langle x \rangle = x_0 \, f\left(\frac{Dt}{x_0^2}\right).$$

Here the scaling function $f(u)$ must approach a constant for small u, so that we recover $x(t) \to x_0$ as $t \to 0$. On the other hand, at long times $Dt \gg x_0^2$, the initial position of the particle should become irrelevant. This occurs only if $f(u) \propto u^{1/2}$ for large u. By using this form for $f(u)$, we see that $\langle x \rangle$ grows as $\sqrt{Dt}$ and that the dependence on x_0 also disappears in the long-time limit. This reproduces the long-time behavior given in Eq. (3.2.11).

3.2.2.2. Biased Diffusion

A nice feature of the image method is that it can be adapted in a simple but elegant manner to biased diffusion. Naively, we might anticipate that if the particle has a drift velocity v, the image particle should move with velocity $-v$. However, because of this opposite velocity, such an image contribution would not satisfy the initial convection–diffusion equation in the domain $x > 0$. The correct solution must therefore consist of an image contribution that moves in the *same* direction as that of the original particle to satisfy the equation of motion. By inspection, the solution is

$$c(x, t) = \frac{1}{\sqrt{4\pi Dt}} \left[e^{-(x-x_0-vt)^2/4Dt} - e^{vx_0/D} e^{-(x+x_0-vt)^2/4Dt} \right], \quad (3.2.12)$$

where the factor $e^{vx_0/D}$ ensures that $c(x = 0, t) = 0$. Note that by transforming to a reference frame that moves at velocity v, we map to the problem of a particle and a static image of magnitude $e^{vx_0/D}$, together with an absorbing boundary that moves with velocity $-v$. This is the just the problem already solved in Eq. (3.2.3)!

Another basic feature of the image solution is that it is manifestly positive for all $x > 0$, as we can see by rewriting $c(x, t)$ as

$$c(x, t) = \frac{1}{\sqrt{4\pi Dt}} \, e^{-(x-x_0-vt)^2/4Dt} (1 - e^{-xx_0/Dt}).$$

Note also that Eq. (3.2.12) can be rewritten as

$$c(x, t) = \frac{1}{\sqrt{4\pi Dt}} \left[e^{-(x-x_0-vt)^2/4Dt} - e^{vx/D} e^{-(x+x_0+vt)^2/4Dt} \right].$$

In this representation, the image appears to move in the opposite direction compared with that of the initial particle. However, the image contribution contains a position-dependent prefactor that decays sufficiently rapidly for large negative x that the true motion of this image contribution is actually in the *same* direction as that of the original particle.

From the concentration profile, we determine the first-passage probability either by computing the flux to the origin, $vc - Dc'|_{x=0}$, or by Taylor expanding Eq. (3.2.12) in a power series in x and identifying the linear term with the first-passage probability. Either approach leads to the simple result

$$F(0, t) = \frac{x_0}{\sqrt{4\pi Dt^3}} e^{-(x_0+vt)^2/4Dt}. \tag{3.2.13}$$

As we might expect, this first-passage probability asymptotically decays exponentially in time because of the bias. For negative bias, the particle is very likely to be trapped quickly, whereas for positive bias the particle very likely escapes to $+\infty$. In either case the first-passage probability should become vanishingly small in the long-time limit.

From the first-passage probability, the survival probability is

$$S(t) = 1 - \int_0^t F(0, t')\, dt,$$

$$= 1 - e^{-vx_0/2D} \int_0^t \frac{x_0}{\sqrt{4\pi Dt'^3}} e^{-x_0^2/4Dt'} e^{-v^2t'/4D}\, dt'. \tag{3.2.14}$$

Using again the substitution $u^2 = x^2/4Dt$ and the Péclet number $Pe = vx_0/2D$, we find that the survival probability is

$$S(t) = 1 - \frac{2}{\sqrt{\pi}} e^{-Pe} \int_{x/\sqrt{4Dt}}^{\infty} e^{-u^2 - Pe^2/4u^2}\, du$$

$$= 1 - \frac{1}{2} e^{-Pe} \left\{ e^{Pe} \left[1 - \mathrm{erf}\left(\frac{x_0}{\sqrt{4Dt}} + \frac{Pe}{2} \frac{\sqrt{4Dt}}{x_0} \right) \right] \right.$$

$$\left. + e^{-Pe} \left[1 - \mathrm{erf}\left(\frac{x_0}{\sqrt{4Dt}} - \frac{Pe}{2} \frac{\sqrt{4Dt}}{x_0} \right) \right] \right\}$$

$$\sim 1 - \frac{1}{2} e^{-Pe-|Pe|} \left[2 - \mathrm{erfc}\left(\frac{|Pe|}{2} \frac{\sqrt{4Dt}}{x_0} \right) \right], \quad t \to \infty. \tag{3.2.15}$$

This last expression has two different behaviors depending on whether the velocity is positive or negative. Using the asymptotic properties of the error function, we find

$$S(t) \sim \begin{cases} 1 - e^{-2Pe} = 1 - e^{-vx_0/D}, & Pe > 0 \\ \dfrac{x_0}{Pe\sqrt{\pi Dt}} e^{-Pe^2 Dt/x_0^2} = \sqrt{\dfrac{4D}{\pi v^2 t}} e^{-v^2 t/4D}, & Pe \le 0 \end{cases}. \quad (3.2.16)$$

From this result, the probability for the particle to be eventually trapped as a function of its starting location, $\mathcal{E}(x_0)$, is simply

$$\mathcal{E}(x_0) = \begin{cases} e^{-vx_0/D}, & \text{for } v > 0 \\ 1, & \text{for } v \le 0 \end{cases}. \quad (3.2.17)$$

Thus for zero or negative bias, a diffusing particle is recurrent, whereas for a positive bias, no matter how small, a diffusing particle is transient. This is a basic extension of the recurrence/transience transition to biased diffusion. Note, finally, that this last result can be obtained more simply from the Laplacian formalism of Chap. 1. For biased diffusion, $\mathcal{E}(x)$ satisfies [see Eq. (1.6.18)]

$$D\mathcal{E}'' + v\mathcal{E}' = 0,$$

subject to the boundary conditions $\mathcal{E}(0) = 1$ and $\mathcal{E}(\infty) = 0$. The first condition corresponds to certain trapping if the particle starts at the origin, and the second merely states that there is zero trapping probability if the particle starts infinitely far away. The solution to this equation is simply Eq. (3.2.17).

3.2.2.3. Long-Range and Generalized Hopping Processes

For long-range hopping, there is a remarkable and not-so-well-known theorem (to physicists at least), which is due to Sparre Andersen, about first passage in the semi-infinite interval [Sparre Andersen (1953); see also Feller (1968) and Spitzer (1976)]. For concreteness, consider a discrete-time random walk that starts at $x_0 > 0$ with each step chosen from a symmetric, continuous, but otherwise *arbitrary* distribution. Then the probability that the random walk first crosses the origin and hits a point on the negative axis at the nth step, $\mathcal{F}(n)$, asymptotically decays as $n^{-3/2}$, just as in the symmetric nearest-neighbor random walk!

The precise statement of the Sparre Andersen theorem is the following: Let $\mathcal{F}(z) = \sum_{n=1}^{\infty} \mathcal{F}(n)z^n$ be the generating function for the first-passage probability. Then for a particle that begins at $x = 0^+$,

$$\mathcal{F}(z) = 1 - \sqrt{1-z}, \quad \text{long-range hopping.}$$

To put this in context, for the symmetric nearest-neighbor random walk, the generating function for the first-passage probability has the nearly identical form [Eq. (1.5.5)]

$$\mathcal{F}(z) = 1 - \sqrt{1 - z^2}, \quad \text{nearest-neighbor hopping.}$$

Clearly these two functions have identical singular behaviors as $z \to 1$, so that their respectively large-n properties are also the same. It is especially intriguing that even for a broad distribution of step lengths in which the mean step length is divergent, the probability of first crossing the origin at the nth still decays as $n^{-3/2}$.

This theorem can be easily applied to more general situations. One instructive example is the first-passage probability of a long-range hopping process in which the time for each step equals the step distance. Suppose that the step lengths ℓ are drawn from a continuous, symmetric distribution $f(\ell) \sim \ell^{-(1+\mu)}$ for large ℓ, with $0 < \mu < 1$ so that the mean step length diverges and the distribution is normalizable. For a particle that starts at $x_0 > 0$, we can now find the probability that the particle first hits the region $x < 0$ at time t by the Sparre Andersen theorem.

After a time t, we may determine the number of steps in the walk by first computing the mean step length in a walk of total duration t. This length is

$$\int_0^t \ell \, f(\ell) \, d\ell \propto t^{1-\mu}.$$

For this finite-duration walk, the mean step length $t^{1-\mu}$ times the number of steps $n(t)$ must clearly equal the total elapsed time t; thus $n(t) \sim t^{\mu}$. Now we may transform from the first-passage probability as a function of n to the first-passage probability as a function of t by means of

$$\mathcal{F}(t) = \mathcal{F}(n)\frac{dn}{dt} \propto n^{-3/2}t^{\mu-1} \sim t^{-(1+\mu/2)}. \tag{3.2.18}$$

As an example, consider the following diffusion process in the two-dimensional half-space $(x > 0, y)$ with absorption when $x = 0$ is reached [Redner & Krapivsky (1996)]. In addition to diffusion, there is a bias $v_x(y) = \text{sign}(y)$. Thus the bias is directed toward the "cliff" for $y < 0$ and away from it for $y > 0$ (Fig. 3.4). How does this "schizophrenic" bias (with mean longitudinal velocity zero) affect the particle survival probability? The key to understanding this problem is to note that successive returns of the vertical displacement to $y = 0$ are governed by the one-dimensional first-passage probability. Therefore the displacement ℓ of the particle in the x direction between each return has a power-law tail $\ell^{-3/2}$, corresponding to the case $\mu = 1/2$ in the previous two paragraphs. Substituting this into Eq. (3.2.18), we conclude that the first-passage probability decays as $t^{-5/4}$, corresponding

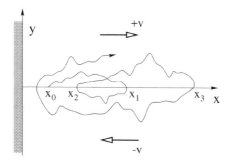

Fig. 3.4. A typical trajectory of a diffusing particle in the flow field $v_x(y) = \text{sign}(y)$. The distribution of distances ℓ between successive crossings of $y = 0$, $\ell = x_n - x_{n-1}$, has a $\ell^{-3/2}$ tail.

to a survival probability that asymptotically decays as $t^{-1/4}$. Consequently this bias leads to an increased survival probability compared to pure diffusion.

Amusingly, this $t^{-5/4}$ decay of the first-passage probability for a Brownian particle in the half-space geometry appears to be relatively ubiquitous. This same decays occurs for free acceleration in the x-direction (Sinai, 1992, Burkhardt, 2000), for unidirectional general shear flows of the form $v_x(y) = -v_x(-y)$ (Redner & Krapivsky (1996)), as well as for a Brownian particle which is subject to a random velocity field in the x-direction, in which $v_x(y)$ is a random zero-mean function of y (Redner (1997)). Perhaps there is a deep relation which connects these apparently disparate systems.

3.2.2.4. Semi-Infinite Slab

We can easily extend our results for the semi-infinite interval to a semi-infinite slab in arbitrary spatial dimension. Although the time dependence of the first-passage probability remains the same as in one dimension, we may also study *where* the particle hits the boundary. Consider the semi-infinite d-dimensional slab where the last coordinate x_d is restricted to positive values by the absorbing boundary condition at $x_d = 0$. A diffusing particle starts at $\vec{h}_0 = (0, 0, \ldots, 0, h_0)$. What is the probability that the particle eventually hits at some point on the $(d - 1)$-dimensional absorbing plane $x_d = 0$?

We may answer this question easily by the image method. First, consider the case of two dimensions. For a particle that starts at $(0, y_0)$ with the line $y = 0$ absorbing, the time-dependent concentration is

$$c(x, y, t) = \frac{1}{4\pi Dt} \left\{ e^{-[x^2+(y-y_0)^2]/4Dt} - e^{-[x^2+(y+y_0)^2]/4Dt} \right\} \quad (3.2.19)$$

The first-passage probability to any point along the absorbing line at time t

is then the magnitude of the diffusive flux at $(x, 0)$:

$$j(x, 0, t) = D \left| \frac{\partial c(x, y, t)}{\partial y} \right|_{y=0}$$

$$= \frac{y_0}{4\pi D t^2} e^{-(x^2 + y_0^2)/4Dt}. \qquad (3.2.20)$$

Finally, the probability that the particle eventually hits $(x, 0)$, $\mathcal{E}(x; 0, y_0)$, is the integral of the first-passage probability at $(x, 0)$ over all time:

$$\mathcal{E}(x; 0, y_0) = \int_0^\infty j(x, 0, t) \, dt.$$

We may evaluate this integral easily by changing variables to $u = (x^2 + y_0^2)/4Dt$ to obtain the *Cauchy distribution* for the hitting probability:

$$\mathcal{E}(x; 0, y_0) = \frac{1}{\pi} \frac{y_0}{x^2 + y_0^2}. \qquad (3.2.21)$$

Because of the x^{-2} decay of this distribution for large x, the mean position of a particle that hits the line, namely, $\langle x \rangle = \int_0^\infty x \mathcal{E}(x; 0, y_0) \, dx$, is infinite.

For general spatial dimension, the time-dependent concentration is

$$c(\vec{x}, t) = \frac{1}{(4\pi D t)^{d/2}} \left\{ e^{-[\vec{x}_\perp^2 + (x_d - h_0)^2]/4Dt} - e^{-[\vec{x}_\perp^2 + (x_d + h_0)^2]/4Dt} \right\}, \qquad (3.2.22)$$

where $\vec{x}_\perp^2 = x_1^2 + x_2^2 + \cdots + x_{d-1}^2$. From this, the first-passage probability to a point on the absorbing hyperplane is

$$j(\vec{x}_\perp, t) = \frac{1}{(4\pi D t)^{d/2}} \frac{h_0}{t} e^{-(\vec{x}_\perp^2 + h_0^2)/4Dt}. \qquad (3.2.23)$$

Integrating over all time and using the same variable change as in two dimensions, we find that the probability of eventually hitting $\vec{x}_\perp$ is simply

$$\mathcal{E}(\vec{x}_\perp; \vec{h}_0) = \frac{h_0}{\pi^{d/2}} \frac{\Gamma(d/2)}{(\vec{x}_\perp^2 + h_0^2)^{d/2}}. \qquad (3.2.24)$$

In greater than two dimensions, the mean lateral position of the distribution of trapped particle density on the absorbing hyperplane is finite. Note, as expected from the equivalence with electrostatics, that this eventual hitting probability is also just the electric field of a point charge located at $\vec{h}_0$ in the presence of a grounded hyperplane at $x_d = 0$. We will return to this equivalence when we discuss first passage in the wedge geometry in Chap. 7.

3.3. Systematic Approach

As a counterpoint to the image method, we now obtain first-passage properties by the systematic Green's function solution that was emphasized in Chap. 2.

3.3.1. The Green's Function Solution

We compute the Green's function for the diffusion equation with the initial condition $c(x, t = 0) = \delta(x - x_0)$ and boundary condition $c(x = 0, t) = 0$. After a Laplace transform, the diffusion equation becomes

$$sc(x, s) - \delta(x - x_0) = Dc''(x, s).$$

In each subdomain, $x < x_0$ and $x > x_0$, the solution is a linear combination of exponential functions. From the boundary condition at $x = 0$, the linear combination must be antisymmetric for $x < x_0$; for $x > x_0$ the boundary condition at $x = \infty$ implies that only the decaying exponential occurs. Thus

$$c_<(x, s) = A \sinh\left(\sqrt{\frac{s}{D}}x\right),$$

$$c_>(x, s) = B \exp\left(-\sqrt{\frac{s}{D}}x\right).$$

Imposing continuity of the concentration at $x = x_0$ and also the joining condition $D(c'_> - c'_<)|_{x=x_0} = -1$, we obtain

$$c(x, s) = \frac{1}{\sqrt{sD}} \sinh\left(\sqrt{\frac{s}{D}}x_<\right)e^{-\sqrt{s/D}\,x_>},$$

with $x_> = \max(x, x_0)$ and $x_< = \min(x, x_0)$. Writing the sinh function in terms of exponentials, we can write this Green's function in the more symmetric and physically revealing form

$$c(x, s) = \frac{1}{4sD}\left[e^{\pm\sqrt{s/D}(x-x_0)} - e^{-\sqrt{s/D}(x+x_0)}\right], \tag{3.3.1}$$

where the positive sign refers to $c_>$ and the negative sign to $c_<$. It is elementary to invert this Laplace transform and recover the image solution form of $c(x, t)$ quoted in Eq. (3.2.1).

For biased diffusion, the Green's function for the Laplace transformed convection–diffusion equation,

$$sc(x, s) - \delta(x - x_0) + vc(x, s)' = Dc(x, s)'',$$

has the general form $c = Ae^{\alpha_+ x} + Be^{\alpha_- x}$, with $\alpha_\pm = (v \pm \sqrt{v^2 + 4Ds})/2D$.

When the boundary conditions are imposed, the Green's function reduces to

$$c(x, s) = \begin{cases} A(e^{\alpha_+ x} - e^{\alpha_- x})e^{\alpha_- x_0}, & x < x_0 \\ A(e^{\alpha_+ x_0} - e^{\alpha_- x_0})e^{\alpha_- x}, & x > x_0 \end{cases} . \qquad (3.3.2)$$

Finally the constant A is obtained from the usual joining condition $D(c'_> - c'_<)|_{x=x_0} = -1$. Thus the Green's function is

$$c(x, s) = \begin{cases} \dfrac{1}{D(\alpha_+ - \alpha_-)}(e^{\alpha_+ x} - e^{\alpha_- x})e^{\alpha_+ x_0}, & x < x_0 \\ \dfrac{1}{D(\alpha_+ - \alpha_-)}(e^{-\alpha_- x_0} - e^{-\alpha_+ x_0})e^{\alpha_- x}, & x > x_0 \end{cases} . \qquad (3.3.3)$$

From this, the flux to the origin is

$$F(0, s) = +Dc'_<|_{x=0} = e^{-\alpha_+ x_0}. \qquad (3.3.4)$$

We now invert this Laplace transform to find the time dependence of the first-passage probability. For this, we use the general theorem that the inverse Laplace transform of the function $f(s + a)$, equals e^{-at} times the inverse Laplace transform of $f(s)$ [Ghez (1988)]. That is, $\mathcal{L}^{-1}[f(s+a)] = e^{-at}\mathcal{L}^{-1}[f(s)]$. For the present example,

$$F(0, s) = e^{-vx_0/2D}e^{-(x_0/\sqrt{D})\sqrt{s+v^2/4D}},$$

so that

$$\begin{aligned} F(0, t) &= e^{-vx_0/2D}[e^{-v^2t/4D}\mathcal{L}^{-1}(e^{-x_0\sqrt{s/D}})] \\ &= \frac{x_0}{\sqrt{4\pi Dt^3}} e^{-x_0^2/4Dt} e^{-vx_0/2D} e^{-v^2t/4D}, \end{aligned} \qquad (3.3.5)$$

which coincides with Eq. (3.2.13). Thus complete results for first-passage on the semi-interval can be obtained either by the Green's function approach or by the intuitive image method.

3.3.2. Constant-Density Initial Condition

We now study the constant-density initial distribution because of its utility for diffusion-controlled reactions in a one-dimensional geometry, in which a chemical reactant is neutralized when it reaches $x = 0$. A simple example is particle–antiparticle annihilation $A + B \rightarrow 0$, where A initially occupies the region $x > 0$ and B occupies the region $x < 0$ (Gálfi & Rácz (1988)). Annihilation takes place within a thin reaction zone that is centered at $x = 0$.

A first step in understanding this kinetics is provided by a study of the simpler system where reactions occur exactly at $x = 0$.

The constant-density initial condition on $(0, \infty)$ can be solved either by the image method or by a direct solution of the diffusion equation; the latter approach is presented. Thus we solve the diffusion equation subject to the initial condition $c(x, t = 0) = c_0$ and the boundary conditions $c(x = 0, t) = 0$ and $c(x = \infty, t) = c_0$. The Laplace transform of the diffusion equation gives $sc(x, s) - c_0 = Dc''(x, s)$, and, after the boundary conditions are applied, the solution is

$$c(x, s) = \frac{c_0}{s}\left(1 - e^{-x\sqrt{\frac{s}{D}}}\right). \tag{3.3.6}$$

When the Laplace transform is inverted, the corresponding solution as a function of time is

$$c(x, t) = c_0 \, \mathrm{erf}\left(\frac{x}{\sqrt{4Dt}}\right). \tag{3.3.7}$$

An important feature of this concentration profile is the gradual depletion of the density near the origin. This is reflected by the fact that the flux to the origin decreases with time as

$$j(x = 0, t) = Dc'(x = 0, t) = \sqrt{\frac{D}{\pi t}} \, c_0. \tag{3.3.8}$$

In the context of diffusion-controlled reactions, this means that the reaction rate at an absorbing boundary also decreases with time.

Note that the concentration profile (see Fig. 3.5) for the uniform initial condition [Eq. (3.3.7)] is identical to the survival probability for the point

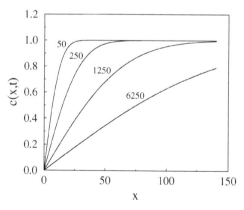

Fig. 3.5. Concentration profile at $Dt = 50, 250, 1250$, and 6250 for an absorbing boundary at $x = 0$ and a constant initial concentration $c_0 = 1$ for $x > 0$.

initial condition [Eq. (3.2.9b)]. This coincidence is no accident! It arises because the underlying calculations for these two results are, in fact, identical. Namely:

- For a particle initially at x_0, we find the survival probability $S(t)$ is found by first calculating the concentration with this initial state, $\delta(x - x_0)$, and then integrating the concentration over all $x > 0$ to find the survival probability.
- For an initial concentration equal to one for $x > 0$, we can view the initial condition of a superposition of delta functions. The concentration for $t > 0$ is found by integrating the time-dependent concentration that arises at each such source over all space.

To show formally that these two computations are equivalent, define $c(x, t; x_0)$ as the Green's function for the diffusion equation on the positive semi-interval for a particle that is initially at x_0. Here we explicitly write the source point x_0 and the observation point x. We now exploit the fact that this Green's function satisfies the reciprocity relation

$$c(x, t; x_0) = c(x_0, t; x) \qquad (3.3.9)$$

under the interchange of the source and the observation points. By definition, the survival probability for a particle that is initially at x_0 equals

$$S(t; x_0) = \int_0^\infty c(x, t; x_0)\, dx. \qquad (3.3.10)$$

On the other hand, for the uniform initial condition, the time-dependent concentration is given by

$$\int_0^\infty c(x, t; x_0)\, dx_0 = \int_0^\infty c(x_0, t; x)\, dx_0 \quad \text{(reciprocity)}$$

$$= \int_0^\infty c(x, t; x_0)\, dx \quad \text{(relabeling)}$$

$$= S(t; x_0), \qquad (3.3.11)$$

thus showing the equivalence of the concentration due to a uniform initial condition and the survival probability due to a point initial condition.

3.4. Discrete Random Walk

3.4.1. The Reflection Principle

We now study first-passage properties of a symmetric nearest-neighbor random walk. Much of this discussion is based on the treatment given in Chap. III in Vol. I of Feller's book [Feller (1968)]. Although the results are equivalent to those of continuum diffusion, the methods of solution are appealing and provide the starting point for treating the intriguing origin crossing properties of random walks. We start by deriving the fundamental *reflection principle* that allows us to derive easily various first-passage properties in terms of the occupation probability.

It is convenient to view a random walk as a directed path in space–time (x, t), with isotropic motion in space and directed motion in time (Fig. 3.6). Let $N(x, t | x_0, 0)$ be the total number of random walks that start at $(x_0, 0)$ and end at $(x > x_0, t)$. Typically, we will not write the starting location dependence unless instructive. This dependence can always be eliminated by translational invariance, $N(x, t | x_0, 0) = N(x - x_0, t | 0, 0) \equiv N(x - x_0, t)$.

Consider a random walk that starts at $(0, 0)$, crosses a point $y > x$, and then ends at (x, t) (Fig. 3.6). Let $X_y(x, t)$ be the number of these "crossing" paths between 0 and x. The reflection principle states that

$$X_y(x, t) = N(2y - x, t). \tag{3.4.1}$$

This relation arises because any crossing path has a unique last visit to y. We may then construct a "reflected" path by reversing the direction of all steps after this last visit (Fig. 3.6). It is clear that there is a one-to-one mapping between the crossing paths that comprise $X_y(x, t)$ and these reflected paths. Moreover the reflected paths are unrestricted because y is necessarily crossed in going from $(0, 0)$ to $(2y - x, t)$. Thus the reflected

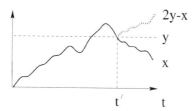

Fig. 3.6. The reflection principle illustrated. The solid curve is a "crossing" path to (x, t) by way of (y, t'), with $t' < t$. The number of such restricted paths is the same as the number of unrestricted paths to $(2y - x, t)$. The reflected portion of the trajectory is shown as a dotted curve.

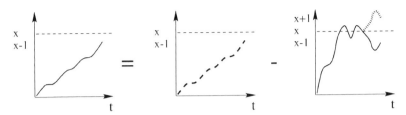

Fig. 3.7. First-passage paths (solid curve) from $(0, 0)$ to $(x - 1, t - 1)$ written as the difference between all paths (dashed curve) and crossing paths. The latter are related to the number of all paths from $(0, 0)$ to $(x + 1, t)$ (dotted curve) by the reflection principle.

paths constitute $N(2y - x, t)$. Parenthetically, this reflection construction could also be applied to the leading segment of the crossing path to give the equality $X_y(x, t) = N(x, t|2y, 0) = N(2y - x, t)$.

3.4.2. Consequences for First Passage

We now use this reflection principle to compute the number of first-passage paths $F(x, t|0, 0)$, that is, paths that start at $(0, 0)$ and arrive at (x, t) *without* previously touching or crossing x. This is a pleasant, graphically based computation (Fig. 3.7). First note that $F(x, t)$ is the same as $F(x - 1, t - 1)$, as there is only one way to continue the first-passage path from $(x - 1, t - 1)$ to (x, t). Next, the number of first-passage paths from $(0, 0)$ to $(x - 1, t - 1)$ can be written as the difference

$$F(x - 1, t - 1) = N(x - 1, t - 1) - X_x(x - 1, t - 1), \qquad (3.4.2)$$

where again $X_x(x - 1, t - 1)$ is the number of paths from $(0, 0)$ to $(x - 1, t - 1)$ which *necessarily* touch or cross x before $t - 1$ steps. By the reflection principle, $X_x(x - 1, t - 1) = N(x + 1, t - 1)$. Consequently, the number of first-passage paths from 0 to x after t steps is

$$\begin{aligned}
F(x, t) &= F(x - 1, t - 1) \\
&= N(x - 1, t - 1) - N(x + 1, t - 1) \\
&= \frac{(t - 1)!}{\left(\frac{t+x-2}{2}\right)! \left(\frac{t-x}{2}\right)!} - \frac{(t - 1)!}{\left(\frac{t+x}{2}\right)! \left(\frac{t-x-2}{2}\right)!} \\
&= \frac{x}{t} \frac{t!}{\left(\frac{t+x}{2}\right)! \left(\frac{t-x}{2}\right)!} \\
&= \frac{x}{t} N(x, t).
\end{aligned} \qquad (3.4.3)$$

Because the total number of paths from $(0, 0)$ to (x, t) is proportional to $e^{-x^2/4Dt}/\sqrt{4\pi Dt}$ in the continuum limit, the number of first-passage paths between these two points therefore scales as $xe^{-x^2/4Dt}/\sqrt{4\pi Dt^3}$, in agreement with our previous derivations of this result.

If the end point is also at $x = 0$, then we have the first return to the origin. In this case, we apply the reflection principle in a slightly different form to obtain the number of first-passage paths. Suppose that the first step is to the right. Then the last step must be to the left, and hence $f(t) \equiv F(0, t) = F(1, t - 1 | 1, 1)$. The reflection principle now gives

$$
\begin{aligned}
f(t) &= F(1, t - 1 | 1, 1) \\
&= N(1, t - 1 | 1, 1) - X_0(1, t - 1 | 1, 1) \\
&= N(1, t - 1 | 1, 1) - N(1, t - 1 | - 1, 1) \\
&= \frac{(t - 2)!}{[(t/2 - 1)!]^2} - \frac{(t - 2)!}{(t/2)! \, (t/2 - 2)!} \\
&= \frac{1}{2t - 1} \frac{t!}{[(t/2)!]^2} \\
&= \frac{1}{2t - 1} N(0, t).
\end{aligned}
\tag{3.4.4}
$$

Note that t must be even because we are considering return to the origin.

One additional amusing result that will also be useful in Subsection 3.4.3 is the equality between the number of paths $\mathcal{N}(t)$ that *never* return to the origin *by* time t and the number of paths $\mathcal{R}(t)$ that return to the origin *at* time t. From the binomial distribution of a random walk, the number of these return paths is

$$
\mathcal{R}(t) = \frac{t!}{[(t/2)!]^2} \sim \frac{1}{\sqrt{\pi t}},
\tag{3.4.5}
$$

where Stirling's approximation is used to obtain the asymptotic result. By their definition, the nonreturning paths are the same as all first-passage paths from the origin to any end point. Therefore, by summing over all end points, we have

$$
\begin{aligned}
\mathcal{N}(t) &= \sum_{x=2}^{t}{}' F(x, t) \\
&= \sum_{x=2}^{t}{}' [N(x - 1, t - 1) - N(x + 1, t - 1)].
\end{aligned}
\tag{3.4.6}
$$

Here the prime on the sum denotes the restriction to the even integers. Successive terms in the second sum cancel in pairs, leaving behind the first series term

$N(1, t - 1) = N(0, t|0, 0)$. Thus we obtain the simple and unexpected result,

$$\mathcal{N}(t) = \mathcal{R}(t). \tag{3.4.7}$$

That is, the number of paths that never return by time t is the same as the number of otherwise unrestricted paths that return exactly at time t! It is striking that these two diametrically different properties are so closely linked.

3.4.3. Origin-Crossing Statistics

3.4.3.1. Qualitative Picture and Basic Questions

Hidden within the survival and the first-passage probabilities of the one-dimensional random walk are fundamental properties about the statistics of crossings of the origin (see Fig. 3.8). Some of these properties are sufficiently counterintuitive that they appear to be "wrong," yet they can be derived simply and directly from the first-passage probability. They dramatically illustrate the profound implications of the power-law tail in the first-passage probability of a random walk.

Consider a symmetric random walk on the infinite line. We have learned that the mean time for such a random walk to return to the origin is infinite [Eq. (3.2.5)]; on the other hand, from Eq. (3.4.4), the probability of return to the origin after a relatively small number of steps is appreciable. For example, the first-passage probabilities $f(t)$ are $\frac{1}{2}, \frac{1}{8}, \frac{1}{16}, \frac{5}{128}$, and $\frac{7}{256}$ for $t = 2, 4, 6, 8$, and 10, respectively. Thus the probability of returning to the starting point within 10 steps is $0.7539\ldots$. This contrast between the relatively large probability of returning after a short time and an infinite mean return time is the source of the intriguing properties of origin crossings by a random walk.

We focus on two basic questions. First:

- What is the average number of returns to the origin in an n-step walk? More generally, what is the probability that there are m returns to the origin in an n-step walk?

Fig. 3.8. Caricature of a one-dimensional random-walk trajectory. There are many quick returns to the origin and fewer long-duration returns, which take most of the elapsed time.

Surprisingly, the outcome with the largest probability is *zero* returns. In fact, the probability for m returns in an n-step walk has the Gaussian form $e^{-m^2/n}$, so that the typical number of returns after n steps is of the order of $\sqrt{n}$ or less. Thus returns are relatively unlikely.

We also study the *lead probability*. To appreciate its meaning, consider again the coin-tossing game in which player A wins $1 from B if a coin toss is heads and gives $1 to B if the result is tails. This is the same as a one-dimensional random walk, in which position x of the walk at time t is the difference in the number of wins for A and B. In a single game with n coin tosses, A will be leading for a time $n\phi$, where ϕ is the *lead fraction*, whereas B will be leading for a time $n(1 - \phi)$. (In a discrete random walk, "ties" can occur; we can eliminate these by defining the leader in the previous step as the current leader.) This gives our second fundamental question:

- For an n-step random walk, what is the lead fraction ϕ? What is the probability distribution for this lead fraction?

Amazingly, the lead fraction is likely to be close to 0 or 1! That is, one player is in the lead most of the time, even though the average amount won is zero! Conversely, the least likely outcome is the naive expectation that each player should be in the lead for approximately one-half the time.

The ultimate expression of these unexpected results is the beautiful arcsine law for the lead probability. To set the stage for this law, we first study the statistics of number of returns to the origin. We then treat the statistics of lead durations, from which the arcsine law follows simply.

3.4.3.2. Number of Returns to the Origin

It is easy to estimate the typical number of returns to the origin in an n-step walk. Because the occupation probability at the origin is $P(n, 0) \propto 1/\sqrt{n}$, after n steps there are of the order of $n/\sqrt{n} = \sqrt{n}$ returns. Accordingly, returns to the origin should be few and far between! In fact, returns to the origin are clustered because the distribution of times between returns has a $t^{-3/2}$ tail. Thus of the $\sqrt{n}$ returns to the origin, most occur after a very short duration between successive returns (Fig. 3.8).

Now let's consider the probability that an n-step random walk contains m returns to the origin. This can be found conveniently in terms of the first- and the mth-passage probabilities of the random walk. Our treatment of multiple returns follows the discussion given in Weiss and Rubin (1983) and Weiss (1994). Let $F(x, n)$ be the usual first-passage probability that a random walk, which starts at the origin, hits x for the first time at the nth step. From Chap. 1, the generating function, $F(x, z) = \sum_n F(x, n)z^n$, is

$$F(x, z) = \frac{P(x, z) - \delta_{x,0}}{P(0, z)}, \tag{3.4.8}$$

where $P(x, z)$ is the generating function for the occupation probability $P(x, n)$.

Now define the mth-passage probability $F^{(m)}(x, n)$ as the probability that a random walk hits x for the mth time *at* the nth step. This mth-passage probability can be expressed recursively in terms of the $(m - 1)$th-passage probability

$$F^{(m)}(x, n) = \sum_{k=0}^{n} F^{(m-1)}(x, k)F(0, n - k). \tag{3.4.9}$$

That is, for the walk to hit x for the mth time at step n, it must hit x for the $(m - 1)$th time at an earlier step k and then return to x one more time exactly $n - k$ steps later. This recursion is conceptually similar to the convolution that relates the first-passage probability to the occupation probability [Eq. (1.2.1)].

As usual, we can solve this recursion easily by substituting the generating function $F^{(m)}(x, z) = \sum_n F^{(m)}(x, n)z^n$ into the above recursion formula to give $F^{(m)}(x, z) = F^{(m-1)}(x, z)F(0, z) = F(x, z)F(0, z)^{m-1}$. Here we use the definition that $F^{(1)}(x, z) = F(x, z)$. We therefore find

$$F^{(m)}(x, z) = \begin{cases} \dfrac{P(x, z)}{P(0, z)} \left[\dfrac{P(x, z) - \delta_{x,0}}{P(0, z)} \right]^{m-1}, & x \neq 0 \\[4mm] \left[1 - \dfrac{1}{P(0, z)} \right]^m, & x = 0 \end{cases} \tag{3.4.10}$$

What we seek, however, is the probability that x is visited m times *sometime* during an n-step walk. This m-visit probability can be obtained from the mth-passage probabilities as follows. First, the probability that x has been visited *at least* m times in an n-step walk is

$$\sum_{j=1}^{n} F^{(m)}(x, j), \qquad x \neq 0,$$

$$\sum_{j=1}^{n} F^{(m-1)}(0, j), \qquad x = 0. \tag{3.4.11}$$

These express the fact that to visit x at least m times, the mth visit must occur in $j \leq n$ steps. Alternatively, we could sum $F^{(m)}(x, n)$ from m to ∞ to get the probability of visiting at least m times, but this formulation is less convenient for our purposes.

The probability that there are m visits to x sometime during an n-step walk, $G^{(m)}(x, n)$, is just the difference in the probability of visiting at least m times and at least $m + 1$ times. Therefore

$$G^{(m)}(x, n) = \begin{cases} \sum_{j=1}^{n} [F^{(m)}(x, n) - F^{(m+1)}(x, n)], & x \neq 0 \\ \sum_{j=1}^{n} [F^{(m-1)}(0, n) - F^{(m)}(0, n)], & x = 0 \end{cases} \tag{3.4.12}$$

In terms of the generating function for G, these relations become

$$G^{(m)}(x, z) = \begin{cases} \dfrac{1}{1 - z} [F^{(m)}(x, z) - F^{(m+1)}(x, z)], & x \neq 0 \\ \dfrac{1}{1 - z} [F^{(m-1)}(0, z) - F^{(m)}(0, z)], & x = 0 \end{cases} \tag{3.4.13}$$

Let us now focus on returns to the origin. Substituting the mth-passage probability [Eq. (3.4.10)] and also the relation between the first-passage and occupation probabilities [Eq. (1.2.1)] into Eq. (3.4.13), we obtain

$$G^{(m)}(0, z) = \frac{1}{1 - z} \left[1 - \frac{1}{P(0, z)} \right]^{m} \frac{1}{P(0, z)}. \tag{3.4.14}$$

It is easy to extract the asymptotic time dependence from this generating function. Using $P(0, z) = (1 - z^2)^{-1/2} \approx (2s)^{-1/2}$ as $z \to 1$ from below, with $s = 1 - z$, we have

$$G^{(m)}(0, s) \approx \frac{1}{s} (1 - \sqrt{2s})^{m} \sqrt{2s}$$

$$\sim \sqrt{\frac{2}{s}} \, e^{-m\sqrt{2s}}. \tag{3.4.15}$$

This is essentially identical to the Laplace transform of the random-walk occupation probability in approximation (1.3.22)! Thus we conclude that the probability for a random walk to make m visits to the origin during the course of an n-step walk is simply

$$G^{(m)}(0, n) \sim \frac{1}{\sqrt{2\pi n}} e^{-m^2/2n}. \tag{3.4.16}$$

We can obtain this same result more simply by noting that Eq. (3.4.14) is essentially the same as the generating function for the occupation probability at displacement $x = m$ [Eq. (1.3.11)].

Equation (3.4.16) has very amusing consequences. First, from among all the possible outcomes for the number of returns to the origin, the most likely

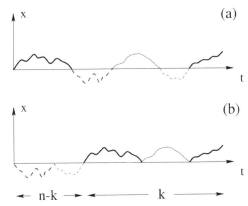

Fig. 3.9. (a) A typical random walk, with k total steps in the region $x > 0$ (solid curves) and $n - k$ steps in $x < 0$ (dashed curves). (b) This configuration is in one-to-one correspondence with a walk in which all the dashed segments are at the beginning and concatenated with all the solid segments. The first element is a return segment of $n - k$ steps whereas the latter is a no-return segment of k steps.

outcome is *zero* returns. In the context of the coin-tossing game, this means that one person would always be in the lead. The probability for $m > 0$ returns to the origin monotonically decreases with m and becomes vanishingly small after m exceeds $\sqrt{n}$. Thus the characteristic number of returns is of the order of $\sqrt{n}$. As a result, ties in the coin-tossing game occur with progressively lower frequency as the game continues, as it is *very* unlikely that the number of returns is of the order of n.

3.4.3.3. *Lead Probability and the Arcsine Law*

We now turn to the lead probability of a random walk. The basic phenomenon is illustrated in Fig. 3.9. For a random walk of n steps, a total of k steps lie in the region $x > 0$, whereas the other $n - k$ steps lie in the region $x < 0$. If x happens to be zero, we again use the tie-breaker rule of Subsection 3.4.3.1 – a walk at the origin is considered as having $x \gtrless 0$ if the location at the previous step was positive (negative). We want to find the probability $Q_{n,k}$ for this division of the lead times in an n-step walk.

We use the fact that the occurrence probabilities for each of the first-passage segments of the walk in Fig. 3.9 are independent. Hence they can be interchanged freely without changing the overall occurrence probability of the walk. We therefore move all the negative segments to the beginning and all the positive segments to the end. This rearranged walk can be viewed as composite of a return segment of length $n - k$ and a no-return segment of length

k. For convenience, we now choose n to be even while k must be even. The occurrence probability of this composite path is just the product of the probabilities for the two constituents. From Eq. (3.4.5), the probability of the return segment is $\mathcal{R}(k)/2^k$, whereas that of the no-return segment is $\mathcal{R}(n-k)/2^{n-k}$. Thus the probability for all paths that are decomposable as in Fig. 3.9 is

$$Q_{n,k} = 2^{-n}\,\mathcal{R}(k)\mathcal{R}(n-k)$$

$$\sim \frac{1}{\pi\sqrt{k(n-k)}}. \tag{3.4.17}$$

As advertised previously, this probability is maximal for $k \to 0$ or $k \to n$ and the minimum value of the probability occurs when $k = n/2$. For large n,

$$Q_{n,k} \to Q(x) = \frac{1}{n\pi}\frac{1}{\sqrt{x(1-x)}}, \tag{3.4.18}$$

where $Q(x)$ is the probability that the lead fraction equals x in an n-step walk.

In addition to the probability that player A is in the lead for exactly k out of n steps, consider the probability $L_{k,n}$ that A is in the lead for *at least* k steps. In the limit $k, n \to \infty$ with $x = k/n$ remaining finite, $L_{n,k} \to L(x)$ becomes

$$L(x) = \int_0^k \frac{dk'}{\pi\sqrt{k'(n-k')}}$$

$$= \frac{2}{\pi}\sin^{-1}\sqrt{x}, \tag{3.4.19}$$

where $x = k/n$. This is the *continuous arcsine distribution*.

The strange consequences of this arcsine distribution arise from the fact that the underlying density $1/\sqrt{x(1-x)}$ has a square-root singularity near the extremes of $x \to 0$ and $x \to 1$, as shown in Fig. 3.10. It is easy to devise paradoxical numerical illustrations of this distribution and many are given in Feller's book [Feller (1968)]. One nice example is our two-person coin-tossing game that is played once every second for 1 year. Then, with probability $1/2$, the unluckier player, which could be either A or B, will be leading in this game for 53.45 days or fewer – that is, less than 0.146 of the total time. With probability $1/10$ (inset of Fig. 3.10), the unluckier player will be leading for 2.24 days or fewer – less than 0.00615 of the total time!

As a final note, when there is a finite bias in the random walk, almost all of the subtleties associated with the tension between certain return and infinite return time disappear. A random walk becomes aware of its bias once the Péclet number becomes large. Beyond this point, all the dynamics is controlled by the bias and a gambler on the wrong side of the odds will always be losing.

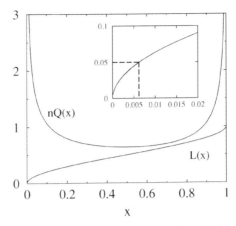

Fig. 3.10. The arcsine distribution. Shown are both the underlying density $nQ(x)$ (upper curve), and the arcsine distribution itself $L(x)$ (lower curve). The inset shows $L(x)$ near $x = 0$. The dashed line highlights the numerical example that, at $x \approx 0.00615$, $L(x)$ has already risen to 0.05.

3.5. Imperfect Absorption

3.5.1. Motivation

We now investigate the role of imperfect boundary absorption on first-passage and related properties of diffusion on the semi-infinite interval. Imperfect absorption means that a random walk may be either reflected or absorbed when it hits the origin. In the continuum limit, this leads naturally to the *radiation boundary condition*, in which the concentration is proportional to the flux at $x = 0$. This interpolates between the limiting cases of perfect absorption and perfect reflection as the microscopic absorption rate is varied. We will discuss basic physical features of this radiation boundary condition and also discuss how imperfect absorption in the semi-infinite interval is equivalent to perfect absorption in the long-time limit.

From a physical perspective, we will also show that a semi-infinite system with a radiation boundary condition is equivalent to an infinite system with free diffusion for $x > 0$ and attenuation for $x < 0$ [Ben-Naim, Redner, & Weiss (1993b)]. The latter provides a simple model for the propagation of light in a turbid medium, such as a concentrated suspension or human tissue. A general discussion of this phenomenon is given in Chap. 7 of Weiss (1994). This equivalence means that we can understand properties of diffusive propagation inside an attenuating medium in terms of the concentration *outside* the medium.

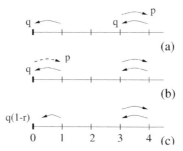

Fig. 3.11. Illustration of (a) absorbing, (b) reflecting, and (c) radiation boundary conditions. The solid arrows denote the bulk hopping probabilities, and the dashed arrow denotes the hopping probability from the origin to site 1.

3.5.2. Radiation Boundary Condition

To see how the radiation boundary condition arises, consider a discrete random walk on $x \geq 0$ with a probability p of hopping to the right and q to the left (see Fig. 3.11). Let $P(i, n)$ be the probability that the particle is at site i at the nth time step. For $i \geq 2$, the occupation probabilities obey the master equations

$$P(i, n + 1) = p P(i - 1, n) + q P(i + 1, n), \qquad (3.5.1)$$

whereas the master equation for site $i = 1$ is distinct because the boundary site $i = 0$ alters the rate at which probability from site 0 reaches 1.

To set the stage for the radiation condition, consider first the absorbing and reflecting boundary conditions. For an absorbing boundary, a particle is absorbed once it reaches site 0 and there is no flux from site 0 to 1. Consequently, the master equation for site 1 is simply

$$P(1, n + 1) = q P(2, n). \qquad (3.5.2)$$

This idiosyncratic equation can be recast in the same form as Eqs. (3.5.1) if the absorbing boundary condition $P(0, n) = 0$ is imposed. In the continuum limit, this clearly translates to zero interface concentration, $c(0, t) = 0$.

For the reflecting boundary, the net probability flux across bond $(i, i + 1)$ at the nth step is simply

$$j_{i,i+1} = p P(i, n) - q P(i + 1, n). \qquad (3.5.3)$$

Zero flux at the boundary means that $p P(0, n) = q P(1, n)$. We can write the flux in a more useful form by defining $p = \frac{1}{2} + \epsilon$ and $q = \frac{1}{2} - \epsilon$, so that

$$j_{i,i+1}(n) = \left(\frac{1}{2} + \epsilon\right) P(i, n) - \left(\frac{1}{2} - \epsilon\right) P(i + 1, n)$$

$$= \frac{1}{2}[P(i, n) - P(i + 1, n)] + \epsilon[P(i, n) + P(i + 1, n)]. \qquad (3.5.4)$$

By Taylor expanding, we obtain the familiar continuum expression for the flux $j(x, t) = -D\frac{\partial c(x,t)}{\partial x} + vc(x, t)$, with diffusion coefficient $D = 1/2$ and velocity $v = 2\epsilon$. For an isotropic random walk, the reflecting boundary condition therefore leads to a zero concentration gradient at the interface.

Finally, we turn to the boundary condition for partial absorption and partial reflection. There is subtlety in formulating the simplest situation that quickly leads to the correct result, as discussed thoroughly in Weiss (1994). We adopt the simple choice that if the random walk hops to the left from site 1, it is reflected and immediately returned to site 1 probability r, whereas with probability $1 - r$ the walk is trapped permanently at site 0. This means that the net flux to site 0 is simply $j_{0,1}(n) = -(1 - r)q P(1, n)$ (Fig. 3.11(c)). On the other hand, comparing this with the general form of the flux in Eq. (3.5.3) leads to $pP(0, n) = rqP(1, n)$. By Taylor expanding this relation to lowest order, we find the radiation boundary condition for isotropic diffusion,

$$\left.\frac{\partial c(x, t)}{\partial x}\right|_{x=0} = \kappa c(0, t), \qquad (3.5.5)$$

where $\kappa = (1 - r)/r\Delta x$ has the units of an inverse length.

To appreciate the physical meaning of the radiation boundary condition, consider diffusion in the one-dimensional domain $x > 0$, subject to the radiation boundary condition at $x = 0$ and the initial condition $c(x, t = 0) = c_0$ for $x > 0$. This problem is easily solved in the Laplace domain, and the resulting time-dependent concentration is

$$c(x, t) = c_0 \operatorname{erf}\left(\frac{x}{\sqrt{4Dt}}\right) + c_0 e^{(\kappa x + \kappa^2 Dt)}\operatorname{erfc}\left(\frac{x}{\sqrt{4Dt}} + \kappa\sqrt{Dt}\right). \quad (3.5.6)$$

In the long-time limit, this solution smoothly approaches that of an absorbing boundary, as illustrated in Fig. 3.12. There are, however, two vestiges of the radiation boundary condition as $t \to \infty$. First, at $x = 0$ the concentration is not zero, but rather approaches zero as $c_0/\kappa\sqrt{Dt}$. Second, the concentration linearly extrapolates to zero at $x = -1/\kappa$. For $\kappa \gg 1$, the interface concentration and the extrapolation length $\ell \equiv 1/\kappa$ are both small and the boundary condition is close to perfectly absorbing. Conversely, for $\kappa \ll 1$, the concentration remains nearly constant and the boundary condition is nearly reflecting.

The finite extrapolation length is suggestive of a "hidden" nonzero concentration in the region $x < 0$. We now discuss how to turn this vague observation into a mapping between the semi-infinite system with a radiating boundary, and an infinite composite medium with free diffusion for $x > 0$ and attenuation for $x < 0$.

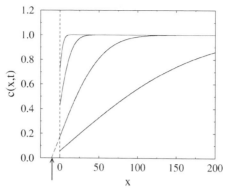

Fig. 3.12. Concentration $c(x, t)$ in Eq. (3.5.6) vs. x for $Dt = 10, 100, 1000$, and $10,000$ (top to bottom), with $\kappa = 0.1$ and $c_0 = 1$. In the range $x \lesssim 20$, the linear extrapolations of these curves to the x axis approaches -10 as t increases (arrow). The dashed line shows this extrapolation for the case $Dt = 1000$.

3.5.3. Connection to a Composite Medium

A practical example of a composite medium with attenuation for $x < 0$ and free diffusion for $x > 0$ is the propagation of light in human tissue. In various diagnostic situations, laser light is incident upon a sample and some fraction of this incident light eventually reemerges. Diffusion with attenuation provides a simple description for propagation in the sample, and useful information about its physical properties may be gained from the intensity and the spatial distribution of the reemitted light. Various practical aspects of this system are reviewed in Weiss (1994); see also Ishimaru (1978) and Bonner et al. (1987) for additional details. Although the real system is three dimensional, we discuss a one-dimensional composite because it illustrates the main features of diffusion with attenuation and it is easily soluble [Ben-Naim, Redner, & Weiss (1993b)].

3.5.3.1. Diffusion–Absorption Equation

For diffusion in an unbounded one-dimensional medium, with attenuation at rate q for $x < 0$ and free propagation for $x > 0$, the concentration is described by a *diffusion–absorption* equation for $x < 0$ and the diffusion equation for $x > 0$ (see Fig. 3.13):

$$\frac{\partial c(x, t)}{\partial t} = D\frac{\partial^2 c(x, t)}{\partial x^2} - qc(x, t), \quad x < 0,$$

$$\frac{\partial c(x, t)}{\partial t} = D\frac{\partial^2 c(x, t)}{\partial x^2}, \qquad x > 0. \qquad (3.5.7)$$

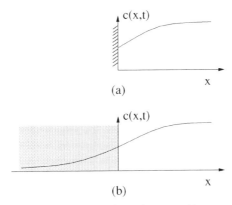

Fig. 3.13. Illustration of the correspondence between (a) a semi-infinite system $x > 0$ with a radiation boundary condition at $x = 0$, and (b) an infinite system with free propagation for $x > 0$ and attenuation for $x < 0$.

In parallel with our earlier discussion of diffusion with an absorbing boundary, we again consider the two natural initial conditions of a single particle and a uniform concentration $c(x, t = 0) = c_0$ for $x > 0$.

Single-Particle Initial Condition. It is simplest to locate the particle initially at the origin because it renders the Laplace transforms of Eqs. (3.5.7) homogeneous. The general solutions to these equations in each subdomain are then the single exponentials $c_>(x, s) = Ae^{-x\sqrt{s/D}}$, for $x > 0$, and $c_<(x, s) = Be^{x\sqrt{(s+q)/D}}$, for $x < 0$. By applying the joining conditions of continuity of the concentration and $D(c'_> - c'_<)|_{x=0} = -1$, we obtain

$$c(x, s) = \begin{cases} \dfrac{1}{\sqrt{Ds}} \dfrac{1}{1 + \alpha(s)} \exp\left(x\sqrt{(s + q)/D}\right), & x < 0 \\[3mm] \dfrac{1}{\sqrt{Ds}} \dfrac{1}{1 + \alpha(s)} \exp\left(-x\sqrt{s/D}\right), & x > 0 \end{cases}, \qquad (3.5.8)$$

where $\alpha(s) = \sqrt{(s + q)/s}$.

To appreciate the meaning of these results, let us focus on the following characteristic quantities of penetration into the attenuating medium:

- The survival probability, $S(t) = \int_{-\infty}^{+\infty} c(x, t)\, dx$.
- The total concentration within the absorbing medium, defined as $C_-(t) = \int_{-\infty}^{0} c(x, t)\, dx$.
- The concentration at the origin.

The first two quantities are simply related. By integrating the first of Eqs. (3.5.7) over the half space $x < 0$, we find

$$\frac{dC_-(t)}{dt} = D\frac{\partial c(x,t)}{\partial x}\bigg|_{x=0} - qC_-(t). \tag{3.5.9}$$

Thus $C_-(t)$ decreases because of absorption within $x < 0$, but is replenished by flux entering at $x = 0$. Additionally, by adding to Eq. (3.5.9) the result of integrating the second of Eqs. (3.5.7) over all $x > 0$, we find $\dot{S}(t) = -qC_-(t)$.

From the concentration in Eq. (3.5.8), we obtain the basic observables in the time domain

$$S(t) = I_0(t/2t_q)\,e^{-t/2t_q},$$

$$C_-(t) = \frac{1}{2}[I_0(t/2t_q) - I_1(t/2t_q)]\,e^{-t/2t_q},$$

$$c(x=0,t) = \frac{t_q}{\sqrt{4\pi Dt^3}}[1 - \exp(-t/t_q)], \tag{3.5.10}$$

where $I_n(x)$ is the modified Bessel function of the first kind of order n [Abramowitz & Stegun (1972)] and $t_q = 1/q$ is the characteristic absorption time of the medium. For short times $t \ll t_q$, the survival probability is close to one, whereas the total probability in the attenuating medium grows linearly with time because of the initial diffusive penetration. Correspondingly, the density at the origin decreases as $(4\pi Dt)^{-1/2}$, as in pure diffusion. In this short-time limit, attenuation does not yet play a role.

Conversely, from the large-z expansion,

$$I_n(z) = \frac{e^z}{\sqrt{2\pi z}}\left(1 - \frac{n-1}{8z} + \cdots\right),$$

we obtain the long-time behaviors for $t/t_q \to \infty$:

$$S(t) \sim (t_q/\pi t)^{1/2},$$

$$C_-(t) \sim \frac{1}{2\sqrt{\pi}}(t_q/t)^{3/2}. \tag{3.5.11}$$

Here the survival probability decays as if there were a perfect absorber at the origin. This arises because a diffusing particle makes arbitrarily long excursions into the attenuating medium so that the attenuation will eventually be total. It is in this sense that an attenuating medium has the same effect as a perfectly absorbing medium in the long-time limit.

Uniform Initial Condition. For the initial condition, $c(x, t = 0) = c_0$ for $x > 0$ and $c(x, t = 0) = 0$ for $x < 0$, the Laplace transform of the concentration is

$$c(x, s) = \begin{cases} \dfrac{c_0}{s} \dfrac{1}{\alpha(s)} \exp(x\sqrt{(s+q)/D}), & x < 0 \\ \dfrac{c_0}{s}\left[1 - \dfrac{1}{1+\alpha(s)^{-1}} \exp(-x\sqrt{s/D})\right], & x > 0 \end{cases}, \quad (3.5.12)$$

from which we find the time dependences

$$c(0, t) = \frac{c_0}{2}[I_0(t/2t_q) + I_1(t/2t_q)] e^{-t/2t_q} \rightarrow \frac{c_0}{\sqrt{\pi t/t_q}}, \quad t \gg t_q$$

$$C_-(t) = \frac{c_0 \ell}{\sqrt{\pi t/t_q}}(1 - e^{-t/t_q}) \qquad \rightarrow \frac{c_0 \ell}{\sqrt{\pi t/t_q}} \quad t \gg t_q. \quad (3.5.13)$$

Here $\ell = \sqrt{D/q} = \sqrt{Dt_q}$ is the typical distance that a particle travels in the attenuating medium before being absorbed. We will show that this distance coincides with the previously introduced extrapolation length. Note also that the concentration at the origin is proportional to the total concentration inside the absorber, $c(x = 0, t) \sim C_-(t)/\ell$.

Finally, by using Eq. (3.5.9) and neglecting the subdominant time-derivative term in this equation, we find that the flux to the origin is, asymptotically,

$$D\frac{\partial c(x, t)}{\partial x}\bigg|_{x=0} \sim qC_-(t) \sim Dc_0/\sqrt{\pi Dt}. \quad (3.5.14)$$

Thus in the long-time limit, the flux to the origin is the same as that to a perfect trap. This is another example of how the region $x < 0$ becomes infinitely attenuating as $t \rightarrow \infty$.

3.5.4. Equivalence to the Radiation Boundary Condition

We now deduce the basic equivalence between an infinite composite medium and a semi-infinite medium with a radiation boundary condition at $x = 0$. From the previous results for $c(0, t)$ and the gradient at the origin, the concentration near the interface has the asymptotic behavior

$$c(x, t) \sim \frac{c_0}{\sqrt{\pi t/t_q}} + \frac{c_0}{\sqrt{\pi Dt}}x, \quad t \gg t_q. \quad (3.5.15)$$

This has the same qualitative form as the concentration for diffusion near a radiating boundary (see Fig. 3.12). In fact, from approximation (3.5.15) the interface concentration and its derivative are simply related by

$$D\frac{\partial c(x, t)}{\partial x}\bigg|_{x=0} = \sqrt{Dq}\, c(0, t). \quad (3.5.16)$$

This is just the radiation boundary condition! By comparing with Eq. (3.5.5), we also infer that $\kappa = \sqrt{q/D} = 1/\ell$, so that the attenuation and the extrapolation lengths are the same.

3.6. The Quasi-Static Approximation

3.6.1. Motivation

Our discussion of first passage in a semi-infinite domain closes by a presentation the *quasi-static approximation* [see e.g., Reiss, Patel, & Jackson (1977)]. This simple approach provides an appealing and versatile way to bypass many technical complications in determining the concentration profile in complex boundary-value problems (see Fig. 3.14). The physical basis of this approximation is that the system can be divided into a "far" region, where the concentration has not yet been influenced by the boundary, and a complementary "near" region, where diffusing particles can explore this zone thoroughly. In this near zone, the concentration should therefore be almost steady, or quasi static, and can be described accurately by the time-independent (and simpler) Laplace equation rather than by the diffusion equation. Then, by matching the near and the far solutions, we obtain the global solution. This approach is now described for both the absorbing and radiation boundary conditions.

3.6.2. Quasi-Static Solution at an Absorbing Boundary

We consider the trivial (and previously solved) concentration profile in the one-dimensional domain $x > 0$ for the initial condition $c(x, t = 0) = 1$ for $x > 0$ and with absorption at $x = 0$. Although the exact solution is nearly as easy to obtain as the quasi-static solution, the latter is much more versatile and can be easily applied to the same geometry in higher dimensions, whereas

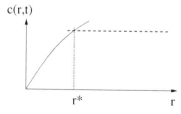

Fig. 3.14. Schematic illustration of the quasi-static approximation for an absorbing boundary condition. The respective near-zone (solid curve) and far-zone (dashed line) concentrations match at $r = r^* \simeq \sqrt{Dt}$.

the exact solution is much more difficult to obtain. A specific illustration of this is given in Subsection 6.5.4.

According to the quasi-static approximation, we first identify the spatial domain $0 < x < \sqrt{Dt}$ as the near zone, where the concentration obeys the Laplace equation $c'' = 0$. Conversely, in the far zone $x > \sqrt{Dt}$, the concentration remains at its initial value. The subtle issue is how to match the solutions from these two zones. The simplest approach is to adjust the amplitude of the near-zone solution so that it equals the far-zone solution at $x^* = \sqrt{Dt}$. Because x^* has a time dependence, this matching induces a time dependence into the Laplacian solution. Thus the near-zone concentration profile is actually quasi static, since it is the solution to the Laplace equation with a time-dependent boundary condition. However, this moving boundary condition is generally easy to apply because of the simplicity of the near-zone solution.

For our example, the near-zone solution is simply $c(x, t)_{\text{near}} = A + Bx$, and we determine A and B by imposing the absorbing boundary condition $c(0, t) = 0$ and the matching condition $c(x = \sqrt{Dt}, t) = 1$. These immediately give

$$c(x, t) \simeq \begin{cases} \dfrac{x}{\sqrt{Dt}} & \text{for } 0 < x < \sqrt{Dt} \\[2mm] 1 & \text{for } x > \sqrt{Dt} \end{cases} \qquad (3.6.1)$$

This simple form reproduces the basic features of the exact solution given in Eq. (3.3.7). In particular, the flux to the origin decays as $\sqrt{D/t}$, compared with the exact result of $\sqrt{D/\pi t}$.

3.6.3. Quasi-Static Solution at a Radiation Boundary

For a radiation boundary condition in one dimension, we again solve the Laplace equation in the near zone $0 < x < \sqrt{Dt}$ and match it to the constant far-zone concentration at $x = \sqrt{Dt}$. The near-zone solution is now subject to the radiation boundary condition $Dc' = \sqrt{Dq}\,c\big|_{x=0}$ (Eq. (3.5.16)). By applying this boundary condition and the matching condition at $x = \sqrt{Dt}$ to the generic near-zone solution, $c(x, t)_{\text{near}} = A + Bx$, we easily find

$$c(x, t)_{\text{near}} = \frac{c_0}{1 + \sqrt{t/t_q}} \left(1 + \frac{x}{\ell}\right). \qquad (3.6.2)$$

It is easy to verify that this very simple result reproduces the main features of the exact concentration profile near a radiating boundary. Namely, the concentration at the interface asymptotically decays as $\sqrt{t_q/t}$, the slope of the concentration at the interface is $1/\sqrt{Dt}$, and the concentration profile extrapolates to zero at $x = -\ell$.

4

Illustrations of First Passage in

Simple Geometries

4.1. First Passage in Real Systems

The first-passage properties of the finite and the semi-infinite interval are relevant to understanding the kinetics of a surprisingly wide variety of physical systems. In many such examples, the key to solving the kinetics is to recognize the existence of such an underlying first-passage process. Once this connection is established, it is often quite simple to obtain the dynamical properties of the system in terms of well-known first-passage properties

This chapter begins with the presentation of several illustrations in this spirit. Our first example concerns the dynamics of integrate-and-fire neurons. This is an extensively investigated problem and a representative sample of past work includes that of Gerstein and Mandelbrot (1964), Fienberg (1974), Tuckwell (1989), Bulsara, Lowen, & Rees (1994), and Bulsara et al. (1996). In many of these models, the firing of neurons is closely related to first passage in the semi-infinite interval. Hence many of the results from the previous chapter can be applied immediately. Then a selection of self-organized critical models [Bak, Tang, & Wiesenfeld (1987) and Bak (1996)] in which the dynamics is driven by extremal events is presented. These include the Bak–Sneppen (BS) model for evolution [Bak & Sneppen (1993), Flyvbjerg, Sneppen, & Bak (1993), de Boer et al. (1994), and de Boer, Jackson, & Wetting (1995)] directed sandpile models [Dhar & Ramaswamy (1989)], traffic jam models [Nagel & Paczuski (1995)], and models for surface evolution [Maslov & Zhang (1995)]. All of these systems turn out to be controlled by an underlying first-passage process in one dimension that leads to the ubiquitous appearance of the exponent $-3/2$ in the description of avalanche dynamics.

We next turn to the kinetics of spin systems, specifically the Ising–Glauber [Glauber (1963)] and the Voter models [Clifford & Sudbury (1973), Holley & Liggett (1975), Liggett (1985), and Cox & Griffeath (1986)]. In both these examples, we will discuss how the evolution of correlations in the system can be viewed abstractly as the propagation of diffusing particles from a point source into an infinite medium. This correspondence is valid in one dimension for the Ising–Glauber model, but applies to the Voter-model in all dimensions. Thus the transition between transience and recurrence of diffusion in spatial dimension $d = 2$ is responsible for the transition in Voter-model dynamics in which there is eventual unanimity for $d < 2$ and perpetual evolution for $d > 2$. This connection to a diffusive system also provides an easy way to understand the asymptotic kinetics of the one-dimensional Ising–Glauber model.

Then substantive extensions of the basic first-passage formalism to more complex processes are presented. These include first passage in composite systems in which, e.g., different bias velocities exist in different subintervals [Frisch et al. (1990), Privman & Frisch (1991), and Le Doussal, Monthus, & Fisher (1999)]. A fundamental equivalence between such a composite and an effective homogeneous system with a uniform renormalized bias velocity is demonstrated. First passage in fluctuating systems [Doering & Gadoua (1992) and Bier & Astumian (1993)] is also discussed, in which the interplay between the internal fluctuations of the system and diffusive fluctuations may lead to resonant phenomena. It is shown how the time-integrated electrostatic formulation of Section 1.6 provides considerable simplification in computing first-passage properties. We solve a generic example and reproduce the intriguing resonant behavior in the mean first-passage time as a function of the frequency of fluctuations. We also discuss the exit from an interval with a spatially varying diffusion coefficient and provide criteria for normal and anomalous behavior of this exit time.

Finally, we study the survival of a diffusing particle in an absorbing interval whose length $L(t)$ grows systematically with time. When $L(t)$ grows either slower or faster than diffusion, simple behavior occurs that we can understand by the correspondingly simple means of the adiabatic [Schiff (1968)] and "free" approximations. In the marginal case in which $L(t)$ grows as $\sqrt{At}$, particularly intriguing behavior arises [Breiman (1966), Uchiyama (1980), Lerche (1986), Salminen (1988), Iglói (1992), Turban (1992), and Krapivsky & Redner (1996b)], in part because of the existence of an dimensionless parameter, namely A/D, where D is the diffusion coefficient. By adapting the previously mentioned simple approaches from the nonmarginal limits to the marginal case, we find that the survival probability $S(t)$ decays as a *nonuniversal* power law in time, $S(t) \sim t^{-\beta}$, with β dependent on A/D. We also derive parallel results for a diffusing particle in the presence of a single moving, absorbing boundary.

These diverse examples provide some hints of the range of issues and physical systems that fall within the rubric of first passage.

4.2. Neuron Dynamics

4.2.1. Some Basic Facts

A neuron is a specialized cell that is a fundamental discrete unit of the central nervous system. In the human body there are of the order of 10^{10} neurons [see, e.g., Amit (1989) for a general perspective]. One important feature of a neuron is that it emits a train of action potentials, or voltage spikes. Considerable effort has been devoted to understanding the distribution of times between voltage spikes, the *interspike interval* (ISI), as this provides basic clues about underlying neurophysiological processes. Useful reviews are given by Fienberg (1974) and Tuckwell (1988, 1989). The basic mechanism for these voltage spikes is that the polarization level in a neuron changes by a small amount because of the influence of excitatory and inhibitory voltage inputs. When the polarization level in the neuron first exceeds a specified level θ, the neuron emits an action potential and the state of the neuron quickly returns to a reference level (Fig. 4.1). Thus the time between action potentials is governed by the first-passage probability for the polarization to reach the threshold level. In the following, we discuss some simple models for this dynamics and its implications for ISI statistics.

4.2.2. Integrate-and-Fire Model

One of the simplest models for neuron dynamics whose predictions can account for the ISI distributions of real neurons is the integrate-and-fire model [Gerstein & Mandelbrot (1964)]. However, the model incorporates few of the

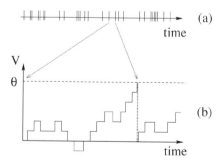

Fig. 4.1. (a) Schematic spike train that is due to a sequence of neuron-firing events. (b) Basic mechanism for the creation of an action potential spike for an integrate-and-fire neuron (finer time scale). The potential of a neuron fluctuates stochastically because of excitatory and inhibitory inputs. When the potential first reaches a threshold level θ, the neuron "fires" and its potential quickly returns to a resting level.

realities of nerve–cell behavior, and the parameters of the model cannot be meaningfully related to physiological variables. In spite of these shortcomings, the model is wonderfully simple and it provides a starting point both for understanding basic features of the ISI distribution and for more realistic modeling.

In the model, excitatory or inhibitory inputs can occur with respective rates λ_E and λ_I. This stimulus causes the polarization of the neuron to correspondingly increase by a_E or decrease by a_I. We use the simplifying assumption that each such input is sufficiently small that we can approximate the discrete evolution of the potential by a diffusive process with bias velocity $v = \lambda_E a_E - \lambda_I a_I$ and diffusivity $D = \sqrt{\lambda_E a_E^2 + \lambda_I a_I^2}$. By this formulation, the potential first reaching a fixed threshold value θ is identical to the first passage to the origin of a biased diffusing particle that starts at the point θ in the semi-infinite interval [Eq. (3.2.13)]. By translating this result to the variables of the present system, the distribution of times between action potentials, or the ISI distribution, is simply

$$F(t) = \frac{\theta}{\sqrt{4\pi D t^3}} e^{-(\theta - vt)^2/4Dt}. \qquad (4.2.1)$$

This distribution is sometimes called the inverse Gaussian density [Chikara & Folks (1989)]. As shown in Fig. 4.2, this distribution rapidly goes to zero in both the short- and the long-time limits and has a long-time power-law tail that eventually gets cut off by an exponential function of time, with the characteristic time for the exponential determined by the relative bias between excitatory and inhibitory inputs.

The basic features of this distribution are quite rich. From the moments of the distribution $\langle t^k \rangle = \int t^k F(t)\, dt$, the mean ISI time and its variance are

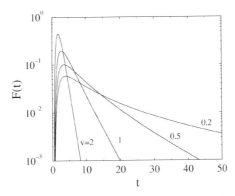

Fig. 4.2. The inverse Gaussian distribution of ISIs on a semilogarithmic scale for $\theta = D = 1$ for the values of v indicated.

given by

$$\langle t \rangle = \frac{\theta}{v},$$

$$\sigma^2 = \langle t \rangle^2 - \langle t \rangle^2 = \frac{D\theta}{v^3} = \frac{\langle t \rangle^2}{Pe}, \qquad (4.2.2)$$

where $Pe = v\theta/D$ is the Péclet number. Although the mean ISI time involves only the threshold and the bias, the variance also depends on the fluctuations in the inputs. Note also that the variance is much larger than $\langle t \rangle^2$ and that the ratio $\sigma^2/\langle t \rangle^2$ diverges as the bias vanishes. This is a manifestation of the power-law tail in $F(t)$ when there is no bias.

Another interesting aspect of the inverse Gaussian density is the dependence of the most probable value of the ISI time, t_{mp}, on basic system parameters. From Fig. 4.2, we see that $\langle t \rangle$ changes significantly as v changes, whereas t_{mp} is much more slowly varying. The location of the peak in $F(t)$ is simply given by

$$t_{mp} = \frac{3D}{v^2}(\sqrt{1 + Pe^2/9} - 1). \qquad (4.2.3)$$

Thus as $Pe \to \infty$, $t_{mp} \to \langle t \rangle$. This merely reflects the relative unimportance of fluctuations in the ISI when the bias is the dominant driving mechanism. When fluctuations are small we should therefore expect that the mean and most probable ISI times become the same. Conversely, as $Pe \to 0$, $t_{mp} \to \theta^2/6D$; that is, t_{mp} becomes roughly of the order of the time for a particle to diffuse a distance θ. On the other hand t_{mp} also approaches $\langle t \rangle Pe/6$. Thus $\langle t \rangle$ becomes arbitrarily large as $Pe \to 0$; this stems from the $t^{-3/2}$ tail in the ISI distribution.

To incorporate more physiological realism, a wide variety of extensions of the basic integrate-and-fire model have been proposed and investigated. The types of processes included in these more refined models include a "leaky" neuron, e.g., an Orstein–Uhlenbeck process [Uhlenbeck & Ornstein (1930)] for the fluctuations of the neuron itself [Tuckwell (1988)], or a periodically varying threshold [Bulsara et al. (1994, 1996); see also Fletcher, Havlin, & Weiss (1988) for a simple realization of this periodic process]. Visually, the effect of a periodically varying threshold is quite dramatic as it leads to complex behavior for the ISI distribution in which the asymptotic decay contains a complex train of modulations.

Because of its simplicity, it is tempting to apply the image method to the ISI problem with a modulated threshold [Bulsara et al. (1994, 1996)]. However, this method represents an uncontrolled approximation as there is no simple distribution of images that can reproduce the periodically varying absorbing boundary condition. Nevertheless, the image solution reproduces many of the qualitative features of the ISI distribution that can be obtained by numerical simulation or by more rigorous calculational methods Bulsara et al. (1996). The general problem of the first passage to a moving boundary with nonmonotonic motion still appears to be relatively open.

As a parenthetical note, the simplest version of the integrate-and-fire model has recently been proposed to account for a simple sliding friction experiment, in which a solid block, initially at rest on a plane with subcritical inclination angle, is subjected to a sudden impulse. The impulse causes the block to slide a finite distance x downslope bfore coming to rest again. Because of local surface heterogeneities, the same initial condition leads to a distribution of stopping distances $P(x)$ which was observed to have a $x^{-3/2}$ tail (Brito & Gomes (1995)). This observation can be accounted for by a positing that the kinetic energy of the block undergoes a biased random walk with drift toward zero because of friction, but with superimposed fluctuations due to the local heterogeneities (Lima et al. (2000)). This leads to precisely the same dynamics as the integrate-and-fire model. From this analogy, the $x^{-3/2}$ distribution of stopping distances follows naturally.

4.3. Self-Organized Criticality

Self-organized criticality has emerged as an attractive principle to describe the dynamical behavior of a wide variety of natural phenomena, such as earthquakes, $1/f$ noise in resistors, traffic flow fluctuations, relaxation in granular materials, etc. [see, e.g., Bak et al. (1987) and Bak (1996)]. In generic models of these systems, simple local dynamical rules give rise to macroscopic dynamics that exhibits fluctuations on all time scales, as well as power-law temporal correlations. The appearance of such power laws without the apparent need to tune microscopic parameters to critical values is the hallmark of self-organized criticality.

The purpose of this section is modest. The basic principles of this phenomenon are not explained nor is an attempt made to justify self-organized critical models. Instead, several examples of these systems are presented that can be naturally described in terms of first-passage processes. In many of these situations, the underlying dynamics is simply that of the first-passage probability of a one-dimensional random walk. Thus the ubiquitous "avalanche" exponent value of $-3/2$ just emerges as the exponent for the decay of the first-passage probability in one dimension. It is striking that many apparently disparate self-organized phenomena seem to have the same basic driving mechanism.

4.3.1. Isotropic and Directed Sandpile Models

The archetypical model of self-organized criticality is the Bak–Tang–Wiesenfeld (BTW) sandpile model [Bak et al. (1987) and Flyvbjerg (1996)]. The basic physical picture that underlies the model is that of a system in which sand grains are added one at a time to a growing pile. Most of the time, the grains simply get incorporated into the pile where they land. However, a point is eventually reached where the slope of the pile is beyond its natural angle of repose. When this occurs, the next grain added will not come to rest at its impact point, but will roll down the slope. In so doing this grain will typically dislodge other nearby downhill grains that then disturb their downhill neighbors, etc. This leads to an avalanche that reduces the slope of the pile to below a critical angle. The understanding of this sporadic cycle of gradual slope buildup followed by avalanches of all scales when the pile is driven by the steady addition of single grains is one of the major goals of self-organized critical models.

The BTW model attempts to describe this dynamics in a clean and minimalist way by the following discrete-lattice model. Each lattice site $\vec{x}$ is endowed with an integer height variable $h(\vec{x})$ that lies between 0 and h_c. A grain of sand is added at random to one of the lattice sites, so that $h(\vec{x}) \to h(\vec{x}) + 1$. Whenever this updated height reaches h_c, the site "topples" and transfers one grain to each of its nearest neighbors. This is described by the following dynamics:

$$h(\vec{x}) \to h(\vec{x}) - z,$$
$$h(\vec{x}') \to h(\vec{x}') + 1.$$

Here z is the lattice coordination number and $\vec{x}'$ denotes the z nearest neighbors of site $\vec{x}$. Without loss of generality we can take $h_c = z$, so that the height at the toppled site is set to zero. For sites that are on the boundary of the system, the dynamics is similar except that grains can leave the system. After the initial toppling event, if the height at one or more of these neighboring sites reaches h_c, each one of these sites is again updated as above. This updating process is repeated at all supercritical sites until all the heights are again less than h_c.

Because of the openness of this system, a statistical steady state is eventually reached at which the input of single grains is matched by the outflow that is due to intermittent avalanches. Basic dynamical characteristics include the number of sites that topple in an avalanche and the time duration of the avalanche. There are many variations on this basic model, and a very rich array of phenomena is still in the process of being discovered, much of which seems to have an underlying description in terms of first passage [see, e.g., Paczuski & Boettcher (1997), Dickman, Vespignani, & Zapperi (1998),

Vergeles, Maritan, & Banavar (1998), De Menech, Stella, & Tebaldi (1998), and Ktitarev et al. (2000) for a sampling of such publications].

To direct our discussion towards first-passage properties, we now restrict the model by imposing a preferred direction for the motion of the toppled grains [Dhar & Ramaswamy (1989)]. Remarkably, this directed version of the original BTW sandpile model is exactly soluble in all dimensions. More relevant for our purposes is that the dynamics of this directed model in two dimensions is equivalent to the first-passage of a one-dimensional random walk. This immediately gives the basic dynamical properties of the directed sandpile model in two dimensions

It is convenient to define the directed sandpile model on a hypercubic lattice. Each site is defined to have a time coordinate that is the sum of the Cartesian components of $\vec{x}$, $T = \sum_{i=1}^{d} x_i$. The preferred direction for the motion of the grains is along this time axis. Grains are added one at a time to random sites at the top of the lattice (Fig. 4.3). Whenever the height or number of grains at one of these sites $\vec{x}$ reaches h_c (which we may generally choose to be equal to d), the same dynamics as that of the isotropic sandpile model occurs, except that grains can move to only the downstream neighbors of $\vec{x}$. These are the sites where the time coordinate is $T(\vec{x}) + 1$. If the height at one or more of these downstream sites reaches h_c, each one of these sites is then

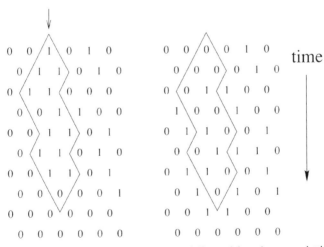

Fig. 4.3. A typical avalanche in the directed sandpile model on the square lattice. The arrow indicates the position where a grain of sand is added, and the numbers are the heights at each site before (left) and after (right) the avalanche. The space–time boundary of the avalanche, enclosing all sites that have toppled, is indicated. This avalanche has a duration of 8 steps, and topplings occur at 14 sites.

updated according to the sandpile dynamics. This process is repeated until all the heights are again less than h_c. Periodic boundary conditions in the transverse direction are imposed, while grains leave at the bottom of the system.

This model has several attractive features that render it exactly soluble [Dhar & Ramaswamy (1989)]. Here we focus on the two-dimensional system that has the additional simplification that all sites that topple on a line of constant T must be contiguous. We can easily appreciate this by following the dynamics of small example systems by hand. Thus the toppled region has no holes, so that the left and the right boundaries of this region each define the space–time trajectory of a nearest-neighbor random walk in one dimension. The distance between these two walks also undergoes a random walk, so that the distribution of times for them to meet has the $t^{-3/2}$ tail of the one-dimensional first-passage probability. Thus one-dimensional first passage naturally applies to directed avalanches in two dimensions, and the probability that an avalanche has a duration t asymptotically decays as $t^{-3/2}$.

4.3.2. Bak–Sneppen Model

Another appealing example of an extremal dynamics system that exhibits self-organized criticality, as well as a connection with first passage, is the *Bak–Sneppen* (BS) model [Bak & Sneppen (1993), Flyvbjery et al. (1993), de Boer et al. (1994), and de Boer, Jackson, & Wettig (1995)]. This was originally introduced as a toy model to describe punctuated equilibrium in the evolution of species, in which long periods of relative stability in the diversity of species are interrupted by short bursts of mass extinctions. The underlying basis of the BS model is to focus on a minimal set of variables that capture the basic features of punctuated equilibrium while ignoring all other details. Although the model thus has the appearance of a "toy," it is phenomenologically rich, generates punctuated equilibrium in an extremely simple fashion, and, for the purposes of this book, its dynamics can be recast as the first-passage of a one-dimensional random walk. By this mapping, many dynamical features of the model can be understood easily.

The BS model describes an ecosystem of N species, each of which is characterized by a single "fitness" variable x_i ($i = 1, 2, \ldots, N$), which, for simplicity, is assumed to be initially distributed uniformly between 0 and 1. Species with small values of x are unfit, whereas those with large values are fit. In a single time step, the weakest species i with the smallest value of x_i becomes extinct, and its ecological niche is replaced with another species whose fitness is randomly drawn from the range [0, 1]. This change modifies the environments of the species at sites $i - 1$ and $i + 1$. Here, we may think of i as indexing trophic levels, so that $i \pm 1$ are the species that would normally

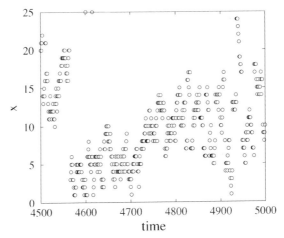

Fig. 4.4. Space–time evolution of avalanches in the BS model in one dimension. Sites with the minimal fitness at any given time are shown by the circles, and sites with subthreshold fitness <0.667 are shown by the lines.

prey on species i or be preyed upon by species i. Accordingly, when species i becomes extinct, the large change in the environments of species $i \pm 1$ leads to their extinction as well. Thus these neighboring species on the food chain are also replaced, with their fitnesses assigned anew uniformly from [0, 1].

With this dynamics, the system self-organizes into a state in which almost all of the species have fitnesses that are above a threshold value, whereas a small fraction of species have subthreshold fitnesses in which sporadic but repeated extinction events occur. An example from a simulation of a small system is shown in Fig. 4.4. It is convenient to use the term avalanche to describe a causally connected sequence of activity that is associated with subthreshold fitness values. For threshold value λ, suppose that at time $t = 0$ all fitnesses happen to be greater than λ. Then the extinction of the least-fit species and its neighbors may lead to the creation of at least one subthreshold species. As these unfit species evolve, a point will eventually be reached at which all fitnesses are again above threshold. This defines a λ avalanche. One basic feature of this avalanche is, How long does it last? As we now discuss, for a mean-field version of the BS model, this property maps directly to a first-passage problem on the semi-infinite interval. This derivation is based on the original presentations in Flyvbjerg et al. (1993) and de Boer et al. (1995).

In the mean-field version of the BS model, all species are considered equivalent and the neighbors of the least-fit species are $K - 1$ randomly chosen species. Thus, in each update, K fitness values are reassigned; for simplicity, we discuss only the case $K = 2$. To characterize the avalanche dynamics, we study $P(n, t)$, the probability that n fitness values are less than λ at time t in a finite ecosystem of N total species. This probability evolves because the number of unfit species can change from n to m with rate $R_{n \to m}$. We now enumerate all the possibilities for the final state m and the events that lead to this final state, from which we will write the master equation for $P(n, t)$.

(i) $R_{n \to n-2}$: Number of unfit species decreases by two.
- Pick the site i with lowest fitness x_i [which is necessarily in the range $(0, \lambda)$] and reassign it a new fitness in $(\lambda, 1)$ – this event occurs with probability $1 - \lambda$.
- Pick the second site j from among the $n - 1$ remaining unfit sites with fitness in $(0, \lambda)$. This occurs with probability $(n - 1)/(N - 1)$. We reassign site j a new fitness in $(\lambda, 1)$ – this occurs with probability $1 - \lambda$.

Adding the rates of these two possibilities gives

$$R_{n \to n-2} = (1 - \lambda)^2 (n - 1)/(N - 1)$$

for $n \geq 2$, whereas the rate is zero if $n = 0$ or 1, as in these cases there is no possibility that the number of unfit species can decrease by two.

(ii) $R_{n \to n-1}$: Number of unfit species decreases by one.
- Reassign the lowest fitness x_i to $(\lambda, 1)$ [with probability $(1 - \lambda)$]. The second site may have fitness either in $(0, \lambda)$ or in $(\lambda, 1)$, and this fitness is reassigned to its original range. The first event occurs with probability $\lambda(n - 1)/(N - 1)$ and the second with probability $(1 - \lambda)[1 - (n - 1)/(N - 1)]$.
- Reassign the lowest fitness x_i to $(0, \lambda)$. Then the second site must have fitness in the range $(0, \lambda)$ and it must be reassigned to $(\lambda, 1)$.

These three processes are illustrated in Fig. 4.5 and give for the rate $R_{n \to n-1}$

$$R_{n \to n-1} = (1 - \lambda) \left[\lambda \left(\frac{n - 1}{N - 1} \right) + (1 - \lambda) \left(1 - \frac{n - 1}{N - 1} \right) \right]$$
$$+ \lambda \left[(1 - \lambda) \frac{n - 1}{N - 1} \right]$$
$$= (1 - \lambda)^2 + \left(\frac{n - 1}{N - 1} \right) (1 - \lambda)(3\lambda - 1).$$

(iii) $R_{n \to n}$: Number of unfit species remains unchanged.
- The lowest fitness is reassigned a new fitness still in $(0, \lambda)$. Then the fitness of the second site – either in $(0, \lambda)$ or $(\lambda, 1)$ – remains in its initial range.

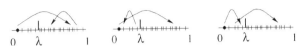

Fig. 4.5. The three basic processes that contribute to $R_{n \to n-1}$. Site fitnesses are indicated by tick marks and the minimal fitnesses by the filled circles. The arrows indicate how the fitnesses are reassigned in a single update event.

- The lowest fitness is reassigned a new fitness in $(\lambda, 1)$ and the second site with fitness in $(\lambda, 1)$ is reassigned to $(0, \lambda)$.

The rates for these events are

$$R_{n \to n} = \lambda(1 - \lambda)\left[1 - \left(\frac{n-1}{N-1}\right)\right]$$
$$+ \lambda\left[\lambda\left(\frac{n-1}{N-1}\right) + (1 - \lambda)\left(1 - \frac{n-1}{N-1}\right)\right]$$
$$= 2\lambda(1 - \lambda) + \lambda(3\lambda - 2)\left(\frac{n-1}{N-1}\right).$$

(iv) $R_{n \to n+1}$: Number of unfit species increases by one.
- The weakest species must be reassigned to $(0, \lambda)$, whereas the second must be initially in $(\lambda, 1)$ and reassigned to $(0, \lambda)$. This occurs with rate λ^2.

Assembling these results, we find that the master equations for $P(n, t)$, for $n \geq 3$, are

$$P(n, t) = \sum_{m=n-1}^{n+2} R_{m \to n} P(m, t). \tag{4.3.1}$$

For $n \leq 3$, the various small-n constraints on the rates lead to the boundary master equations

$$P(2, t) = \lambda^2[P(0, t) + P(1, t)] + 2\lambda(1 - \lambda)P(2, t)$$
$$+ (1 - \lambda)^2 P(3, t),$$
$$P(1, t) = 2\lambda(1 - \lambda)[P(0, t) + P(1, t)] + (1 - \lambda)^2 P(2, t),$$
$$P(0, t) = (1 - \lambda)^2[P(0, t) + P(1, t)]. \tag{4.3.2}$$

These equations simplify considerably in the thermodynamic limit $N \to \infty$. For $n \geq 3$, the master equations reduce to

$$P(n, t) = \lambda^2 P(n - 1, t) + 2\lambda(1 - \lambda)P(n, t) + (1 - \lambda)^2 P(n + 1, t). \tag{4.3.3}$$

After all this bookkeeping, it is striking that the master equation reduces to that for the probability distribution of a biased random walk in one dimension with a reflecting boundary condition at the origin! This boundary condition reflects the fact that if the number of unfit species reaches zero, the random-walk process starts anew. For avalanche dynamics, however, the number of unfit species reaching zero signals the end of the avalanche and we want to track the time for this event. Accordingly, we define $Q(n, t)$ as the probability that there are n subthreshold fitnesses at time t, given that an avalanche started at $t = 0$. The master equation for $Q(n, t)$ is essentially the same as that for $P(n, t)$ except that the rates of going from $0 \to 1$ or $0 \to 2$ are zero. That is, once n reaches zero, the avalanche has finished and Q does not evolve further. Because the avalanche starts at $t = 0$, $P(0, 0) = 1$, and this induces the initial conditions $Q(1, 1) = 2\lambda(1 - \lambda)$, $Q(2, 1) = \lambda^2$, and $Q(n, 1) = 0$ for $n \geq 3$. The solution for $Q(n, t)$ then is [Flyvbjerg et al. (1993) and de Boer et al. (1995)]

$$Q(n, t) = \frac{2n}{t + n + 1} \binom{2t + 1}{t + n} \lambda^{t+n+1}(1 - \lambda)^{t-n+1}. \qquad (4.3.4)$$

Finally, from basic probabilistic grounds, the distribution $q(t)$ that an avalanche has duration t satisfies the relation

$$q(t + 1) = (1 - \lambda)^2 \left[Q(1, t) + \frac{1}{N - 1} Q(2, t) \right]. \qquad (4.3.5)$$

These two terms account for the respective rates at which a system with one and with two unfit species evolves to a state with no unfit species. Using Eq. (4.3.4), we then find that, in the $N \to \infty$ limit,

$$q(t) = \frac{1}{t + 1} \binom{2t}{t} \lambda^{t-1}(1 - \lambda)^{t+1}. \qquad (4.3.6)$$

This expression coincides with the first-passage probability of a one-dimensional biased random walk! For large time, the asymptotic behavior of $q(t)$ is given by

$$q(t) \sim \frac{(1 - \lambda)}{\lambda} \frac{1}{\pi^{1/2}t^{3/2}} [4\lambda(1 - \lambda)]^t, \qquad (4.3.7)$$

so that distribution of avalanche lifetimes has a $t^{-3/2}$ long-time tail for $\lambda \to 1/2$ and an exponential cutoff otherwise.

During the evolution of the system in which $K = 2$ sites are updated in each event, if a point is reached where all the fitness values are greater than $1/2$, then on average we must wait an infinite time until all fitnesses are again greater than $1/2$. More generally, for the mean-field BS model in which K fitness values are updated in each event, the threshold value of λ is $1/K$, at which point a $t^{-3/2}$ tail in the distribution of avalanche durations universally occurs [Flyvbjerg et al. (1993) and de Boer et al. (1995)]. This shows the close connection between the avalanche dynamics of the BS model and first passage in one dimension.

It is worth mentioning a very recent and intriguing extension of the BS model is to an "economy" in one dimension [Nørrelykke & Bak (2000)]. Here traders buy a single good from their right neighbors and sell a single product to their left neighbors in a single trading day. Each individual tries to maximize a generic form for his or her utility function, while satisfying the constraint that the total value of the goods bought and sold by an individual is equal. After a single transaction, the individual with the smallest profit lowers the price of the good he or she produces and the process begins anew. This reassignment of the parameter of the least-fit individual is akin to the extinction event of the original BS model. Once again, we can define the notion of avalanche as a causally connected sequence of events associated with individuals whose profits are less than a specified threshold. Just as in the original BS model, the distribution of avalanche durations has a $t^{-3/2}$ tail.

Perhaps it will soon be possible to see clearly if there is an unambiguous connection between the plethora of self-organized models that exhibit a $t^{-3/2}$ distribution of avalanche times and first-passage of diffusion in one dimension.

4.3.3. Related Systems

The basic connection between the avalanche dynamics in systems that are driven by extremal dynamics and the first-passage of a random walk in the infinite semi-interval appears to be ubiquitous. Two additional examples are described that illustrate this ubiquity.

4.3.3.1. One-Dimensional Traffic Jams

The understanding of traffic flow is clearly a basic problem in modern-day life. In addition to classic models of traffic based on car-following theories [Prigogine & Herman (1971)], a wide variety of simple discrete models, which are partially inspired by ideas from self-organized criticality, has been recently proposed to describe various aspects of traffic flow [see, e.g., Wolf & Schreckenberg (1998)]. Here we discuss a particularly simple one-dimensional model [Nagel & Paczuski (1995)] that leads to intermittent jams and periods of free flow and that is qualitatively analogous to the punctuated equilibrium of the BS model. The dynamics of these jams can also be described in terms of the first-passage of a random walk on the infinite semi-interval.

The system is a special limit of the Nagel–Schreckenberg (NS) model [Nagel & Schreckenberg (1992)] that, loosely speaking, describes aggressive drivers on a one-dimensional road. In the NS model, if there is a large open space in front of a car (headway), a driver is likely to accelerate until a fixed maximum speed v_{max} is reached. Conversely, if the headway is such that a collision would occur in the next time step, the driver decelerates to prevent the collision. Thus each driver generally tries to move as fast as possible at all times consistent with not hitting the preceding car. The precise rules of the model are the following:

- Velocity update steps:
    - (i) A *free* car that is moving at v_{max} continues at this speed if the length n of the empty space in front (the headway) satisfies $n \geq v_{max}$.
    - (ii) With probability $1/2$, a car with speed $v < v_{max}$ and headway $n > v$ accelerates to speed $v + 1$. Otherwise, the speed is maintained. The acceleration ends when v_{max} is reached.
    - (iii) A car with $v > n$ slows down to the maximum safe speed $v = n$ with probability $1/2$ or to the larger of $n - 1$ and 0 otherwise. The last two events mean that, with probability $1/2$, the driver overreacts and slows down more than needed.
- Movement step: Each car advances by $v \Delta t$.

A traffic avalanche can be generated in this model by taking a large system in which all cars are freely flowing and by momentarily reducing the speed of a single car to zero. This perturbed car will eventually accelerate to v_{max}; however, the next car may approach too closely and be forced to decelerate. This leads to a chain reaction of subsequent deceleration events and can result in a substantial traffic jam. Eventually, however, the jam gets resolved and all cars are flowing freely again. Numerically, it is found that the distribution of jam lifetimes $P(t)$ has a $t^{-3/2}$ decay at long times [Nagel & Paczuski (1995)]. Once again, the appearance of the exponent $-3/2$ suggests a connection between jam lifetimes and first passage.

We now construct a random-walk argument, following Nagel and Paczuski (1995), to show this correspondence in the limiting case of $v_{max} = 1$; that is, cars may be either at rest or moving with speed 1. Thus a jam consists of a line of cars at rest that occupy consecutive lattice points. Cars leave the front end of the jam at a rate r_{out}, which is determined by the probabilistic acceleration rule of the NS model. Similarly, cars add to the back end of the jam at a rate r_{in}, which depends on the density of free-flowing cars behind the jam. Consequently, the probability that the jam contains n cars at time t obeys the random-walk master equation

$$P(n, t + 1) = r_{out} P(n + 1, t) + r_{in} P(n - 1, t) + (1 - r_{out} - r_{in}) P(n, t).$$

In the continuum limit, this master equation becomes the one-dimensional convection–diffusion equation,

$$\frac{\partial P}{\partial t} + (r_{in} - r_{out})\frac{\partial P}{\partial n} = \frac{1}{2}(r_{in} + r_{out})\frac{\partial P}{\partial n},$$

subject to an absorbing boundary condition at $n = 0$, which signals the end of the jam. The interesting situation is that of a stationary jam where $r_{in} = r_{out}$. Because this corresponds exactly to one-dimensional diffusion with an absorbing boundary condition at the origin, the distribution of traffic jam lifetimes has an asymptotic $t^{-3/2}$ power-law decay.

4.3.3.2. Anisotropic Interface Depinning

Another simple self-organized critical system that exhibits a $t^{-3/2}$ distribution of avalanche durations is an anisotropic interface depinning model. The presentation here follows the basic ideas of Maslov and Zhang (1995). Roughly speaking, the model is a idealization of the motion of a driven interface in the presence of random pinning centers. At each time step, the interface is released from the pinning center where the difference between the local force and the pinning threshold is maximal. These forces are then transferred to nearby pinning centers and the release process occurs again. The interface continues to move until there are no above-threshold pinning centers remaining.

At a simple-minded level, much of this dynamics is captured by an anisotropic version of the process in one dimension in which each site i is endowed with a force variable $f(i)$. At each time step, the site with the largest force is located, and an amount r of this force, which is uniformly distributed between 0 and 1, is transferred to the left to site $i - 1$; that is,

$$f(i) \rightarrow f(i) - r,$$
$$f(i - 1) \rightarrow f(i - 1) + r.$$

Once again, we can construct the notion of an avalanche. Suppose that at some time the maximal force in the system is smaller than a value λ and that in the next update a force $f(i) > \lambda$ is created. This avalanche will then propagate in a causally connected fashion until the largest force first drops below λ again. We term this event a λ-avalanche.

An important feature of the model dynamics is that the size of an avalanche does not depend on the order at which sites are activated. This *Abelian* property means that we can reorder the activation events to simplify the determination of the duration of an avalanche. This motivates the following variation. Once a site exceeds threshold, it is activated as many times as necessary to drive it below threshold. Only after this occurs can the left neighbor be activated. Thus a typical λ avalanche proceeds as follows: Initially the force $f(i)$ at some site i equals λ and $f(i)$ is reduced by a random number r, whereas $f(i-1)$ is increased by r. If $f(i-1) > \lambda$, we subtract random numbers from $i-1$ until $f(i-1) < \lambda$ and transfer the excess force $\Delta f(i-1)$ to site $i-2$.

At any stage of the process, let $\Delta f(i)$ denote the force excess (above λ) at site i immediately after the transfer from $i+1$. Then the force transferred from i to $i-1$ will equal $\Delta f(i) + r_1$, where r_1 is a random number uniformly distributed on $[0, 1]$. Because the initial force on $i-1$ will be smaller than λ by an arbitrary amount r_2, $\Delta f(i-1) = \Delta f(i) + r_1 - r_2$. Thus the force excess performs a random walk as a function of site index i. The parameter λ determines the relative bias of this hopping process. If $\lambda > \lambda_c$, $\langle r_2 \rangle > \langle r_1 \rangle$ and avalanches will tend to die out. Conversely for, $\lambda < \lambda_c$, an avalanche will grow with time. At the critical point λ_c, $\langle r_1 \rangle = \langle r_2 \rangle$ and the force excess performs an isotropic random walk subject to an absorbing boundary condition at the origin, corresponding to the end of the avalanche. Because of this correspondence, we again conclude that the distribution of avalanche durations again has a $t^{-3/2}$ decay at long times.

4.4. Kinetics of Spin Systems

4.4.1. Background

From basic results about first passage in the semi-infinite interval, we will now show how to understand, in a simple way, the long-time kinetics of two important interacting spin systems, namely, the Ising–Glauber model and the Voter model. We begin by presenting some pertinent facts about these two systems [see, e.g., Huang (1987) for general information about the Ising model and see Liggett (1985) for the Voter model], and then give the solution for their kinetics.

The Ising model is an equilibrium statistical–mechanical model for ferromagnetism in which each site of a regular lattice is endowed with a spin σ_i that can assume one of two discrete states $\sigma_i = \pm 1$. There is a nearest-neighbor interaction between spins that favors ferromagnetic order. The resulting Hamiltonian of the system is

$$\mathcal{H} = -J \sum_{\langle i, j \rangle} \sigma_i \sigma_j, \tag{4.4.1}$$

where $\langle i, j \rangle$ denotes a sum over all nearest-neighbor spin pairs. Thus aligned nearest-neighbor pairs contribute a factor $-J$ to the total energy and misaligned pairs a factor $+J$. The goal of equilibrium statistical mechanics is to compute the partition function $\mathcal{Z} = \sum e^{-\mathcal{H}/kT}$, where the sum is over all spin states, T is the absolute temperature, and k is the Boltzmann constant. In spatial dimensions $d > 1$, this system undergoes a continuous phase transition as a function of temperature, between a high-temperature disordered phase to a low-temperature ordered, or ferromagnetic, phase. This ordering is defined by the existence of a nonzero average spin value, $s \equiv \langle \sigma_i \rangle \equiv \sum \sigma_i e^{-\mathcal{H}/kT} / \sum e^{-\mathcal{H}/kT} \neq 0$, as well as long-range correlations between spins, $c(i, j) \equiv \langle \sigma_i \sigma_j \rangle > 0$ for large $|i - j|$.

The Ising–Glauber model [Glauber (1963)] is an extension of the equilibrium Ising model in which each spin is also endowed with a kinetics so that the system can make transitions between different microscopic states. The kinetics is defined by allowing each spin to flip at a rate that is chosen to give simple behavior and that also ensures that the system reaches the thermodynamic equilibrium in the long-time limit. This extension of the Ising model provides a fundamental and physically appealing way to model the time evolution of ferromagnetic systems.

The Voter model is an even simpler interacting two-state spin system, in which we can equivalently view each site as being a "voter" with only two opinions. The dynamics of the system involves picking a voter at random and updating it according to the following rule: A randomly selected voter merely assumes the state of a randomly chosen nearest neighbor. We can think of each voter as an individual with zero self-confidence who blindly assumes the state of its neighbor in an update event. This is a purely kinetic model, so that there is no notion of an equilibrium for this system. In the long-time limit, the Voter model either reaches complete unanimity for spatial dimension $d \leq 2$, or it reaches a steady state in which voters change their opinion ad infinitum, for $d > 2$.

The crucial feature for the solvability of both these systems is that the equations for correlation function are closed. That is, the equation for the n-body correlation function does not involve any higher-order correlations. The exact solutions for the kinetics of these two models are now presented, following closely the original literature.

4.4.2. Solution to the One-Dimensional Ising–Glauber Model

The fundamental quantity is the probability distribution $P(\{\sigma\}, t)$ that the spins of the system are in a particular state, denoted collectively as $\{\sigma\}$. This probability obeys the master equation

$$\frac{d}{dt}P(\{\sigma\},t) = \sum_i w(\{\sigma'\}_i \to \{\sigma\})P(\{\sigma'\}_i,t)$$

$$- \sum_i w(\{\sigma\} \to \{\sigma'\}_i)P(\{\sigma\},t). \qquad (4.4.2)$$

Here $\{\sigma'\}_i$ denotes the state in which only the ith spin is reversed compared with the state $\{\sigma\}$. This equation thus accounts for the gain in $P(\{\sigma\})$ that is due to transitions into this spin state at rate $w(\{\sigma'\}_i \to \{\sigma\})$ and for the loss in $P(\{\sigma\})$ that is due to transitions out of this state by the flip of a single spin σ_i, with rate $w(\{\sigma\} \to \{\sigma'\}_i)$. On physical grounds, Glauber found the following a particularly convenient choice for the transition rates:

$$w(\{\sigma\} \to \{\sigma'\}_i) = \frac{\alpha}{2}\left[1 - \frac{\gamma}{2}\sigma_i(\sigma_{i-1} + \sigma_{i+1})\right],$$

$$= \begin{cases} \dfrac{\alpha}{2}(1-\gamma) & \text{if } \uparrow\uparrow\uparrow \longrightarrow \uparrow\downarrow\uparrow \\[2mm] \dfrac{\alpha}{2} & \text{if } \uparrow\uparrow\downarrow \longrightarrow \uparrow\downarrow\downarrow, \\[2mm] \dfrac{\alpha}{2}(1+\gamma) & \text{if } \uparrow\downarrow\uparrow \longrightarrow \uparrow\uparrow\uparrow \end{cases}$$

with the same rates applying under an overall sign change of all these spins.

As constructed, these flip rates favor spin alignment. These rates must also satisfy the *detailed balance condition* to ensure that the correct thermodynamic equilibrium is reproduced [Glauber (1963)]. This detailed balance condition is

$$\frac{w(\{\sigma\} \to \{\sigma'\}_i)}{w(\{\sigma'\}_i \to \{\sigma\})} = \frac{P(\{\sigma'\}_i)}{P(\{\sigma\})},$$

which translates to

$$\frac{1 - \frac{\gamma}{2}\sigma_i(\sigma_{i-1} + \sigma_{i+1})}{1 + \frac{\gamma}{2}\sigma_i(\sigma_{i-1} + \sigma_{i+1})} = \frac{e^{-\beta J\sigma_i(\sigma_{i-1} + \sigma_{i+1})}}{e^{+\beta J\sigma_i(\sigma_{i-1} + \sigma_{i+1})}}.$$

This merely expresses the fact that in equilibrium the probability current from one state to another is the same as the current in the reverse direction for any pair of states.

By writing $e^x = \cosh x + \sinh x$ and using the property that $\sinh(\epsilon x) = \epsilon \sinh(x)$ for $\epsilon = \pm 1$, we find that the transition rate reduces to, after several simple steps [Glauber (1963)],

$$w(\{\sigma\} \to \{\sigma'\}_i) = \frac{\alpha}{2}\left(1 - \frac{\gamma}{2}E_i\right), \qquad (4.4.3)$$

where $\gamma = \tan(2\beta J)$ and $E_i = -\sigma_i(\sigma_{i-1} + \sigma_{i+1})$ is defined as the local spin energy. Henceforth we take the rate $\alpha = 1$ because it only sets the overall time scale.

We now solve for the mean spin and the mean-spin correlation function in one dimension with these flip rates. The mean spin at site j is given by

$$s_j \equiv \langle \sigma_j \rangle = \sum_{\{\sigma\}} \sigma_j P(\{\sigma\}, t),$$

where the sum is over all spin configurations of the system. From the master equation for the probability distribution, s_j obeys the equation of motion

$$\frac{ds_j}{dt} = \sum_{\{\sigma\}} \sigma_j \frac{d}{dt} P(\{\sigma\}, t)$$

$$= \sum_{\{\sigma\}} \sigma_j \sum_i \{(w(\{\sigma'\}_i \rightarrow \{\sigma\})P(\{\sigma'\}_i, t) - w(\{\sigma\} \rightarrow \{\sigma'\}_i)P(\{\sigma\}, t)\}$$

$$= \sum_{\{\sigma\}} \sigma_j \sum_i \left[\frac{1}{2}\left(1 + \frac{\gamma}{2}E_i\right) P(\{\sigma'\}_i, t) - \frac{1}{2}\left(1 - \frac{\gamma}{2}E_i\right) P(\{\sigma\}, t) \right].$$

$$(4.4.4)$$

We can simplify these sums by exploiting the following basic facts:

$$\sum_{\{\sigma\}} \sigma_j P(\{\sigma\}) = +\langle \sigma_j \rangle \quad \text{definition},$$

$$\sum_{\{\sigma\}} \sigma_j P(\{\sigma'\}_j) = -\langle \sigma_j \rangle,$$

$$\sum_{\{\sigma\}} \sigma_j P(\{\sigma'\}_{k \neq j}) = +\langle \sigma_j \rangle.$$

These identities reduce the above equation of motion (4.4.4) into a simple closed equation for the mean spin [Glauber (1963)]:

$$\frac{ds_j}{dt} = -s_j + \frac{\gamma}{2}(s_{j-1} + s_{j+1}). \qquad (4.4.5)$$

We see that s_j obeys the same type of master equation as the one-dimensional random walk, as discussed previously in Subsection 1.3.2. From this equivalence, the solution for $s_n(t)$ for the initial condition $s_n(t = 0) = \delta_{n,0}$ is simply

$$s_n(t) = e^{-t} I_n(\gamma t), \qquad (4.4.6)$$

where again $I_n(x)$ is the modified Bessel function of the first kind of order n. Using the asymptotic properties of this Bessel function [Abramowitz &

Stegun (1972)], we find that the mean value of the spin at any site asymptotically decays as

$$s_n(t) \sim \frac{1}{\sqrt{2\pi\gamma t}} e^{-(1-\gamma)t}. \tag{4.4.7}$$

This is exponential decay in time for $T > 0$ and power-law decay for $T = 0$. Although the mean value of each spin decays to zero, note that the average magnetization $m \equiv \sum_j s_j / N$, where N is the total number of spins, is strictly conserved at $T = 0$! This seemingly paradoxical feature follows directly from equation of motion (4.4.5). As a result, the mean magnetization is not a suitable measure for characterizing the degree of spin alignment in the system.

For this purpose, the two-spin correlation function, $c(i, j) \equiv \langle \sigma_i \sigma_j \rangle$, is much more useful. Because $\sigma_i \sigma_j = +1$ when the two spins are parallel and $\sigma_i \sigma_j = -1$ when the spins are antiparallel, the correlation function measures the extent to which these two spins are aligned. By translational invariance and isotropy, this function may be written as $c(i, j) = c_n$ for large n, where $n = |i - j|$ is the distance between the two spins. If c_n goes to zero rapidly as n increases, then there is no long-range spin alignment, whereas if c_n decays as a power law in n as $n \to \infty$, long-range alignment exists.

The evolution of the correlation function is described by a rate equation that is very similar in structure to Eq. (4.4.4) for the mean spin. Following exactly the same approach as that used to solve for the mean spin (all the details are in Glauber's original paper), we find that the correlation function obeys the master equation

$$\frac{dc_n}{dt} = -2c_n + \gamma(c_{n-1} + c_{n+1}). \tag{4.4.8}$$

This is identical to the master equation for the mean spin (up to an overall factor of 2), but with the crucial difference that the boundary condition for the correlation function is $c_0(t) = \langle \sigma_i^2 \rangle = 1$, because $\sigma_i = \pm 1$.

The most interesting situation is the limit of zero temperature, in which the master equation for the correlation function reduces to

$$\frac{dc_n}{dt} = c_{n-1} - 2c_n + c_{n+1} \equiv \Delta^{(2)} c_n, \tag{4.4.9}$$

subject to the boundary condition is $c_0(t) = 1$. Here $\Delta^{(2)}$ is the discrete second derivative. For simplicity, we also assume that the spins are initially uncorrelated, that is, $c_n = 0$ for $n > 0$. This restriction is immaterial for our main conclusions, however. In the continuum limit, Eq. (4.4.9) describes diffusion in the half-line $x > 0$, with the concentration at the origin fixed to be one and the initial concentration equal to zero for all $x > 0$. By linearity of the diffusion equation, this is trivially related to diffusion in the same geometry with the initial concentration equal to one and the concentration at origin fixed to be zero. Thus from our discussion on diffusion in the semi-infinite interval (Section 3.3), we conclude that the correlation function is

$$c_n(t) = 1 - \text{erf}\left(\frac{n}{\sqrt{4t}}\right) \sim 1 - \frac{n}{\sqrt{4t}}. \tag{4.4.10}$$

Therefore the correlation function decays to one as $t^{-1/2}$ for any n. This simply means that the correlation between spins extends to a range that is of the order of $\sqrt{t}$. This gives the physical picture that the spins organize into a coarsening mosaic of aligned domains of typical length which grows as $\sqrt{t}$. From the correspondence to diffusion in one dimension, this continuous coarsening arises from the recurrence of the diffusion process.

On the other hand, for nonzero temperature $\gamma < 1$. In this case, the continuum limit of master equation Eq. (4.4.8) becomes

$$\frac{\partial}{\partial t} c(x, t) = \gamma \frac{\partial^2}{\partial x^2} c(x, t) - 2(1 - \gamma)c(x, t), \tag{4.4.11}$$

where $c(x, t)$ is the continuum limit of the correlation function $c_n(t)$. This equation has the steady-state solution $c(x) = e^{-x\sqrt{2(1-\gamma)/\gamma}}$. Thus there is no long-range spin correlations at any nonzero temperature in the one-dimensional Ising model.

Part of the reason for highlighting the one-dimensional Ising-Glauber model is that it closely related to diffusion-controlled single-species annihilation $A + A \rightarrow 0$ in one dimension, in which diffusing particles on the line annihilate whenever a pair of particles meet. As we will discuss in detail in Section 8.4, the particle density in this reaction is intimately related to the spin correlation function in the Ising–Glauber model.

4.4.3. Solution to the Voter Model in all Dimensions

A classic approach to solving the Voter model in the mathematical literature is based on a duality relation to coalescing random walks [Bramson & Griffeath (1980b) and Liggett (1985)]. If the time is reversed in the Voter model, then the trajectories of different opinions are equivalent coalescing random walks, and one can exploit known results about the latter system to solve the Voter model. Here we follow a different approach, which is based on writing the master equation for the Voter model in a form similar to that of the Ising–Glauber model. In the mathematics literature, this construction was apparently discovered by Spitzer (1981); see also Bhattacharya and Waymire (1990) for a general discussion of this method. In the physics literature, this approach was independently found by Scheucher and Spohn (1988), Krapivsky (1992a, 1992b), and Frachebourg and Krapivsky (1996), and was also extended to treat higher-order correlation functions.

According to the dynamics of the Voter model, the transition rate between a state $\{\sigma\}$ and a state $\{\sigma'\}_i$ in which only the spin (or voter) σ_i has been flipped is

$$w(\{\sigma\} \rightarrow \{\sigma'\}_i) = \frac{1}{2}\left(1 - \frac{\sigma_i}{z}\sum_{k \text{ n.n.} i} \sigma_k\right),$$

where the sum is over the z nearest neighbors of site i on a lattice of co-ordination number z. The overall factor of $1/2$ is for convenience only. With these rates, a spin σ_i on the square lattice changes from -1 to $+1$ with rates $1, \frac{3}{4}, \frac{1}{2}, \frac{1}{4}$, and 0 when the number of neighboring up spins are 4, 3, 2, 1, and 0, respectively. Substituting these transition rates into master equation Eq. (4.4.2) and following the same steps as those developed for the Ising–Glauber model, we find that the mean voter spin (opinion) at site j evolves according to

$$
\begin{aligned}
\frac{ds_j}{dt} &= \sum_{\{\sigma\}} \sigma_j \frac{d}{dt} P(\{\sigma\}, t) \\
&= \sum_{\{\sigma\}} \sigma_j \sum_i \{w(\{\sigma'\}_i \rightarrow \{\sigma\})P(\{\sigma'\}_i, t) - w(\{\sigma\} \rightarrow \{\sigma'\}_i)P(\{\sigma\}, t)\} \\
&= \frac{1}{2}\sum_{\{\sigma\}} \sigma_j \sum_i \left[\left(1 + \frac{\sigma_i}{z}\sum_k \sigma_k\right)P(\{\sigma'\}_i, t) - \left(1 - \frac{\sigma_i}{z}\sum_k \sigma_k\right)P(\{\sigma\}, t)\right].
\end{aligned}
$$

$$(4.4.12)$$

By evaluating the sums in this equation we eventually find

$$\frac{ds_j}{dt} = -s_j + \frac{1}{z}\sum_k s_k. \qquad (4.4.13)$$

This equation has a structure similar to that of the master equation for the mean spin in the Ising–Glauber model. However, the linearly interpolating form of the transition rates as a function of the local magnetization allows us to factorize this equation along different coordinate axes. This gives an equation of motion that is just discrete diffusion in d dimensions with the boundary condition of unit concentration at the origin. This is the mechanism by which the Voter model is exactly soluble in all dimensions.

We now rewrite Eq. (4.4.13) in a form that exposes this factorizability. When each site is labeled by its vector location, the master equation becomes

$$\frac{ds_{\vec{r}}}{dt} = \frac{1}{z}\left[(s_{\vec{r}+\hat{e}_1} + s_{\vec{r}-\hat{e}_1} - 2s_{\vec{r}}) + \cdots + (s_{\vec{r}+\hat{e}_d} + s_{\vec{r}-\hat{e}_d} - 2s_{\vec{r}})\right]$$

$$\equiv \frac{1}{z}\vec{\Delta}^{(2)}s_{\vec{r}}, \tag{4.4.14}$$

where $\hat{e}_k$ denotes a unit vector in the kth coordinate direction. The last line defines the vector discrete second-difference operator. Following exactly the same steps as in solving the master equation for the Ising–Glauber model, we obtain, for the initial condition $s_n(0) = \delta_{n,0}$ (e.g., one "Democrat" in a sea of uncommitted voters),

$$s_{\vec{r}}(t) = e^{-2t} I_{\vec{r}}(4t/z), \tag{4.4.15}$$

where

$$I_{\vec{r}}(z) = \prod_{i=1}^{d} I_{x_i}(z),$$

and where x_i is the ith Cartesian coordinate of $\vec{r}$.

By very similar considerations [Frachebourg & Krapivsky (1996)], the master equation for the correlation function is

$$\frac{dc(\vec{r})}{dt} = \frac{2}{z}\vec{\Delta}^{(2)}c(\vec{r}), \tag{4.4.16}$$

subject again to the initial condition that $c(0) = 1$. This similarity of the equation for the one- and the two-spin correlation functions is exactly as in the Ising–Glauber model in one dimension. Once again, the exact solution for the discrete system has been obtained by Frachebourg and Krapivsky (1996). However, it is much simpler to consider the continuum limit to obtain the asymptotic properties of the correlation function. In the continuum, the correlation function $c(\vec{r}, t)$ obeys the diffusion equation with the boundary condition $c(0, t) = 1$, and we may choose the initial condition to be $c(\vec{r}, 0) = 0$. The long-time behavior follows directly from what we already know about the first-passage probability for diffusion to an absorbing point in arbitrary spatial dimension.

In one dimension, the Voter–model behavior is identical to that of the Ising–Glauber model in which the correlation function asymptotically goes to one. In two dimensions, the survival probability of a diffusing particle in the presence of an excluded point decays as $1/\ln t$ [approximation (1.5.8)]. This implies that the correlation function for any r asymptotically approaches one with a correction that vanishes as $const/\ln t$. Thus unanimity is eventually achieved, albeit very slowly. In three dimensions and greater, there is a finite probability for a diffusing particle to never reach the origin; this leads to a steady-state concentration profile about the origin that has a $1/r^{d-2}$ deviation from the value one at large r. Correspondingly, the steady-state correlation function decays to zero as $1/r^{d-2}$ at large distances – distant voters become uncorrelated! Thus the high-dimensional Voter model reaches a steady state in which voters change opinions ad infinitum, whereas the average correlation function goes to a steady-state value with vanishing correlations at large distances.

4.5. First Passage in Composite and Fluctuating Systems

4.5.1. Motivation

Much effort has been devoted to investigating the role of spatial disorder on on the probability distribution and the first-passage properties of random-walk processes [Alexander et al. (1981), Havlin & ben-Avraham (1987), Bouchaud & Georges (1990), and ben-Avraham & Havlin (2000)]. Considerable progress in developing exact solutions has been made in one dimension both for the probability distribution [typical examples include those of Azbel (1982) and Derrida (1983)] and for first-passage properties [Noskowicz & I. Goldhirsch (1988), Murthy & Kehr (1989), Le Doussal (1989), Kehr & Murthy (1990), Raykin (1993), and Le Doussal et al. (1999)]. In higher dimensions, exact analysis does not appear to be feasible, but there is a good understanding of long-time transport properties, based on a variety of approximate methods [see, e.g., Havlin & ben-Avraham (1987), Bouchaud & Georges (1990), and ben-Avraham & Havlin (2000)].

This section is devoted to treating the first-passage properties of one useful class of such systems, namely, composite media, which are constructed from different segments of homogeneous material (Fig. 4.6). A natural question for a composite medium is whether it can be replaced with an effective homogeneous system whose transport properties accurately describe

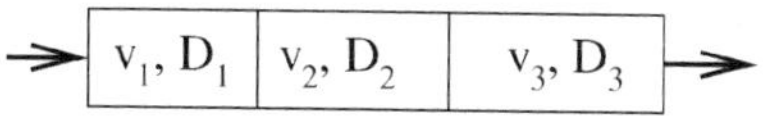

Fig. 4.6. A composite medium that consists of homogeneous segments, with distinct values of the diffusion coefficient D_i and/or bias velocity v_i in the ith segment. A diffusing particle is injected into the first segment and is absorbed when it reaches the end of the last segment.

those of the composite. As we shall discuss in the two generic examples of
(i) an isotropic composite in which each element has a distinct diffusivity, and
(ii) a composite with a different bias in each segment, such an averaging does
not hold for a purely diffusive system, but does apply for media with different bias velocities in each segment. This greatly simplifies the description of
long-distance transport properties for the latter systems.

4.5.2. Segments with Different Diffusivities

Consider first the series composition of two homogeneous blocks of lengths
L_1 and L_2 and diffusion coefficients D_1 and D_2, respectively. A unit current
is injected at $x = 0$, and we shall compute the first-passage probability of
leaving the right edge of the system at $x = L_1 + L_2$. In each segment, we
need to solve the diffusion equation

$$\frac{\partial}{\partial t} c_i(x, t) = D_i \frac{\partial^2}{\partial x^2} c_i(x, t),$$

with $i = 1, 2$, subject to the absorbing boundary condition $c_2(x, t) = 0$ at $x =
L_1 + L_2$ and continuity of the concentration and the flux at $x = L_1$. The initial
condition is the injection of a unit pulse of current at $t = 0$, $j(0, t) = \delta(t)$. The
Laplace transform of the concentration in each segment has the usual general
form $c_i(x, s) = A_i e^{x\sqrt{s/D_i}} + B_i e^{-x\sqrt{s/D_i}}$, and, by imposing the boundary and
initial conditions and performing some straightforward algebra, we find that
the Laplace transform of the output flux is

$$j_2(L_1 + L_2, s) =$$
$$\left(\sqrt{\frac{D_1}{D_2}} \sinh \sqrt{\frac{sL_1^2}{D_1}} \sinh \sqrt{\frac{sL_2^2}{D_2}} + \cosh \sqrt{\frac{sL_1^2}{D_1}} \cosh \sqrt{\frac{sL_2^2}{D_2}} \right)^{-1}.$$

$$(4.5.1)$$

This expression just fails to have the form of a single cosh function of a
suitable composite argument. Thus it is not possible to replace this two-
segment composite with an effective homogeneous medium.

From this first-passage probability, the corresponding mean first-passage
time is

$$\langle t \rangle = \frac{L_1^2}{2D_1} + \frac{L_2^2}{2D_2} + \frac{L_1 L_2}{D_2}. \qquad (4.5.2)$$

Again, this is almost, but not quite, equal to

$$\left(\frac{L_1}{\sqrt{D_1}} + \frac{L_2}{\sqrt{D_2}} \right)^2 .$$

Because of this feature and the lack of symmetry between indices 1 and 2, it is clear that we cannot, in general, write $\langle t \rangle$ in the form $\langle t \rangle = L_{\text{eff}}^2 / D_{\text{eff}}$.

For a linear chain of N blocks, each of length L_i and diffusion coefficient D_i, we find, after a more involved calculation, that the average first-passage time through the composite is

$$\langle t \rangle = \frac{1}{2} \sum_{i=1}^{N} \frac{L_i^2}{D_i} + \sum_{i<j}^{N} \frac{L_i L_j}{D_j}. \qquad (4.5.3)$$

For simple special cases, such as an alternating array of segments (L_1, D_1) and (L_2, D_2), the above expression can be expressed as $\langle t \rangle \sim L_{\text{eff}}^2 / D_{\text{eff}}$ when the number of blocks is large. However, general power-law distributions of block lengths and diffusion coefficients can easily give nondiffusive behavior. For disorder in the diffusivities only, it is well known that transport is sub-diffusive when the distribution is sufficiently singular in the small diffusivity limit [Alexander et al. (1981)]. This slower transport arises because regions of progressively weaker hopping rates are explored as time increases, leading to scale-dependent transport properties. For the complementary situation of a power-law distribution of block lengths (with a nonsingular distribution of block diffusion coefficients), a similar scale dependence in effective macroscopic properties arises.

4.5.3. *Segments with Different Bias Velocities*

Much richer behavior occurs when each segment of the composite has a distinct bias velocity. When the bias fields are all in the same direction, a very simple form of a homogenization principle holds – the composite can be represented by an effective homogeneous medium. When the bias is arranged so there is a local potential-energy minimum in the system, then transport properties are controlled *only* by the highest barrier in the local potential energy in going from the input to the output. This then determines the properties of the equivalent homogeneous system. We discuss these features explicitly for two- and three-segment system; this can be generalized to an arbitrary number of segments [Le Doussal et al. (1999)].

4.5.3.1. Two Segments

To illustrate the basic ideas, consider first the two-component slab, with bias velocities v_1 in the segment $0 \leq x \leq x_1$ and v_2 for $x_1 \leq x \leq 1$. To simplify computational details, we take the diffusion coefficients of each segment to be equal throughout. This does not affect long-time properties, as the role of diffusion is subdominant with respect to the bias. As in Subsection 4.5.2, we determine the first-passage probability to the output at $x = 1$ when a unit current is injected at $x = 0$. Thus we solve the convection–diffusion equation

$$
\frac{\partial c_i}{\partial t} + v_i \frac{\partial c_i}{\partial x} = D \frac{\partial^2 c_i}{\partial x^2}, \quad i = 1, 2,
$$

with the initial condition $j_1(0, t) = \delta(t)$, the boundary condition $c(1, t) = 0$, and continuity of the concentration and the flux $j = vc - Dc'$ at $x = x_1$.

The Laplace transform of the concentration has the general form $c_i(x, s) = A_i e^{\alpha_i x} + B_i e^{\beta_i x}$, where $\alpha_i, \beta_i = (v_i \pm \sqrt{v_i^2 + 4Ds})/2D$. The four constraints (two at the end points and two at $x = x_1$) determine A_i and B_i, from which the Laplace transform of the outgoing flux is, after the usual tedious algebra,

$$
j_2(x = 1, s) = D(\alpha_2 - \beta_2) \left[\frac{f - g}{p - q} \right], \tag{4.5.4}
$$

where

$$
f = \frac{1}{D\alpha_1} \left(\frac{e^{\beta_1 x_1}}{\alpha_1 e^{\alpha_1 x_1} - \beta_1 e^{\beta_1 x_1}} \right), \quad g = \frac{1}{D} \left[\frac{e^{\beta_1 x_1}}{\alpha_1 \beta_1 (e^{\alpha_1 x_1} - e^{\beta_1 x_1})} \right],
$$

$$
p = \frac{e^{\alpha_2(x_1-1)} - e^{\beta_2(x_1-1)}}{\alpha_1 e^{\alpha_1 x_1} - \beta_1 e^{\beta_1 x_1}}, \quad q = \frac{\beta_2 e^{\alpha_2(x_1-1)} - \alpha_2 e^{\beta_2(x_1-1)}}{\alpha_1 \beta_1 (e^{\alpha_1 x_1} - e^{\beta_1 x_1})}.
$$

This expression is unwieldy, and many of the remaining computations in this section were performed with the help of Mathematica [Wolfram (1991)]. By expanding this first-passage probability in a power series in s, we formally have $j_2(1, s) = 1 - s \langle t \rangle + \ldots$, from which the mean first-passage time is

$$
\langle t \rangle = \frac{1}{v_1 v_2} \left\{ (1 - e^{-v_1 x})\left[1 - e^{v_2(x-1)} \right] + v_2 x \right\}
$$

$$
+ \frac{1}{v_1^2 v_2^2} \left\{ v_2^2 (e^{-v_1 x} - 1) + v_1^2 \left[e^{v_2(x_1-1)} + v_2(1 - x_1) - 1 \right] \right\}. \tag{4.5.5}
$$

When the biases in both segments are large and in the same direction, this

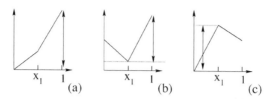

Fig. 4.7. Dependence of the potential energy on x for a two-segment interval for (a) $v_1, v_2 < 0$, (b) $v_1 > 0$, $v_2 < 0$, and (c) $v_1 < 0$, $v_2 > 0$. Also shown for each case is the relevant energy barrier that controls the mean first-passage time to $x = 1$.

first-passage time reduces to

$$\langle t \rangle \to \begin{cases} \dfrac{x}{v_1} + \dfrac{1-x}{v_2}, & v_1, v_2 > 0 \\[2mm] \dfrac{1}{|v_1 v_2|} e^{|v_1|x + |v_2|(1-x)}, & v_1, v_2 < 0 \end{cases} . \tag{4.5.6}$$

The former corresponds to simple constant-velocity motion in each segment, with $\langle t \rangle$ simply equal to $t_1 + t_2$, where t_i is the convection time for each segment. The latter shows that the mean first-passage time is proportional to

$$e^{+(\text{barrier height})}, \tag{4.5.7}$$

where the relevant barrier is indicated in Fig. 4.7(a). When the biases are oppositely oriented, the limiting behavior of Eq. (4.5.5) is determined by the barriers shown in Figs. 4.7(b) and (c).

4.5.3.2. Three Segments

The above treatment can be readily extended to three segments. This system is particularly interesting because it provides a conceptually simple and soluble example of escape from a metastable state to a stable state by surmounting an energetic barrier. This is a problem of fundamental importance in many stochastic processes, which started with the original work of Kramers (1940) and is still continuing [see Hänggi, Talkner, & Borkovec (1990) for a fairly recent review]. We consider the generic three-segment geometries depicted in Fig. 4.8. The restriction to piecewise linear potential, as first studied by Frisch et al. (1990), Privman and Frisch (1991), and then by Le Doussal et al. (1999), provides an enormous simplification over the original Kramers' problem. We can obtain a complete solution for the first-passage probability by applying conceptually simple, although labor-intensive, methods.

As in the case of the two-segment system, we find the first-passage probability by solving the convection–diffusion equation for each segment of the composite, with injection of unit current at $x = 0$, particle absorption at $x = 1$,

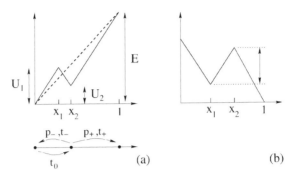

Fig. 4.8. Dependence of the potential energy on x and the associated energy barriers for a three-segment interval with different biases in each segment. Shown are the cases in which the bias signs are (a) $-+-$ and (b) $+-+$. In (a), the effective bias is indicated by the dashed line. Also shown below is the renormalized three-site random walk corresponding to the barrier-crossing problem. In (b), the first-passage time is controlled by the indicated barrier between x_1 and x_2.

and continuity of the concentration and the flux at segment junctions. Because the velocity v_i in each segment always appears in the combination $v_i x/D$ in the first-passage probability, we can achieve the effect of different speeds in different segments without loss of generality by using equal magnitudes for the speeds and adjusting the segment lengths appropriately.

There are two particularly interesting situations that we shall focus on. The first is that in which the bias velocities in each segment have the signs $-+-$, corresponding to the potential-energy profile sketched in Fig. 4.8(a). Using Mathematica, we obtain a very unwieldy expression for the Laplace transform of the first-passage probability. From this, the mean first-passage time is the linear term in s in the power-series expansion of this passage probability. This gives

$$\langle t \rangle = \frac{1}{v^2}\left[e^{v(1+2x_1-2x_2)} + 2e^{v(1-x_2)} + 4e^{v(x_1-x_2)} + 2e^{vx_1} - 2e^{v(2x_1-x_2)}\right.$$
$$\left. - 2e^{v(1+x_1-2x_2)} + v(2x_2 - 2x_1 - 1) - 5\right]. \tag{4.5.8}$$

In the limit $v \to \infty$, this result has a simple physical interpretation. Keeping only the dominant terms in the mean first-passage time as $v \to \infty$, we obtain

$$\langle t \rangle = \frac{1}{v^2}\left[e^{v(1+2x_1-2x_2)} + 2e^{vx_1} + 2e^{v(1-x_2)}\right] + \cdots. \tag{4.5.9}$$

From Fig. 4.8(a), we see that the exponents in each of the terms in Eq. (4.5.9) represent the total energy barrier $E = v(1 + 2x_1 - 2x_2)$, the first energy barrier $U_1 = vx_1$, and the second energy barrier $E - U_2 = v(1 - x_2)$. Thus the asymptotic behavior of the mean first-passage time is determined *solely* by the largest of these three barriers. For the case in which the total energy barrier E is the largest, the three-segment system can therefore be replaced with an effective homogeneous system with a spatially uniform bias velocity $v_{\text{eff}} = 1 + 2x_1 - 2x_2$, as indicated by the dashed line in Fig. 4.8(a). Conversely, when U_1 is the largest barrier, then its height determines the first-passage time.

When E is the largest barrier, we may also give a simple physical estimate for the mean first-passage time in terms of a "renormalized" random walk on a three-site chain that consists of the points 0, x_2, and 1. This walk starts at $x = 0$ and reaches x_2 in a time of the order of $t_0 \sim e^{U_1}$. From our discussion of the two-segment system, we know that, starting from x_2, the particle surmounts the barrier at $x = x_1$ and returns to the origin with probability $p_- \propto e^{-(U_1 - U_2)}$, whereas with very small probability $p_+ \propto e^{-(E - U_2)}$, the particle reaches $x = 1$ and the process ends. Because these probabilities must add to one, $p_+ \approx e^{-(E-U_2)+(U_1-U_2)} = e^{-(E-U_1)}$ and $p_- = 1 - p_+$. The respective times for these steps to the left and to the right are $t_- \sim e^{(U_1 - U_2)}$ and $t_+ \sim e^{(E - U_2)}$. Then the first-passage time obeys the recursion formula

$$\langle t \rangle = t_0 + p_-(t_- + \langle t \rangle) + p_+ t_+,$$

from which we find

$$\langle t \rangle = \frac{t_0}{p_+} + \frac{p_- t_-}{p_+} + t_+$$

$$\sim \frac{t_0}{p_+} = e^E, \tag{4.5.10}$$

which agrees asymptotically with the exact result.

In the converse case, in which the bias velocities have the signs $+ - +$, the mean first-passage time is asymptotically dominated by the factor $4e^{v(x_2 - x_1)}/v^2$, which is the only energy barrier in the system.

4.5.4. Resonant First Passage in a Fluctuating Medium

An important feature that arises in various physical realizations of first-passage processes is that the medium itself or the boundary conditions may be stochastically fluctuating. The interplay between these external fluctuations and the intrinsic fluctuations of diffusive motion can lead to stochastic resonant phenomena, in which the first-passage probability exhibits nonmonotonic behavior as a function of the frequency of the external fluctuations. A practical example of such external fluctuations is ligand binding to myoglobin [Austin et al. (1975)]. There are a wide variety of examples of this genre,

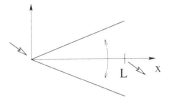

Fig. 4.9. Schematic of the fluctuating-barrier system. The arrows denote particle input at $x = 0$ and output at $x = L$.

such as first passage through a fluctuating bottleneck [Zwanzig (1990, 1992)] or a stochastic gate [Eizenberg & Klafter (1995)], gated chemical reactions [Spouge, Szabo, & Weiss (1996)], or escape from a fluctuating environment [Doering & Gadoua (1992), Bier & Astumian (1993), and Bar-Haim & Klafter (1999)]. To begin to understand the complex dynamics of such systems, we study a model introduced by Doering and Gadoua (1992) in which the external fluctuations are sufficiently simple in character that resonant behavior in first-passage characteristic time can be computed analytically.

Specifically, we study the mean first-passage time of a Brownian particle in the one-dimensional interval $[0, L]$, with injection of a unit flux and subsequent reflection at $x = 0$, and absorption at $x = L$, in the presence of a fluctuating barrier. This is created by having a spatially uniform bias velocity in the interval that stochastically switches between two values (Fig. 4.9). For simplicity, we assume that the velocity switches from $+v$ to $-v$ according to a Poisson process at a rate γ. Thus the bias alternately aids and hinders the particle in reaching $x = L$. We wish to understand the dependence of the mean first-passage time $\langle t \rangle$ on the frequency at which the bias changes.

It is easy to infer the behavior of $\langle t \rangle$ in the extreme limits of $\gamma \to 0$ and $\gamma \to \infty$. In the former case, $\langle t \rangle$ is dominated by the configuration in which the bias velocity is negative, and, from Eqs. (2.2.31),

$$\langle t \rangle \approx \frac{1}{2}\frac{L^2}{D}\left[\left(\frac{D}{vL}\right)^2 (e^{vL/D} - 1) - \frac{D}{vL}\right]$$
$$+ \frac{1}{2}\frac{L^2}{D}\left[\frac{D}{vL} - \left(\frac{D}{vL}\right)^2 (1 - e^{-vL/D})\right]$$
$$\sim \frac{D}{2v^2} e^{vL/D}. \tag{4.5.11}$$

Conversely, for $\gamma \to \infty$, the Brownian particle "sees" only the average, zero-bias environment, so that

$$\langle t \rangle \to \frac{1}{2}\frac{L^2}{D}.$$

The intriguing feature of this fluctuating system is that $\langle t \rangle$ has a minimum at an intermediate frequency value. Such stochastic resonant behavior may be responsible for facilitating chemical reactions in fluctuating environments. A variety of examples are discussed by Zwanzig (1990).

The standard approach to determine $\langle t \rangle$ in this system is conceptually straightforward but computationally formidable. We would need to solve the coupled convection–diffusion equations for the concentration with the medium in its two different states. These are

$$\frac{\partial c_+}{\partial t} + v\frac{\partial c_+}{\partial x} = \frac{\partial^2 c_+}{\partial x^2} + \gamma(c_- - c_+),$$

$$\frac{\partial c_-}{\partial t} - v\frac{\partial c_-}{\partial x} = \frac{\partial^2 c_-}{\partial x^2} + \gamma(c_+ - c_-). \tag{4.5.12}$$

where $c_\pm \equiv c_\pm(x, t)$ is the probability that the particle is at position x at time t when the bias velocity is $\pm v$. The appropriate boundary conditions are $c_\pm = 0$ at $x = L$ and $j_\pm = \pm v c_\pm - Dc'_\pm = \frac{1}{2}\delta(t)$ at $x = 0$. The latter corresponds to a unit input flux at $t = 0$, where the barrier is in each of its two states with probability $1/2$. From $c_\pm(x, t)$, we would then compute the flux to $x = L$, from which we can determine $\langle t \rangle$. Such a computation is carried out, for example, in Doering and Gadoua (1992) and Bier and Astumian (1993).

However, if we are interested in *only* the mean first-passage time, then we can simplify the problem considerably by exploiting the time-integrated formalism of Section 1.6. Once again, the basic point is that the conventional approach involves solving a straightforward but relatively cumbersome time-dependent problem and then integrating the first-passage probability over all time to obtain $\langle t \rangle$. Instead, we first integrate the original equation of motion over all time to obtain a simpler time-independent system. The mean first-passage time can then be obtained by the spatial integral of the time-integrated concentration. This calculation is much simpler than the full time-dependent solution. A similar approach was applied in Bier and Astumian (1993) and Zürcher and Doering (1993).

We now provide the essential steps to find $\langle t \rangle$ by this approach. First we integrate Eqs. (4.5.12) over all time to give the coupled ordinary differential equations for $C_\pm(x) \equiv \int_0^\infty c_\pm(x, t)\,dt$:

$$\frac{d^2 C_+}{dx^2} - v\frac{dC_+}{dx} + \gamma(C_- - C_+) = 0,$$

$$\frac{d^2 C_-}{dx^2} + v\frac{dC_-}{dx} + \gamma(C_+ - C_-) = 0. \tag{4.5.13}$$

Note that there is no end-point contribution from the time integral of $\partial c/\partial t$ because the system is initially empty. These equations are also subject to the time-integrated boundary conditions $\mathcal{J}_\pm = \pm v\mathcal{C}_\pm - D\mathcal{C}'_\pm = \frac{1}{2}$ at $x = 0$ and $\mathcal{C}_\pm = 0$ at $x = L$. From this, the mean first-passage time is

$$\langle t \rangle = \int_0^L [\mathcal{C}_+(x) + \mathcal{C}_-(x)]\,dx. \qquad (4.5.14)$$

To solve Eqs. (4.5.13) it is helpful to define $\sigma = \mathcal{C}_+ + \mathcal{C}_-$ and $\delta = \mathcal{C}_+ - \mathcal{C}_-$. These quantities satisfy

$$D\sigma'' - v\delta' = 0,$$
$$D\delta'' - v\sigma' - 2\gamma\delta = 0. \qquad (4.5.15)$$

These can be recast as the single equation

$$\sigma^{(4)} - \left(\frac{v^2}{D^2} + \frac{2\gamma}{D}\right)\sigma'' = 0, \qquad (4.5.16)$$

where the superscript (4) denotes the fourth derivative, and with the boundary conditions $\sigma = \delta = 0$ at $x = L$, and $\mathcal{J}_\sigma \equiv \mathcal{J}_+ + \mathcal{J}_- = 1$ and $\mathcal{J}_\delta \equiv \mathcal{J}_+ - \mathcal{J}_- = 0$ at $x = 0$. Substituting the exponential solution $\sigma = Ae^{\alpha x}$ into Eq. (4.5.16) gives the four solutions $\alpha = 0$ (doubly degenerate) and $\alpha = \pm\sqrt{v^2 + 2D\gamma}/D$. Finally, from the first of Eqs. (4.5.15), the general solution to these equations is

$$\sigma = A_1 + A_2 x + A_3 e^{\alpha x} + A_4 e^{-\alpha x},$$

$$\delta = -\frac{vA_2}{2\gamma} + \frac{D\alpha A_3}{v}e^{\alpha x} - \frac{D\alpha A_4}{v}e^{-\alpha x}.$$

Applying the boundary conditions gives four linear equations for $\sigma(L)$, $\delta(L)$, $\mathcal{J}_\sigma(0)$, and $\mathcal{J}_\delta(0)$. Then it is relatively simple to solve for the A_i, with the result

$$A_3 = \frac{1}{2v}\left\{e^{-k}\left[\frac{2Pe\Gamma}{1+\Gamma} - \frac{1}{(1+\Gamma)^{3/2}}\right] - \frac{\Gamma}{(1+\Gamma)^{3/2}}\right\}\bigg/(1 + \Gamma\cosh k),$$

$$A_4 = \frac{1}{2v}\left\{e^{k}\left[\frac{2Pe\Gamma}{1+\Gamma} + \frac{1}{(1+\Gamma)^{3/2}}\right] + \frac{\Gamma}{(1+\Gamma)^{3/2}}\right\}\bigg/(1 + \Gamma\cosh k),$$

$$A_1 = \Gamma(A_3 + A_4),$$

$$A_2 = -\frac{1}{D}\frac{\Gamma}{1+\Gamma},$$

where $Pe = vL/2D$ is the Péclet number, and $k = \alpha L$. We have also defined the dimensionless frequency $\Gamma = 2D\gamma/v^2$ that reduces the ubiquitous variable combination $(D\alpha)^2$ to $v^2(1 + \Gamma)$. For numerical purposes, it is better to reexpress Γ in terms of the manifestly dimensionless combination $\Gamma = \gamma \tau_D / Pe^2$.

In terms of the A_i, the mean first-passage time is

$$
\langle t \rangle = \int_0^L \sigma \, dx
$$

$$
= A_1 L + \frac{1}{2} A_2 L^2 + \frac{A_3}{\alpha}(e^{\alpha L} - 1) + \frac{A_4}{\alpha}(1 - e^{-\alpha L}),
$$

and, after some straightforward algebra, we finally obtain

$$
\frac{\langle t \rangle}{\tau_c} = \frac{1}{\Delta} \left\{ \left[\frac{4Pe\Gamma^2}{1+\Gamma} + \frac{1}{Pe(1+\Gamma)^2} \right] \cosh k + \frac{4\Gamma}{(1+\Gamma)^{3/2}} \sinh k \right.
$$
$$
\left. + \frac{\Gamma}{Pe(1+\Gamma)^2} \right\} - \frac{1}{2Pe} \frac{1}{(1+\Gamma)^2} - \frac{Pe\Gamma}{1+\Gamma}, \tag{4.5.17}
$$

where $\Delta = (2 + 2\Gamma \cosh k)$ and $\tau_c = L/v$ is the time to convect across the interval.

Although the final result for $\langle t \rangle$ appears opaque, it is easy to extract numerical results from this expression. In particular, there is a robust minimum in the mean first-passage time as a function of the driving frequency (Fig. 4.10). This same resonant behavior occurs in more general situations, such as a barrier that fluctuates between two different positive values [Bier & Astumian (1993)] and for other forms of the switching rates between internal states of the system [Van den Broeck (1993)].

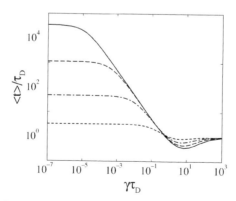

Fig. 4.10. Dependence of the scaled mean first-passage time $\langle t \rangle / \tau_D$ on the scaled frequency $\gamma \tau_D$. Shown are the cases of Péclet number $Pe = 2, 4, 6,$ and 8 (bottom to top).

4.6. Interval with Spatially Variable Diffusion

4.6.1. Basic Examples

What are the first-passage properties in an interval in which the diffusion coefficient depends on position? Here we consider the symmetric interval $[-N, N]$ with absorbing boundaries and with the diffusion coefficients (i) $D(x) = 1 - (x/N)^2$, and (ii) $D(x) = 1 - |x|/N$, where x is the position in the interval. We will study how these spatially varying diffusion coefficients affect the mean exit time, and contrast with the constant diffusivity case, in which the exit time starting from x is $t(x) = (N^2/2D)(1 - x^2)$ [Eq. (2.2.17)].

There are two simple realizations of variable diffusivity systems. The first is the Voter model in the mean-field limit [Liggett (1985) and Krapivsky (1992a, 1992b)]. Suppose that there are n_A and n_B voters of type A and B, respectively, in a finite system of $N = n_A + n_B$ sites. In the mean-field limit, an update involves picking a voter at random and having it assume the state of another randomly chosen voter. The rate at which n_A and $n_B = N - n_A$ changes is then proportional to $n_A n_B$. As either n_A or $n_B \rightarrow N$, the density of opposite pairs goes to zero and the evolution slows down. In terms of the number difference $x = n_A - n_B$, an update occurs with a rate proportional to $1 - (x/N)^2$ in which x changes by ± 2 equiprobably. This process corresponds to diffusion on the interval with a diffusivity proportional to $1 - (x/N)^2$.

A diffusivity $1 - |x|/N$ corresponds to the mean-field limit of the monomer–monomer surface-reaction model [ben-Avraham et al. (1990b)]. Here, equal fluxes of A and B particles impinge on a surface and the unoccupied sites are filled at a rate proportional to their density. There is also an immediate surface reaction in which one A and one B are removed whenever an A adsorbs onto a system that already contains Bs (or vice versa). Thus the surface contains only As or only Bs, and the quantity x now equals n_A or n_B, whichever is nonzero. Again, in each reaction event x changes by ± 2 equiprobably. The overall process corresponds to diffusion on the interval with diffusivity proportional to $1 - |x|/N$.

4.6.2. Diffusivity $1 - (x/N)^2$

Let $P(n_A, n_B, t)$ be the probability that an N-site system contains n_A and n_B voters of type A and B at time t. According to the Voter model rules, this probability obeys the master equation

$$P_{n_A}(t + \Delta t) = \left(\frac{n_A - 1}{N}\right)\left(1 - \frac{n_A - 1}{N}\right) P_{n_A-1}(t)$$

$$+ \left(\frac{n_A + 1}{N}\right)\left(1 - \frac{n_A + 1}{N}\right) P_{n_A+1}(t)$$

$$- \left[1 - 2\left(\frac{n_A}{N}\right)\left(1 - \frac{n_A}{N}\right)\right] P_{n_A}(t), \qquad (4.6.1)$$

for $0 < n_A < N$, and

$$P_0(t + \Delta t) = \left(\frac{1}{N}\right)\left(1 - \frac{1}{N}\right) P_1(t) + P_0(t),$$

$$P_N(t + \Delta t) = \left(\frac{1}{N}\right)\left(1 - \frac{1}{N}\right) P_{N-1}(t) + P_N(t), \qquad (4.6.2)$$

for $n_A = 0, N$. In terms of $x = (n_A - n_B)$, the continuum limit of this master equation is

$$\frac{\partial P(x,t)}{\partial t} = \frac{\partial^2}{\partial x^2}\left\{\left[1 - \left(\frac{x}{N}\right)^2\right] P(x,t)\right\}, \qquad (4.6.3)$$

which is a diffusion process with progressively slower motion near the interval extremeties.

The probability density can be written as an eigenfunction expansion in terms of the Gegenbauer polynomials [ben-Avraham et al. (1990b)]. Amusingly, the lowest eigenmode is a constant, in contrast to the half-sine-wave lowest eigenmode of the probability distribution for constant diffusivity. This feature is a manifestation of the slower diffusion near the interval ends, leading to a slower exit rate and a concomitant local enhancement of the density.

In a similar spirit, the adjoint equation for mean exit time [Eq. (1.6.22)], when starting from the state (n_A, n_B), obeys the recursion relation,

$$t(n_A) = \left(\frac{n_A}{N}\right)\left(1 - \frac{n_A}{N}\right)\{[t(n_A - 1) + \Delta t] + [t(n_A + 1) + \Delta t]\}$$

$$+ \left[1 - 2\left(\frac{n_A}{N}\right)\left(1 - \frac{n_A}{N}\right)\right][t(n_A) + \Delta t], \qquad (4.6.4)$$

which becomes, in the continuum limit,

$$\frac{d^2 t(x)}{dx^2} = -\frac{4}{1 - (x/N)^2}. \qquad (4.6.5)$$

Solving this equation with the boundary conditions $t(N) = t(-N) = 0$ gives

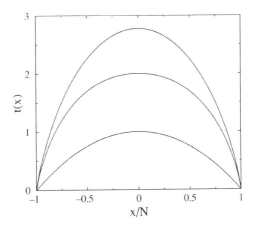

Fig. 4.11. Exit time from an absorbing interval for (i) $D(x) = 1$ (free diffusion), (ii) $D(x) = (1 - |x|/N)$, and (iii) $D(x) = [1 - (x/N)^2]$ (bottom to top).

(see Fig. 4.11)

$$t(x) = -2N^2 \left[\left(1 - \frac{x}{N}\right) \ln\left(\frac{1}{2} - \frac{x}{2N}\right) + \left(1 + \frac{x}{N}\right) \ln\left(\frac{1}{2} + \frac{x}{2N}\right) \right].$$

$$(4.6.6)$$

An amusing feature of this result is that for a particle that starts one step away from the boundary, $x = \pm(N - 1)$, the exit time is $t(x) \approx 2N \ln(2N)$, compared with an exit time proportional to N for the same initial condition in a unrestricted diffusion (Subsection 2.2.2). This increased trapping time is again due to the suppression of the diffusion process close to the system boundaries.

4.6.3. Diffusivity $1 - |x|/N$

In analogy with the discussion of Subsection 4.6.2, in the continuum limit the master equation for the probability density becomes

$$\frac{\partial P(x, t)}{\partial t} = \frac{1}{2} \frac{\partial^2}{\partial x^2} \left[\left(1 - \frac{|x|}{N}\right) P(n, t) \right], \qquad (4.6.7)$$

whose solution can be written as an eigenfunction expansion over Bessel functions of order 1 [ben-Avraham et al. (1990b)].

The recursion formula for the mean exit time when starting at x is

$$t(x) = \frac{1}{2} \left(1 - \frac{x}{N}\right) [t(x - 1) + t(x + 1) + 2\Delta t] + \frac{x}{N} [t(x) + \Delta t],$$

$$(4.6.8)$$

whose continuum limit is

$$\frac{d^2t(x)}{dx^2} = -\frac{2}{1 - \frac{x}{N}}, \qquad (4.6.9)$$

again subject to the boundary conditions $t(\pm N) = 0$. The solution is

$$t(x) = 2N^2 \left(1 - \frac{x}{N}\right)\left[1 - \ln\left(1 - \frac{x}{N}\right)\right]. \qquad (4.6.10)$$

If a particle starts one step from the boundary, $x = \pm(N - 1)$, the exit time becomes $t(x) \approx 2N \ln N$, once again showing the effect of suppressed diffusion near the boundary.

4.6.4. Diffusivity $(1 - |x|/N)^\mu$

We have seen that for a diffusivity that vanishes linearly in the distance to the ends of the interval, there is a logarithmic enhancement in the exit time compared with free diffusion. This logarithmic factor suggests the possibility of new behavior if $D(x) = (1 - |x|/N)^\mu$, with $\mu \neq 1$. We briefly study this more general situation.

The adjoint equation for the exit time is

$$\frac{d^2t(x)}{dx^2} = -\left(1 - \frac{x}{N}\right)^{-\mu}. \qquad (4.6.11)$$

The solution that is symmetric in x and that obeys the boundary conditions is

$$t(x) = \frac{N^2}{(1 - \mu)}\left[\left(1 - \frac{x}{N}\right) - \frac{\left(1 - \frac{x}{N}\right)^{2-\mu}}{(2 - \mu)}\right]. \qquad (4.6.12)$$

If the particle starts at the origin, the mean exit time is still of the order of N^2. However, if the particle starts close to the exit, $x = \pm(N - 1)$, then the exit time varies as

$$t[x = \pm(N - 1)] \propto \begin{cases} N, & 0 \leq \mu \leq 1 \\ N \ln N, & \mu = 1 \\ N^\mu, & 1 < \mu < 2 \end{cases}. \qquad (4.6.13)$$

This result shows how suppressed diffusion leads to an increased exit time when starting near the boundary. If the suppression is stronger than quadratic, however, then the exit time is infinite for any starting location.

4.7. The Expanding Cage

4.7.1. General Considerations

For a fixed-length interval, we have learned that the survival probability of a diffusing particle $S(t)$ asymptotically decays as $e^{-\pi^2 Dt/L^2}$. What happens if the interval length $L(t)$ (the "cage") expands with time? This simple question raises a variety of intriguing issues about the relative effects of diffusion and the motion of the boundary on first-passage properties. This is a classic problem in the first-passage literature, especially when the boundary motion matches that of diffusion. Solutions to this problem have been obtained by a variety of methods [see, e.g., Breiman (1966), Uchiyama (1980), Salminen (1988), Iglói (1992), and Turban (1992)]. The discussion in this section is closely based on a physically motivated approach originally given in Krapivsky and Redner (1996b).

It is easy to infer the survival probability for a slowly expanding and a rapidly expanding cage:

- For $L(t) \propto t^\alpha$ with $\alpha < 1/2$, a diffusing particle eventually overtakes the ends of the cage, after which the survival probability decays rapidly in time. We may expediently treat this slowly moving boundary-value problem by the *adiabatic approximation* [Schiff (1968)]. Here, the asymptotic probability is "engineered" to have the same functional form as in the fixed-size cage, but also to satisfy the time-dependent boundary condition. This leads to a stretched exponential decay, $S(t) \propto \exp[-At^{(1-2\alpha)}]$, where A is a constant.
- For $\alpha > 1/2$, the particle is unlikely reach the boundaries and the probability distribution is close to that for free diffusion. This is the basis of the *free approximation* that leads to a nonzero limiting value for $S(t)$ as $t \to \infty$.

The existence of these two simple approaches stems from the closeness of the respective probability distributions to exactly soluble limits – a fixed-length interval for $\alpha < 1/2$ or a freely diffusing particle for $\alpha > 1/2$.

The most interesting case is when the cage expands at the same rate as diffusion. Here a new dimensionless parameter appears – the ratio of the diffusion length to the cage length. As will be shown, this leads to $S(t)$ decaying as a nonuniversal power law in time. Within the rubric of a marginally expanding cage, there are still two distinct situations:

- The boundaries of the cage each diffuse.
- The cage expands as a power law in time, $L(t) = (At)^\alpha$.

We may solve the former by mapping the trajectories of the particle and the two absorbing boundary points to a single diffusing particle in three dimensions with static absorbing boundaries. For continuity of presentation, this discussion is deferred to Section 8.3, where a complete treatment of the reactions of three diffusing particles on the line is given. For the latter situation, we will show that $S(t)$ decays as $t^{-\beta}$, with $\beta = \beta(A, D)$ diverging as D/A for $A/D \ll 1$ and approaching zero as $\exp(-A/D)$ for $A/D \gg 1$. We focus primarily on the adiabatic and the free approximations because they are easy to use and can be readily adapted to other moving boundary-value problems. These methods also work much better than we might anticipate, given their simple-minded nature.

4.7.2. Slowly Expanding Cage: Adiabatic Approximation

For $L(t) \ll \sqrt{Dt}$, we invoke the adiabatic approximation, in which the spatial dependence of the concentration for an interval of length $L(t)$ is assumed to be identical to that of the static diffusion equation at the instantaneous value of L. This is based on the expectation that the concentration in a slowly expanding cage is always close to that of a fixed-size interval. Thus we write

$$c(x, t) \approx f(t) \cos\left(\frac{\pi x}{2L(t)}\right) \equiv c_{\text{ad}}(x, t), \qquad (4.7.1)$$

with $f(t)$ to be determined. The corresponding survival probability is

$$S(t) \approx \int_{-L(t)}^{L(t)} c_{\text{ad}}(x, t)\, dx = \frac{4}{\pi}\, f(t)L(t). \qquad (4.7.2)$$

For convenience, we have defined the cage to be $[-L(t), L(t)]$.

To obtain $f(t)$, we substitute approximation (4.7.1) into the diffusion equation, as in separation of variables, to give

$$\frac{df}{dt} = -\left(\frac{D\pi^2}{4L^2}\right) f - \left(\frac{\pi x}{2L^2}\right) \frac{dL}{dt} \tan\left(\frac{\pi x}{2L}\right) f. \qquad (4.7.3)$$

Note that variable separation does not strictly hold, as the equation for $f(t)$ also involves x. However, when $L(t)$ increases as $(At)^\alpha$ with $\alpha < 1/2$, the second term on the right-hand side is negligible. Then the controlling factor of $f(t)$ is given by

$$f(t) \propto \exp\left[-\frac{D\pi^2}{4} \int_0^t \frac{dt'}{L^2(t')}\right]$$
$$= \exp\left[-\frac{D\pi^2}{4(1 - 2\alpha)A^{2\alpha}}\, t^{1-2\alpha}\right]. \qquad (4.7.4)$$

We can also obtain this dependence (but without the crucial factor $1 - 2\alpha$ inside the exponential) merely by substituting $L(t) \propto t^\alpha$ into the survival probability for the fixed-size interval [Eq. (2.2.4)]. We see that $f(t)$ reduces to exponential decay for a fixed-size interval, whereas for $\alpha \to 1/2$, the nature of the controlling factor suggests a more slowly decaying functional form for $S(t)$.

4.7.3. Rapidly Expanding Cage: Free Approximation

For a rapidly expanding cage, the escape rate from the system is small and the absorbing boundaries should eventually become irrelevant. We therefore expect that the concentration profile remains close to the Gaussian distribution of free diffusion. We may then account for the slow decay of the survival probability by augmenting the Gaussian with an overall decaying amplitude. This free approximation is another example in which the existence of widely separated time scales, $\sqrt{Dt}$ and $L(t)$, suggests the nature of the approximation itself.

According to the free approximation, we write

$$c(x, t) \approx \frac{S(t)}{\sqrt{4\pi Dt}} \, e^{-x^2/4Dt} \equiv c_{\text{free}}(x, t).$$

Although this concentration does not satisfy the absorbing boundary condition, the inconsistency is negligible at large times because the density is exponentially small at the cage boundaries. We may now find the decay of the survival probability by equating the probability flux to the cage boundaries, $2D|\frac{\partial c}{\partial x}|$, to the loss of probability within the cage. For $L(t) = (At)^\alpha$, this flux is

$$\frac{S(t)A^\alpha}{\sqrt{4\pi D}} \, t^{\alpha-3/2} \, \exp\left(-\frac{A^{2\alpha}}{4D} t^{2\alpha-1}\right), \qquad (4.7.5)$$

which rapidly goes to zero for $\alpha > 1/2$. Because this flux equals $-(dS/dt)$, it follows that the survival probability approaches a nonzero limiting value for $\alpha > 1/2$ and that this limiting value goes to zero as $\alpha \to 1/2$.

4.7.4. Marginally Expanding Cage

4.7.4.1. Heuristics

For the marginal case of $\alpha = 1/2$, the adiabatic and the free approximations are ostensibly no longer appropriate, as $L(t) = \sqrt{At}$ and $\sqrt{Dt}$ have a fixed ratio. However, for $A/D \ll 1$ and $A/D \gg 1$, we might expect that these methods could still be useful. Thus we will continue to apply these heuristic approximations and check their accuracy a posteriori. In fact, the survival probability exponents predicted by these two approximations are each quite close to the exact result except for $A/D \approx 1$.

When the adiabatic approximation is applied, the second term in Eq. (4.7.3) is, in principle, nonnegligible for $\alpha = 1/2$. However, if $A \ll D$, the cage still expands more slowly (in amplitude) than free diffusion and the error made by neglect of the second term in Eq. (4.7.3) may still be small. The solution to this truncated equation immediately gives $f(t)$, which, when substituted into approximation (4.7.2) leads to $S_{ad}(t) \approx t^{-\beta_{ad}}$, with

$$\beta_{ad} \approx \frac{D\pi^2}{4A} - \frac{1}{2}. \tag{4.7.6}$$

The trailing factor of $-1/2$ should not be taken completely seriously, as the neglected term in Eq. (4.7.3) leads to additional corrections which are also of the order of 1.

Similarly for $A \gg D$, the free approximation gives

$$\frac{dS}{dt} \approx -2D \left| \frac{\partial c(x,t)}{\partial x} \right|_{x=\sqrt{At}} \tag{4.7.7}$$

$$= -\frac{S(t)}{t} \sqrt{\frac{A}{4\pi D}} e^{-A/4D}. \tag{4.7.8}$$

This again leads to the nonuniversal power law for the survival probability, $S_{free} \sim t^{-\beta_{free}}$, with

$$\beta_{free} = \sqrt{\frac{A}{4\pi D}} e^{-A/4D}. \tag{4.7.9}$$

As shown in Fig. 4.12, these approximations are surprisingly accurate over much of the range of A/D.

4.7.4.2. Asymptotics

To complete our discussion, a first-principles analysis is outlined for the survival probability of a diffusing particle in a marginally expanding cage, following Krapivsky and Redner (1996b). Here β is determined implicitly from the eigenvalue of a differential equation, from which the exponent may be evaluated numerically. The necessity of resorting to numerics to complete the exact solution is another reason for the attractiveness of the heuristic approaches.

When $L(t) = \sqrt{At}$, a natural scaling hypothesis is to write the density in terms of the two dimensionless variables:

$$\xi \equiv \frac{x}{L(t)}, \quad \sigma \equiv \frac{x}{\sqrt{Dt}}.$$

It is more convenient to use the equivalent choices of ξ and $\rho = \xi/\sigma = \sqrt{A/D}$

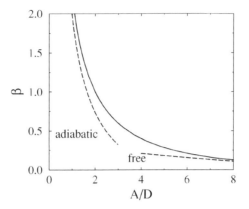

Fig. 4.12. The survival probability exponent β for the marginal case $L(t) = (At)^{1/2}$. Shown is the exact value of β (solid curve) from the numerical solution of Eq. (4.7.12), together with the predictions from the adiabatic and the free approximations, Eqs. (4.7.6) and (4.7.9) respectively (both dashed curves).

as the basic scaling variables. We now seek solutions for the concentration in the form

$$c(x, t) = t^{-\beta - 1/2} \mathcal{C}_\rho(\xi), \qquad (4.7.10)$$

where $\mathcal{C}_\rho(\xi)$ is a two-variable scaling function that encodes the spatial dependence. The power-law prefactor ensures that the survival probability, namely, the spatial integral of $c(x, t)$, decays as $t^{-\beta}$, as defined in Subsection 4.7.4.1.

When Eq. (4.7.10) is substituted into the diffusion equation, the scaling function satisfies the ordinary differential equation

$$\frac{1}{\rho^2} \frac{d^2 \mathcal{C}}{d\xi^2} + \frac{\xi}{2} \frac{d\mathcal{C}}{d\xi} + \left(\beta + \frac{1}{2} \right) \mathcal{C} = 0.$$

Then, by introducing $\eta = \xi \sqrt{\rho/2}$ and $\mathcal{C}(\xi) = e^{-\eta^2/4} \mathcal{D}(\eta)$, we transform this into the parabolic cylinder equation [Abramowitz & Stegun (1972)]

$$\frac{d^2 \mathcal{D}}{d\eta^2} + \left[2\beta + \frac{1}{2} - \frac{\eta^2}{4} \right] \mathcal{D} = 0. \qquad (4.7.11)$$

For an unbounded range of η, this equation has solutions for quantized values of the energy eigenvalue $E = 2\beta + \frac{1}{2} = \frac{1}{2}, \frac{3}{2}, \frac{5}{2}, \ldots$ [Schiff (1968)].

For our cage problem, the range of η is restricted to $|\eta| \leq \sqrt{A/2D}$. In the equivalent-quantum mechanical system, this corresponds to a particle in a harmonic-oscillator potential for $|\eta| < \sqrt{A/2D}$ and an infinite potential for $|\eta| > \sqrt{A/2D}$. For this geometry, a spatially symmetric solution to Eq. (4.7.11), appropriate for the long-time limit for an arbitrary starting point, is

$$\mathcal{D}(\eta) \equiv \frac{1}{2}[\mathcal{D}_{2\beta}(\eta) + \mathcal{D}_{2\beta}(-\eta)],$$

where $\mathcal{D}_\nu(\eta)$ is the parabolic cylinder function of order ν. Finally, the relation between the decay exponent β and $\sqrt{A/D}$ is determined implicitly by the absorbing boundary condition, namely,

$$\mathcal{D}_{2\beta}(\sqrt{A/2D}) + \mathcal{D}_{2\beta}(-\sqrt{A/2D}) = 0. \qquad (4.7.12)$$

This condition for $\beta = \beta(A/D)$ simplifies in the limiting cases $A/D \ll 1$ and $A/D \gg 1$. In the former, the exponent β is large and the second two terms in the brackets in Eq. (4.7.11) can be neglected. Equivalently, the physical range of η is small, so that the potential plays a negligible role. The solution to this limiting free-particle equation is just the cosine function, and the boundary condition immediately gives the limiting expression of Eq. (4.7.6), but without the subdominant term of $-1/2$. In the latter case of $A \gg D$, $\beta \to 0$ and Eq. (4.7.11) approaches the Schrödinger equation for the ground state of the harmonic oscillator. In this case, a detailed analysis of the differential equation reproduces the limiting exponent of Eq. (4.7.9) [Krapivsky & Redner (1996b)]. These provide rigorous justification for the limiting values of the decay exponent β that we obtained by heuristic means.

Finally, it is amusing to consider the mean lifetime of a diffusing particle in the expanding cage. Because this lifetime can be written as $\langle t \rangle = \int_0^\infty S(t)\,dt$, $\langle t \rangle$ is finite for $\beta > 1$ and infinite for $\beta < 1$. In the borderline case of $\beta = 1$, Eq. (4.7.11) describes the second excited state of the simple harmonic oscillator, with wave function $\mathcal{D}(\eta) = (1 - \eta^2)e^{-\eta^2/4}$. The absorbing boundary condition of Eq. (4.7.12) clearly requires that $\eta^2 = 1$ or $A = 2D$. Thus the transition between a finite and an infinite survival time occurs at $A = 2D$.

4.7.5. *Iterated Logarithm Law for Ultimate Survival*

In the marginal situation of $L(t) = (At)^{1/2}$, we have seen that the survival probability $S(t)$ decays as a power law $t^{-\beta(A,D)}$, with $\beta \to 0$ as $A/D \to \infty$. This decay becomes progressively slower as A becomes large. On the other hand, when $L(t) \propto t^\alpha$, with α strictly greater than $1/2$, the survival probability at infinite time is greater than zero. This naturally leads to the following question: What is the nature of the transition between certain death, defined as $S(t \to \infty) = 0$, and finite survival, defined as $S(t \to \infty) > 0$?

The answer to this question is surprisingly rich. There is an infinite sequence of transitions, in which $L(t)$ involves additional iterated logarithmic time dependences that define regimes in which $S(t)$ assumes progressively slower functional forms. Each of these regimes is roughly analogous to a successive layer of an onion. The limit of this sequence is given by $L^*(t) = \sqrt{4Dt(\ln\ln t + \frac{3}{2}\ln\ln\ln t + \ldots)}$. The first term in this series is known as the Khintchine iterated logarithm law [Khintchine (1924) and Feller (1968)]. Although the Khintchine law can be obtained by rigorous methods, we can also obtain this intriguing result, as well as the infinite sequence of transitions, with relatively little computation by the free approximation.

Because we anticipate that the transition between life and death occurs when the amplitude A diverges, we make the hypothesis $A \propto u(t)$, with $u(t)$ diverging slower than a power law as $t \to \infty$. Because $L(t)$ now increases more rapidly than the diffusion length $(Dt)^{1/2}$, the free approximation should be asymptotically exact as it already works extremely well when $L(t) = (At)^{1/2}$ with A large. Within this approximation, we rewrite Eq. (4.7.7) as

$$\ln S(t) \sim -\int^t \frac{dt'}{t'}\sqrt{\frac{A}{4\pi D}}\,e^{-A/4D}. \tag{4.7.13}$$

Here we neglect the lower limit because the free approximation is valid only as $t \to \infty$, where the short-time behavior is irrelevant. In this form, it is clear that, if $L = (At)^{1/2}$, $\ln S$ decreases by an infinite amount because of the divergent integral as $t \to \infty$ and $S(t) \to 0$. To make the integral converge, the other factors in the integral must somehow cancel the logarithmic divergence that arises from the factor dt/t. Accordingly, let us substitute $L(t) = \sqrt{4Dt\,u(t)}$ into approximation (4.7.13). This gives

$$\ln S(t) \sim -\frac{1}{\sqrt{\pi}}\int^t \frac{dt'}{t'}\sqrt{u(t')}\,e^{-u(t')}. \tag{4.7.14}$$

To simplify this integral, it is helpful to define $x = \ln t$ so that

$$S(x) \sim -\frac{1}{\sqrt{\pi}}\int^{\ln t} dx\,\sqrt{u(x)}\,e^{-u(x)}. \tag{4.7.15}$$

To lowest order, it is clear that if we choose $u(x) = \lambda \ln x$ with $\lambda > 1$, the integral converges as $t \to \infty$ and thus the asymptotic survival probability is positive. Conversely, for $\lambda \leq 1$, the integral diverges and the particle surely dies. In this latter case, we evaluate the integral to lowest order to give

$$S(t) \sim \exp\left[-\frac{(\ln t)^{1-\lambda}\sqrt{\lambda\ln\ln t}}{\sqrt{\pi}(1-\lambda)}\right] \quad \lambda < 1. \tag{4.7.16}$$

This decay is slower than any power law, but faster than any power of logarithm, that is, $t^{-\beta} < S(t) < (\ln t)^{-\gamma}$ for $\beta \to 0$ and $\gamma \to \infty$.

What happens in the marginal case of $\lambda = 1$? Here we can refine the criterion between life and death still further by incorporating into $u(x)$ an additional correction that effectively cancels the subdominant factor $\sqrt{u(x)}$ in approximation (4.7.15). We therefore define $u(x)$ such that $e^{-u(x)} = 1/x(\ln x)^{\mu}$. Then, in terms of $y = \ln x$, approximation (4.7.15) becomes

$$S(y) \sim -\frac{1}{\sqrt{\pi}} \int^{\ln \ln t} \frac{dy}{y^{\mu - 1/2}} \left(1 + \frac{\mu \ln y}{y} \right). \qquad (4.7.17)$$

This integral now converges for $\mu > 3/2$ and diverges for $\mu \leq 3/2$. In the latter case, the survival probability now lies between the bounds $(\ln t)^{-\gamma} < S(t) < (\ln \ln t)^{-\delta}$ for $\gamma \to 0$ and $\delta \to \infty$. At this level of approximation, we conclude that when the cage length grows faster than

$$L^*(t) = \sqrt{4Dt \left(\ln \ln t + \frac{3}{2} \ln \ln \ln t + \cdots \right)} \qquad (4.7.18)$$

a diffusing particle has a nonzero asymptotic survival probability, whereas for a cage that expands as L^*, there is an extremely slow decay of the survival probability.

Clearly, we can generate an infinite series of correction terms in the expression for $L^*(t)$ by successively constructing more accurate life–death criteria along the lines of approximation (4.7.17). The ultimate life–death transition corresponds to an ultraslow decay in which $S(t)$ has the form $S(t) \sim \lim_{n \to \infty} 1/\ln_n t$, where $\ln_2 t = \ln \ln t$ and $\ln_n t = \ln \ln_{n-1} t$.

4.8. The Moving Cliff

As a complement to the expanding cage, we study the first-passage properties to an absorbing, moving boundary in a semi-infinite domain (a "cliff"). Once again, the asymptotic behavior of the survival probability depends on the relative time dependences of the cliff position $x_{cl}(t)$ and the diffusion length $\sqrt{Dt}$. When $x_{cl}(t) \sim t^\alpha$ with $\alpha > 1/2$, the particle motion is subdominant. Consequently $S(t)$ approaches a nonzero asymptotic value for a receding cliff and quickly goes to zero for an advancing cliff. For $\alpha < 1/2$, the cliff motion is asymptotically irrelevant and the survival probability decays as $t^{-1/2}$, as in the case of a static boundary.

The interesting situation is again $x_{cl}(t) \sim \sqrt{At}$, in which we find that $S(t)$ decays as $t^{-\beta}$, with β dependent on A/D. As a counterpoint, we also discuss a *diffusing* cliff with diffusivity D_{cl}. Here, the motion of the cliff does not affect the exponent of $S(t)$ but only its overall amplitude, leading to $S(t) \propto 1/\sqrt{(D + D_{cl})t}$. Nevertheless, the spatial probability distribution exhibits interesting features that are dependent on the particle and cliff diffusivities.

In addition to the intriguing behavior of the survival probability, the marginal case of $x_{cl}(t) \sim \sqrt{At}$ corresponds to a *Stefan problem* [Crank (1987)]. This typically involves the motion of an interface between two phases during a phase transition. A familiar example is the growth of a layer of ice on a body of water that remains at $T = 0\,°C$ when the air temperature is $-\Delta T$. The temperature in the ice obeys the diffusion equation, subject to the moving boundary conditions that $T = 0\,°C$ at the ice–water interface and that the temperature equals $-\Delta T$ at the air–ice interface. For this idealized situation, the ice thickness grows as $\sqrt{t}$, and the heat flux, which plays the role of the first-passage probability, determines the interface motion. The idealized problem considered in Subsection 4.8.2 therefore provides a useful approach for studying these phase-change problems.

4.8.1. Rapidly Moving Cliff

Consider a particle initially at $x = x_0$ and the cliff initially at $x_{cl} = 0$. For a quickly receding cliff, $x_{cl}(t) \sim -At^\alpha$ with $\alpha > 1/2$, the particle concentration profile is barely affected by the cliff, and we may approximate this profile by the profile of free diffusion. We then estimate the probability of the particle being absorbed at the cliff as the flux of this Gaussian distribution evaluated at $x = x_{cl}$. Because $|x_{cl}| > \sqrt{Dt}$, this flux decays rapidly with time [see expression (4.7.5)], and the asymptotic survival probability goes to a nonzero constant at long times.

For an approaching cliff, $x_{cl} \sim +At^\alpha$, we may obtain a rough, but asymptotically correct, approximation for $S(t)$ by simply using the survival probability of biased diffusion in the semi-infinite interval [the latter expression in Eq. (3.2.16), with v replaced with $\dot{x}_{cl} \propto t^{\alpha-1}$]. This immediately leads to

$$S(t) \propto \frac{1}{t^{\alpha-\frac{1}{2}}} e^{-(At^{2\alpha-1})}. \tag{4.8.1}$$

Thus the survival probability decays faster than any power law in time.

4.8.2. Marginally Moving Cliff

The interesting situation is again the case in which the cliff motion matches that of diffusion. We analyze this system by the methods used for the marginally expanding interval. It is convenient to change variables from (x, t) to $[x' = x - x_0(t), t]$ to fix the absorbing cliff at the origin. The initial diffusion equation is then transformed to the convection–diffusion equation (in which the prime is now dropped):

$$\frac{\partial c}{\partial t} - \frac{x_{cl}}{2t}\frac{\partial c}{\partial x} = D\frac{\partial^2 c}{\partial x^2}, \quad \text{for} \quad 0 \le x < \infty, \tag{4.8.2}$$

with the boundary condition $c(x = 0, t) = 0$. This describes biased diffusion with velocity $-x_{cl}/2t$ and absorption at the origin. If the cliff in the initial system is approaching, then in the transformed problem the bias velocity is toward a fixed cliff and vice versa for the receding cliff.

We then introduce the dimensionless length $\xi = \pm x/x_{cl}$, where $\pm$ refers to the approaching and the receding cliffs, respectively, and then we make the scaling ansatz for the concentration profile

$$c(x, t) = t^{-\beta - 1/2} C_\rho(\xi),$$

with $\rho = \sqrt{A/D}$ and the power-law prefactor fixed by the condition that $S(t)$ decays as $t^{-\beta}$. Substituting this ansatz into Eq. (4.8.2) gives

$$\frac{D}{A}\frac{d^2 C}{d\xi^2} + \frac{1}{2}(\xi \pm 1)\frac{dC}{d\xi} + \left(\beta + \frac{1}{2}\right) C = 0.$$

It is now expedient to transform variables yet again by

$$\eta = \sqrt{\frac{A}{2D}}(\xi \pm 1), \quad \mathcal{D}(\eta) = \mathcal{C}(\xi)\, e^{\eta^2/4}.$$

With these definitions, $\mathcal{D}(\eta)$ satisfies the same parabolic cylinder equation as the interval problem [Eq. (4.7.11)], but with the boundary conditions

$$\mathcal{D}_{2\beta}(\pm\sqrt{A/2D}) = 0, \tag{4.8.3a}$$

which correspond to absorption of the particle when it hits the cliff (+ sign for approaching cliff and − sign for receding), and

$$\mathcal{D}_{2\beta}(\eta = \infty) = 0, \tag{4.8.3b}$$

which ensures that $S(t) = \int_0^\infty dx\, c(x, t)$ remains bounded.

The determination of the survival exponent β and the concentration profile $\mathcal{D}(\eta)$ is therefore equivalent to finding the ground-state energy and wave function of a quantum particle in a potential composed of an infinite barrier at $\eta = \pm\sqrt{A/2D}$ and the harmonic-oscillator potential for $\eta > \pm\sqrt{A/2D}$. Higher excited states do not contribute in the long-time limit. This prescription provides a general, but implicit, solution for $\beta = \beta(A/D)$, which must be solved numerically (see Fig. 4.13). However, we can obtain some of this limiting behavior for β for small and large values of A/D directly from

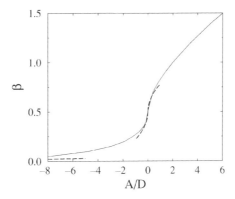

Fig. 4.13. Survival probability exponent β versus A/D for the cliff geometry when its position is $L(t) = \pm(At)^{1/2}$. Positive (negative) A/D corresponds to an approaching (receding) cliff. Shown are the exact value of β from the numerical solution of Eq. (4.8.3a) and the approximate predictions (both dashed) for $|A|/D \ll 1$ [approximation (4.8.4a)] and for $-A/D \gg 1$ [approximation (4.8.4b)].

the parabolic cylinder equation [Eq. (4.7.11)] together with elementary facts about the quantum mechanics of the harmonic oscillator.

In the limit $A \ll D$, the infinite barrier in the equivalent quantum problem is close to $\eta = 0$. When the wall is exactly at the origin, the ground state of this truncated harmonic-oscillator potential is obviously the first excited state of the pure harmonic oscillator, namely, $\mathcal{D}(\eta) = \eta e^{-\eta^2/4}$ and correspondingly $\beta = 1/2$. For $A \ll D$, it is possible to construct a perturbative solution for the concentration profile, which ultimately gives [Krapivsky & Redner (1996b)]

$$\beta \approx \frac{1}{2} \pm \sqrt{\frac{A}{4\pi D}} \quad A \ll D. \tag{4.8.4a}$$

When $A \gg D$ and the cliff is receding, the position of the absorbing boundary goes to $-\infty$. Hence the ground state for this system is close to the harmonic-oscillator ground state, namely, $\mathcal{D}(\eta) = e^{-\eta^2/4}$ and $\beta \approx 0$. For this case, we can also apply the free approximation, but with the obvious modification that there is only a single boundary at which flux leaves the system. By this approach, the survival probability exponent is

$$\beta \approx \sqrt{\frac{A}{16\pi D}} \, e^{-A/2D}, \quad A \gg D \text{ and receding cliff.} \tag{4.8.4b}$$

For the advancing cliff, the equivalent quantum system is the harmonic-oscillator potential for $\eta > \sqrt{A/2D}$ and an infinite barrier at $\eta = \sqrt{A/2D}$.

Here, the most naive estimate for the ground-state energy is simply the minimum value of the potential energy, that is, $E_0 \approx \eta^2/4 = A/8D$. This immediately gives

$$\beta \approx \frac{A}{16D}, \quad A \gg D \text{ and advancing cliff.} \qquad (4.8.4c)$$

This crude estimate turns out to be relatively inaccurate, however, even though it gives the correct asymptotic result. Part of the inaccuracy stems from the neglect of the zero-point fluctuations in the corresponding determination of the ground-state energy of a quantum particle in the truncated oscillator potential. Using the Heisenberg uncertainty principle, we obtain relatively large corrections to approximation (4.8.4c), which are of the order of $(A/D)^{1/3}$. Finally, note that when $A/D = 2$, the survival exponent equals one. Thus the mean survival time of a diffusing particle near an advancing cliff is finite for $A/D > 2$ and infinite for $A/D \leq 2$.

4.8.3. Diffusing Cliff

When both the particle and the cliff diffuse, with respective diffusivities D and D_{cl}, their separation diffuses with diffusivity $D + D_{cl}$. Therefore the particle survival probability asymptotically decays as $S(t) \sim x_0/\sqrt{\pi(D + D_{cl})t}$, where x_0 is the initial distance between the particle and the cliff.

Although this time dependence is trivial, the spatial probability distribution of the particle is more interesting. We can obtain this distribution easily by mapping the cliff and the diffusing particle in one dimension to an effective single-particle system in two dimensions and then solving the latter by the image method (see Fig. 4.14). To construct this mapping, we

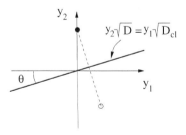

Fig. 4.14. Mapping of the particle and the boundary coordinates in one dimension to the planar coordinates $y_1 = x_{cl}/\sqrt{D_{cl}}$ and $y_2 = x/\sqrt{D}$. The initial $\vec{y}$ coordinates of the particle–boundary pair $(0, \sqrt{D})$ and its image are indicated by the solid and open circles, respectively. The particle survives if it remains above the absorbing line $y_2\sqrt{D} = y_1\sqrt{D_{cl}}$.

introduce the scaled coordinates $y_1 = x_{cl}/\sqrt{D_{cl}}$ and $y_2 = x/\sqrt{D}$ to render the two-dimensional diffusive trajectory (y_1, y_2) isotropic. The probability density in the plane $p(y_1, y_2, t)$ satisfies an absorbing boundary condition when $y_2\sqrt{D} = y_{y1}\sqrt{D_{cl}}$, corresponding to absorption of the particle when it hits the cliff. For simplicity and without loss of generality, we assume that the cliff and the particle are initially at $x_{cl}(0) = 0$ and $x(0) = 1$, respectively, or $[y_1(0), y_2(0)] = (0, \sqrt{D})$.

The probability density in the plane is therefore the sum of a Gaussian centered at $(0, \sqrt{D})$ and an image anti-Gaussian. From the orientation of the absorbing boundary (Fig. 4.14), this image is centered at $(\sqrt{D}\sin 2\theta, -\sqrt{D}\cos 2\theta)$, where $\theta = \tan^{-1}\sqrt{D_{cl}/D}$. In this representation, the probability density in two dimensions is simply

$$p(y_1, y_2, t) = \frac{1}{4\pi t}\left\{e^{-[y_1^2+(y_2-\sqrt{D})^2]/4t} - e^{-[(y_1-\sqrt{D}\sin 2\theta)^2+(y_2+\sqrt{D}\cos 2\theta)^2]/4t}\right\}.$$

(4.8.5)

The probability density for the particle to be at y_2 in one dimension is then the integral of the two-dimensional density over the accessible range y_1 of the boundary coordinate

$$p(y_2, t) = \int_{-\infty}^{y_2\cot\theta} p(y_1, y_2, t)\, dy_1.$$

(4.8.6)

By substituting Eq. (4.8.5) for $p(y_1, y_2, t)$, we can express the integral in Eq. (4.8.6) in terms of error functions. We then transform back to the original

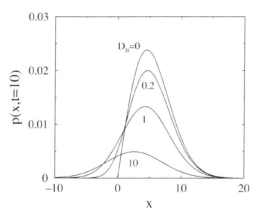

Fig. 4.15. Probability distribution of the particle in one dimension at time $t = 10$ [Eq. (4.8.7)] when the cliff and the particle are initially at $x_{cl} = 0$ and $x = 1$, respectively. Here D is fixed to be one and D_{cl} is varied as indicated.

particle coordinate $x = y_2\sqrt{D}$ by using $p(x, t)\,dx = p(y_2, t)\,dy_2$ to obtain

$$p(x, t) = \frac{1}{\sqrt{16\pi Dt}}\left[e^{-(x-1)^2/4Dt}\,\text{erfc}\left(-\frac{x\cot\theta}{\sqrt{4Dt}}\right)\right.$$
$$\left. - e^{-(x+\cos 2\theta)^2/4Dt}\,\text{erfc}\left(\frac{\sin 2\theta - x\cot\theta}{\sqrt{4Dt}}\right)\right], \qquad (4.8.7)$$

where $\text{erfc}(z) = 1 - \text{erf}(z)$ is the complementary error function. A plot of $p(x, t)$ is shown in Fig. 4.15 for various values of the cliff diffusivity D_{cl}. As the cliff becomes more mobile, the particle survival probability at fixed time rapidly decreases. However, there is also a greater penetration of the particle concentration into the region $x < 0$. Finally, when the cliff is stationary, $\theta = 0$ and Eq. (4.8.7) reduces to Eq. (3.2.1).

This mapping of the two-particle system to diffusion in the plane with an absorbing boundary condition is conceptually simple and extremely convenient. We will apply this same approach to solve the first-passage properties of diffusing three-particle systems in one dimension in Section 8.3.

5

Fractal and Nonfractal Networks

5.1. Beyond One Dimension

This chapter is devoted to the first-passage properties of fractal and non-fractal networks including the Cayley tree, hierarchically branched trees, regular and hierarchical combs, hierarchical blob structures, and other networks. One basic motivation for extending our study of first passage to these geometries is that many physical realizations of diffusive transport, such as hopping conductivity in amorphous semiconductors, gel chromatography, and hydrodynamic dispersion, occur in spatially disordered media. For general references see, e.g., Havlin and ben-Avraham (1987), Bouchaud and Georges (1990), and ben-Avraham and Havlin (2000). Judiciously constructed idealized networks can offer simple descriptions of these complex systems and their first-passage properties are often solvable by exact renormalization of the master equations.

In the spirit of simplicity, we study first passage on hierarchical trees, combs, and blobs. The hierarchical tree is an iterative construction in which one bond is replaced with three identical bonds at each stage; this represents a minimalist branched structure. The comb and the blob structures consist of a main backbone and an array of sidebranches or blob regions where the flow rate is vanishingly small. By varying the relative geometrical importance of the sidebranches (or blobs) to the backbone, we can fundamentally alter the first-passage characteristics of these systems.

When transport along the backbone predominates, first-passage properties are essentially one dimensional in character. For hierarchical trees, the role of sidebranches and the backbone are comparable, leading to a mean first-passage time that grows more quickly than the square of the system length. As might be expected, this can be viewed as the effective spatial dimension of such structures being greater than one. For more extensively branched structures, such as hierarchical combs, exploration along sidebranches can give the main contribution to the first-passage time through the system.

The first-passage characteristics of such systems are now characterized by a multiplicity of time scales.

Another important theme of this chapter is the role of bias on first-passage properties. Here we consider the situation of "hydrodynamic" bias, in which an external pressure difference is imposed across a network, leading to heterogeneous local pressure differences across each bond. Each bond is considered as a linear "resistance" element in which the local flow equals the local pressure drop times the bond permeability. This contrasts with the situation of a spatially uniform bias, in which other interesting effects, such as "pileups" in dead ends or backflow regions, can occur. One striking feature of hydrodynamic transport in networks, which perhaps is not as widely appreciated as it should be, is the *equal-time theorem*. This states that in the limit of large bias, equal subvolumes of the network give equal contributions to the mean first-passage time across the system. As a consequence: *the mean first-passage time across a network equals the system volume divided by the total external current flow*. This basic theorem was first elucidated for continuous-flow systems in the 1950s [Levenspiel & Smith (1957), Danckwerts (1958a, 1958b), and Spalding (1958)] and then more recently for network models of porous media [Koplik, Redner, & Wilkinson (1988)].

In the following sections, examples are presented to illustrate the basic features of first passage in these networks.

5.2. Cayley Tree

As our first example, an exact solution is given for a random walk on the Cayley tree of coordination number q (Fig. 5.1). Related approaches were given by Hughes and Sahimi (1982) and Monthus and Texier (1996). This is an infinite tree in which $q - 1$ "children" are created from each "parent." Because the number of sites in the nth generation grows exponentially in n and there are no closed loops, the Cayley tree can be viewed as equivalent to an infinite dimensional lattice. Part of the appeal of the Cayley tree is that it is often easy to solve physical processes on this structure and the solution corresponds to the mean-field limit.

From the connection with infinite spatial dimension, we expect that a random walk on the Cayley tree is transient, that is, the probability of its return to the starting point is less than one. We can understand this very simply by lumping the sites of the Cayley tree according to their generation number. This leads to an effective biased one-dimensional random walk in the generation coordinate, with a hopping probability $(q - 1)/q$ toward larger

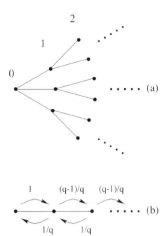

Fig. 5.1. (a) A Cayley tree of coordination number $q = 3$ organized by generation number. (b) The rates of the effective one-dimensional random walk that are obtained when all sites in a given generation are lumped together.

generation number and $1/q$ toward smaller generation number (Fig. 5.1(b)). This effective bias makes return to the origin uncertain.

To compute the first-passage probability, we first write the master equations for the occupation probability. Let $P_n(t)$ be the probability that the random walk is at any site in the nth generation at time t. By the isotropy of the hopping process at each site, the hopping rate along any bond is identical, which, for convenience, we take to be $1/q$. Then the master equations for $P_n(t)$ are

$$\frac{\partial P_0(t)}{\partial t} = \frac{1}{q} P_1(t) - P_0(t),$$

$$\frac{\partial P_1(t)}{\partial t} = \frac{1}{q} P_2(t) - P_1(t) + P_0(t),$$

$$\frac{\partial P_n(t)}{\partial t} = \frac{1}{q} P_{n+1}(t) - P_n(t) + \frac{q-1}{q} P_{n-1}(t), \quad n \geq 2. \quad (5.2.1)$$

We solve these equations for the initial condition of a particle initially at site 0, $P_n(0) = \delta_{n,0}$. Laplace transforming Eqs. (5.2.1) leads to the recursion formulas for $P_n(s)$:

$$(s+1)P_0(s) = \frac{1}{q} P_1(s) + 1,$$

$$(s+1)P_1(s) = \frac{1}{q} P_2(s) + P_0(s),$$

$$(s+1)P_n(s) = \frac{1}{q} P_{n+1}(s) + \frac{q-1}{q} P_{n-1}(s), \quad n \geq 2. \quad (5.2.2)$$

For $n \geq 2$, the general solution to this constant-coefficient linear system of equation is $P_n \propto \lambda^n$ for $n \geq 1$. By inspection of the equations for P_0 and P_1, the exponential solution continues to hold for for P_1, while P_0 is distinct. Therefore substituting $P_n \propto \lambda^n$ into Eqs. (5.2.2) for $n \geq 2$ leads to a characteristic equation for λ whose roots are

$$\lambda_{\pm}(s) = \frac{(s+1)q \pm \sqrt{(s+1)^2 q^2 - 4(q-1)}}{2},$$

with $\lambda_{\pm} \gtrless 1$ for $q > 2$. The general solution is $P_n = A_+ \lambda_+^n + A_- \lambda_-^n$, but the coefficient A_+ must be set to zero to have P_n finite at $n = \infty$. We may now obtain the coefficient A_-, as well as P_0, by substituting $P_1 = A_- \lambda_-$ and $P_2 = A_- \lambda_-^2$ into the first two recursion formulas to yield the solution

$$A_- = \frac{q}{(q-1)(s+1) - \lambda_-},$$

$$P_0 = \frac{q-1}{q} A_-,$$

$$P_n = \frac{q}{(q-1)(s+1) - \lambda_-} \lambda_-^n. \tag{5.2.3}$$

We now determine the first-passage characteristics. Let $F_n(t)$ be the probability for a random walk to reach a site in the nth generation for the first time at time t. Then $F_n(t)$ may be found in terms of $P_n(t)$ by the basic convolution relation between $P_n(t)$ and $F_n(t)$ given in Chap. 1 [Eq. (1.2.1)]:

$$P_n(t) = \int_0^t F_n(t') P_0(t - t') \, dt' + \delta_{n,0} \delta(t). \tag{5.2.4}$$

The corresponding relation in terms of Laplace transforms is

$$F_n(s) = \frac{P_n(s) - \delta_{n,0}}{P_0(s)}. \tag{5.2.5}$$

Thus for a random walk that starts at site 0, the Laplace transform of the first-return probability is

$$F_0(s) = 1 - \frac{1}{P_0(s)}. \tag{5.2.6}$$

For $s \to 0$, Eqs. (5.2.3) give $P_0(s = 0) = (q-2)/(q-1)$, and we obtain the fundamental result

$$F_0(s = 0) = \frac{1}{q-1} \tag{5.2.7}$$

for the Laplace transform of the first-return probability at $s = 0$. This coincides with the time integral of the first-passage probability, namely, the eventual return probability to the origin. Because this quantity is less than one, a random walk on the Cayley tree is transient.

5.3. Hierarchical Three-Tree

We now study the first-passage properties of a hierarchically branched three-coordinated tree. This self-similar or fractal object is generated by replacing iteratively a single bond with a first-order tree (Fig. 5.2). This object is minimal in the sense that iterating with three bonds is the smallest number that generates a branched, self-similar structure. The three-tree provides a simple laboratory to understand the role of branching and self-similarity on first-passage characteristics. The master equations for a random walk on the tree exhibit an algebraic self-similarity that mirrors the geometrical self-similarity. We exploit this feature to solve the master equations [Kahng & Redner (1989)]; this basic approach works generally for random walks on arbitrary self-similar systems [see also, e.g., Maritan & Stella (1986), Keirstead & Huberman (1987), Hoffmann & Sibani (1988), and Van den Broeck (1989)].

There are two generic transport modes on the tree – transmission and reflection. In the former, a random walk is injected at the left edge and exits at the right edge. Here, an absorbing boundary condition is imposed at the exit site, with all other sites at the end of cul-de-sacs reflecting. The first-passage probability remains zero until sufficient time has passed for the particle to cross the system. At long times, the first-passage probability decays exponentially with time on a finite-order tree because of its finiteness. In reflection mode, a particle is injected at one end and leaves the system only when it returns

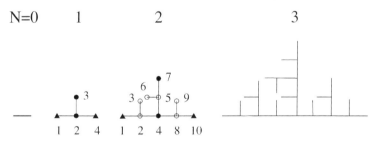

Fig. 5.2. First four iterations of the hierarchical three-tree. The solid symbols – triangles for inlet and outlet sites and circles for interior sites – indicate first-order sites, and the open circles denote second-order sites. The site labels used in the master equations are written for the first- and the second-order trees.

to its starting point. This leads to richer first-passage properties than those in transmission. Now there is an appreciable probability for a walk to return to the starting point almost immediately. This leads to an intermediate-time power-law tail in the first-passage probability before the ultimate exponential decay, because of the finiteness of the system, sets in. This is analogous to the crossover between the $t^{-3/2}$ and the exponential decay of the first-passage probability in one dimension with absorption at $x = 0$ and reflection at $x = L$, when a particle starts close to $x = 0$ (Subsection 2.2.2).

5.3.1. Transmission in the First-Order Tree

Consider a random walk on the first-order tree that starts at site 1 and is absorbed at site 4 (Fig. 5.2) and hops equiprobably to any of its nearest-neighbor sites at each discrete-time step. There are three such neighbors for sites in the interior of the tree, whereas at branch end points there is only the single neighboring site. Let $P_i(t)$ be the probability for a random walk to be at site i at time t. We seek the first-passage probability $F(t)$ that the walk reaches output site 4 at time t. Note that if we view the output site as sticky, then the first-passage probability is related to the occupation probability at the output site by $P_4(t) = P_4(t - 1) + F(t)$. For the first-order tree, the master equations are

$$P_1(t + 1) = \frac{1}{3} P_2(t),$$

$$P_2(t + 1) = P_1(t) + P_3(t),$$

$$P_3(t + 1) = \frac{1}{3} P_2(t),$$

$$F(t + 1) = \frac{1}{3} P_2(t).$$

For a particle that is initially at site 1, the corresponding master equations for the generating functions, $P_i(z) = \sum_{t=0}^{\infty} P_i(t) z^t$ and $F(z) = \sum_{t=0}^{\infty} F(t) z^t$, are

$$P_1(z) = a P_2(z) + 1,$$

$$P_2(z) = 3a P_1(z) + 3a P_3(z),$$

$$P_3(z) = a P_2(z),$$

$$F(z) = a P_2(z), \tag{5.3.1}$$

where $a = z/3$ and the factor of 1 on the right-hand side of the first equation arises from the initial condition. Solving for $F(z)$ yields

$$F(z) = \frac{3a^2}{1 - 6a^2} = \frac{z^2/3}{(1 - 2z^2/3)}. \qquad (5.3.2)$$

From this generating function, it is easy to obtain all first-passage properties. The series expansion of the generating function in powers of z is

$$F(z) = \frac{z^2}{3}\left[1 + \frac{2}{3}z^2 + \frac{4}{9}z^4 + \cdots + \left(\frac{2}{3}\right)^n z^{2n} + \cdots\right],$$

from which the first-passage probability at time t is

$$F(t) = \frac{1}{3}\left(\frac{2}{3}\right)^{(n-2)/2}. \qquad (5.3.3)$$

Finally, the mean first-passage time is given by

$$\langle t \rangle = \sum t F(t) = z\frac{d}{dz}F(z)\Big|_{z=1} = 6, \qquad (5.3.4)$$

and higher moments can be obtained similarly. This will be done explicitly below for the Nth-order tree.

5.3.2. Exact Renormalization for the Nth-Order Tree

In principle, we can calculate the first-passage probability in higher-order trees by solving the master equations for all the site probabilities and then computing the flux leaving the system. This is tedious and does not take advantage of the hierarchical organization. It is much better to perform the calculation iteratively by eliminating the master equations associated with occupation probabilities of the Nth-order sites in the Nth-order tree to give renormalized equations for a tree of one lower order. This process preserves the form of the master equations, and therefore can be iterated to the lowest level to give the exact solution. This is the basic idea of the renormalization group approach [see, e.g., Goldenfeld (1992) or Binney et al. (1992) for a general discussion of the method] that can be implemented exactly for diffusion on hierarchical structures.

For the second-order tree ($N = 2$ in Fig. 5.2), we first eliminate the variables associated with the second-order sites in the tree (open circles) to obtain renormalized master equations for the first-order sites (filled symbols) and then solve these renormalized equations. The master equations for the generating functions on the second-order tree are

$$P_1(z) = a P_2(z) + 1,$$
$$P_2(z) = 3a P_1(z) + 3a P_3(z) + a P_4(z),$$
$$P_3(z) = a P_2(z),$$
$$P_4(z) = a P_2(z) + a P_5(z) + a P_8(z),$$
$$P_5(z) = a P_4(z) + 3a P_6(z) + 3a P_7(z),$$
$$P_6(z) = a P_5(z),$$
$$P_7(z) = a P_5(z),$$
$$P_8(z) = a P_4(z) + 3a P_9(z),$$
$$P_9(z) = a P_8(z),$$
$$F(z) = a P_8(z). \tag{5.3.5}$$

We now eliminate the equations for the second-order sites, namely, 2, 3, 5, 6, 8, and 9, and identify the remaining sites, 1, 4, 7, and 10, as the renormalized first-order sites $1'$, $2'$, $3'$, and $4'$, respectively. With some straightforward algebra, we obtain the renormalized master equations

$$P_{1'} = a' P_{2'} + c,$$
$$P_{2'} = 3a' P_{1'} + 3a' P_{3'},$$
$$P_{3'} = a' P_{2'},$$
$$F' = \frac{a'}{c} P_{2'}, \tag{5.3.6}$$

with $a' = a^2/(1 - 6a^2)$ and $c = (1 - 3a^2)/(1 - 6a^2)$. These have the same form as the first-order master equations, Eqs. (5.3.1), except that the hopping rate has been renormalized and the quantity c appears as both the initial condition factor in the equation for $P_{1'}$ and as an overall factor in the equation for F'. These two factors cancel in solving for the first-passage probability, and we find

$$F' = 3a'^2/(1 - 6a'^2). \tag{5.3.7}$$

This has the same form as Eq. (5.3.2) for the first-passage probability on the first-order tree! Thus the master equations on the tree renormalize exactly. Now we may iteratively eliminate all lower-order sites to yield the first-passage probability on an Nth-order tree:

$$F^{(N)}(z) = \frac{3\left[a^{(N)}\right]^2}{1 - 6\left[a^{(N)}\right]^2}, \tag{5.3.8}$$

where $a^{(N)}$ is recursively defined by

$$a^{(N)} = \frac{\left[a^{(N-1)}\right]^2}{1 - 6\left[a^{(N-1)}\right]^2} \qquad (5.3.9)$$

and $a^{(0)} = z/3$. These two equations may be combined to give a renormalization equation in terms of F only:

$$F^{(N)}(z) = \frac{\frac{1}{3}\left[F^{(N-1)}\right]^2}{1 - \frac{2}{3}\left[F^{(N-1)}\right]^2}. \qquad (5.3.10)$$

Because a single-parameter renormalization gives the first-passage probability, a single time scale accounts for all the moments of the first-passage time.

We now study the asymptotics of this first-passage probability. We start with basic facts about the relation between $F(t)$ and $F(z)$. By definition, $F^{(N)}(z)$ increases monotonically in z, with $F^{(N)}(z=1) = 1$ (see Fig. 5.3), because the probability of eventually hitting the output site equals one. Because of the finiteness of the tree, $F^{(N)}(t)$ must decay exponentially with time as $t \to \infty$, with a decay time $\tau^{(N)}$ that grows with N. We write this as

$$F^{(N)} \sim e^{-t/\tau^{(N)}} \sim \left[\mu^{(N)}\right]^t, \qquad (5.3.11)$$

with $\mu^{(N)} \sim 1 - (1/\tau^{(N)})$ approaching one from below as N increases.

The corresponding generating function is

$$F^{(N)}(z) = \sum_{t=0}^{\infty} F^{(N)}(t)z^t$$

$$\sim \frac{A}{\left[1 - \mu^{(N)}z\right]}$$

$$\equiv \frac{A}{\left[1 - z/z_c(N)\right]}, \qquad (5.3.12)$$

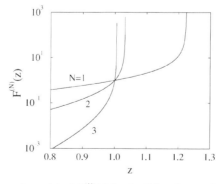

Fig. 5.3. Dependence (to scale) of $F^{(N)}(z)$ for the Nth-order tree for $N = 1, 2$, and 3.

where A is a constant as $\mu \to 1$ from below. Thus the location of the closest pole to $z = 1$ in the generating function $F^{(N)}(z)$ determines the decay time. Substituting this asymptotic form into Eq. (5.3.10), we find that this pole is located at $\sqrt{3/2} = 1.2247\ldots$, $\sqrt{27/(2 + \sqrt{24})} = 1.0320\ldots$, and $1.0052\ldots$, for $N = 1$, 2, and 3, respectively. In general, $z_c^{(N)}$ is determined by the recursion formula

$$z_c^{(N)} = \sqrt{\frac{3z_c^{(N-1)}}{1 + 2z_c^{(N-1)}}}. \tag{5.3.13}$$

Because $z_c^{(N)}$ is close to one, we define $z_c^{(N)} \equiv 1 + \epsilon_N$, with $\epsilon_N \ll 1$, and substitute this into Eq. (5.3.13) to find $\epsilon_N = \frac{1}{6}\epsilon_{N-1}$. Thus we conclude that $F^{(N)}(z)$ has a simple pole at $z_c^{(N)}$ that asymptotically approaches one from above as 6^{-N}. This corresponds to the characteristic time $\tau^{(N)}$ in the exponential decay of $F^{(N)}(t)$ increasing as 6^N.

In a similar spirit, we may determine higher moments of the first-passage time. Formally, the kth moment on an Nth-order tree is

$$\langle [t^{(N)}]^k \rangle = \left(z\frac{\partial}{\partial z}\right)^k F^{(N)}(z)\Big|_{z=1}. \tag{5.3.14}$$

Because $F^{(N)}(z = 1) = 1$, it is expedient to expand the generating function in a power series about $z = 1$. With $z = 1 - \delta$, this is

$$F^{(N)}(z) = 1 + \sigma_1^{(N)}\delta + \sigma_2^{(N)}\delta^2 + \ldots \tag{5.3.15}$$

By applying Eq. (5.3.14) to this series and comparing terms with the same powers of δ, we identify the coefficients $\sigma_i(N)$

$$\sigma_1^{(N)} = -\langle t^{(N)} \rangle,$$
$$\sigma_2^{(N)} = \frac{1}{2}\{\langle [t^{(N)}]^2 \rangle - \langle t^{(N)} \rangle\},$$

etc. The renormalization properties of these coefficients can now be obtained from the recursion relation for $F^{(N)}$, Eq. (5.3.10). By expressing the right-hand side of this equation as a power series in δ, we find

$$F^{(N)} = 1 + 6\sigma_1^{(N-1)}\delta + \delta^2\big[6\sigma_2^{(N-1)} + 27\big(\sigma_1^{(N-1)}\big)^2\big] + \ldots,$$

from which

$$\sigma_1^{(N)} = 6\sigma_1^{(N-1)},$$
$$\sigma_2^{(N)} = 6\sigma_2^{(N-1)} + 27\big(\sigma_1^{(N-1)}\big)^2.$$

The former immediately yields

$$\langle t^{(N)} \rangle = 6^N, \tag{5.3.16}$$

whereas the recursion relation for $\sigma_2^{(N)}$ gives, when written in terms of $\langle t \rangle$ and $\langle t^2 \rangle$,

$$\langle t^{(N)} \rangle = 6 \langle t^{(N-1)} \rangle,$$
$$\langle [t^{(N)}]^2 \rangle = 6 \langle [t^{(N-1)}]^2 \rangle + 54 \langle t^{(N-1)} \rangle^2.$$

To solve for the second moment, we diagonalize these two relations. We accomplish this by first defining $u_N \equiv \langle [t^{(N)}]^2 \rangle + \lambda \langle t^{(N)} \rangle^2$. Then, by choosing $\lambda = -9/5$, we find that the recursion formula for u_N reduces to $u_N = 6u_{N-1}$. Because the initial condition becomes $u_0 = -\frac{4}{5}$, we find $u_N = -\frac{4}{5} \times 6^N$, or

$$\langle [t^{(N)}]^2 \rangle = \frac{9}{5}(6^{2N}) - \frac{4}{5}(6^N) \to \frac{9}{5} \langle t^{(N)} \rangle^2. \tag{5.3.17}$$

Thus the second moment scales as the square of the first moment. This approach can be extended to arbitrary-order moments and gives $\langle [t^{(N)}]^k \rangle \sim \langle t^{(N)} \rangle^k$. Thus all moments of the first-passage time are described by a single time scale that grows by a factor of 6 at each iteration. This same overall approach can be applied to transmission in any self-similar structure and will generally lead to a qualitatively similar scaling picture.

5.3.3. Reflection in the Three-Tree

We now consider the reflection process, in which a random walk starts at the left edge of the tree (site 1 of Fig. 5.2) and is absorbed on return to this starting point. All other end points of the tree are reflecting. The analysis of the master equations is similar to that given for transmission and we discuss only on the new features associated with reflection. In our calculation, we use the simplifying device of first solving for the occupation probabilities and then exploiting the basic relation between occupation and first-passage probabilities [Eq. (1.2.3)] to extract the latter. The master equations on the reflecting tree are nearly identical to those in the transmission problem, Eqs. (5.3.5), except that we must now include an equation for the rightmost (reflecting) site of the tree (Fig. 5.2). For the second-order tree, for example, the rightmost site is #10 and its master equation is $P_{10}(z) = a P_8(z)$.

Following the same renormalization as in transmission and after some algebra, we find that the generating function for the occupation probability at the origin on an Nth-order tree is given by

$$P_1^{(N)}(z) = \frac{1 - 6[a^{(N-1)}]^2}{1 - 9[a^{(N-1)}]^2} \prod_{k=0}^{N-2} \frac{1 - 3[a^{(k)}]^2}{1 - 6[a^{(k)}]^2},$$

$$\equiv \frac{1}{\mathcal{Q}^{(N)}\{1 - 9[a^{(N-1)}]^2\}}, \qquad (5.3.18)$$

with $a^{(N)}$ defined in Eq. (5.3.9) and $a^{(0)} \equiv a = z/3$. The reason for the peculiar definition in the second line will become clear momentarily. We now determine first-passage properties from this formal solution. To find the moments of the return time to the origin we start with basic relation between the first-passage and occupation probabilities:

$$F^{(N)}(z) = 1 - \frac{1}{P_1^{(N)}(z)} \qquad \text{from Eq. (1.2.3)}$$

$$\equiv 1 - \mathcal{Q}^{(N)}\left(1 - 9\left(a^{(N-1)}\right)^2\right), \qquad (5.3.19)$$

and then we apply Eq. (5.3.14). For the mean return time, the computation of the first derivative simplifies considerably because the factor $1 - 9(a^{(N-1)})^2$ vanishes at $z = 1$. Thus we need to consider only

$$\langle t^{(N)} \rangle = z \frac{dF^{(N)}(z)}{dz}\Big|_{z=1}$$

$$= \left(1 - \{1 - 9[a^{(N-1)}]^2\}\mathcal{Q}^{(N)}\right)'\Big|_{z=1}$$

$$= +18\mathcal{Q}^{(N)} a^{(N-1)} a^{(N-1)'}\Big|_{z=1}. \qquad (5.3.20)$$

Using the chain rule repeatedly, together with the fact that $a^{(k)} = 1/3$ at $z = 1$ for all k, we may easily compute the derivative of $a^{(k)}$, from which we find

$$\langle t^{(N)} \rangle = 2 \times 3^N. \qquad (5.3.21)$$

The second moment $\langle [t^{(N)}]^2 \rangle$ may be computed similarly (but more tediously) and gives asymptotically

$$\langle [t^{(N)}]^2 \rangle \sim A \times 2^N \times 3^{2N}, \qquad (5.3.22)$$

where A is a constant of the order of 1. The important feature is that the second moment is much larger than the first moment squared, that is,

$$\langle [t^{(N)}]^2 \rangle \propto 2^N \langle t^{(N)} \rangle^2 \gg \langle t^{(N)} \rangle^2. \qquad (5.3.23)$$

In general, this same approach gives $\langle [t^{(N)}]^k \rangle \propto 2^{(k-1)N} \times 3^{kN}$. Thus each moment of the return time scales independently. This property arises because of the intermediate-time power-law decay in the distribution of return times.

We can infer the exponent of this power law in the first-return probability from the disparity between the first and the second moments, together with simple reasoning. On physical grounds, the first-return probability has the form

$$F^{(N)}(t) \sim t^{-\mu} e^{-t/\tau^{(N)}},$$

where μ must be greater than one for the distribution to be normalizable when $N \to \infty$ and $\tau^{(N)}$ correspondingly diverges. This fact means that the exponential plays a subdominant role in the normalization integral, $\int_0^\infty F^{(N)}(t)\,dt = 1$, so that there is no power-law prefactor that involves $\tau^{(N)}$ in $F^{(N)}(t)$. With this form for $F^{(N)}(t)$, we then deduce

$$\langle t^{(N)} \rangle \sim \int_0^\infty t^{1-\mu} e^{-t/\tau^{(N)}}\,dt \sim \left[\tau^{(N)}\right]^{2-\mu},$$

$$\langle [t^{(N)}]^2 \rangle \sim \int_0^\infty t^{2-\mu} e^{-t/\tau^{(N)}}\,dt \sim \left[\tau^{(N)}\right]^{3-\mu}. \tag{5.3.24}$$

In these formulas, the cutoff time $\tau^{(N)}$ for an Nth-order tree is 6^N, as determined by our results in Subsection 5.3.2 about transmission mode. Comparing this with the previously derived moments $\langle t^{(N)} \rangle \propto 3^N$ and $\langle [t^{(N)}]^2 \rangle \propto 2^N \times 3^N$, we infer that the decay exponent μ is given by

$$\mu = 2 - \frac{\ln 3}{\ln 6} \approx 1.3868\ldots . \tag{5.3.25}$$

5.3.4. Conclusion

The first-passage properties of a random walk on self-similar structures are qualitatively similar to those on a one-dimensional chain, with geometric features of the structure reflected primarily in quantitative details of first passage. As long as there is exact self-similarity, a single time scale will account for all moments of the first-passage time in transmission mode. By renormalization, this mean first-passage time on an Nth-order tree is $\langle t^{(N)} \rangle_N = 6^N$. For comparison, the first-passage time on a linear chain is related to the length by $\langle t^{(N)} \rangle \sim L^2 \equiv L^\zeta$, thus defining a dynamical exponent $\zeta = 2$. The length of an Nth-order tree is simply 2^N, which then gives $\zeta = \ln 6/\ln 2 \approx 2.585$. This larger exponent value indicates how branching retards the progress of a random walk to the output point. For an arbitrary self-similar structure, the dynamical exponent will depend on the microscopic geometry but invariably will be larger than 2.

Reflection mode is more subtle, as the return-time distribution has an intermediate power-law decay before a long-time exponential cutoff sets in.

For the Nth-order three-tree, we found that $\langle [t^{(N)}]^2 \rangle / \langle t^{(N)} \rangle^2 \propto 2^N$. Together with the result that the finite-size cutoff time scales as 6^N, the probability of first returning to the starting point at time t scales as $t^{-\mu}$, with $\mu = 2 - \ln 3 / \ln 6 \approx 1.3868 \ldots$. Although the values of the exponent μ and those that describe the scaling of the moments themselves will depend on the geometry of the underlying network, the qualitative features of these results should again hold on any self-similar structure.

5.4. Comb Structures

5.4.1. Introduction

We now treat first passage on comb structures (Fig. 5.4), in which the relative influence of the backbone and sidebranches can be easily controlled by varying the sidebranch lengths. This model was proposed by Goldhirsch and Gefen (1986) and by Weiss and Havlin (1986). Because of its relative simplicity and because the model captures some aspects of transport in disordered networks, this model was investigated thoroughly [Weiss & Havlin (1987), Havlin, Kiefer, & Weiss (1987), Kahng & Redner (1989), Aslangul, Pottier, & Chvosta (1994), and Aslangul & Chvosta (1995)]. Such geometries have been used to model chromatography or hydrodynamic dispersion [Coats & Smith (1964)], in which a particle diffuses freely along a main backbone but

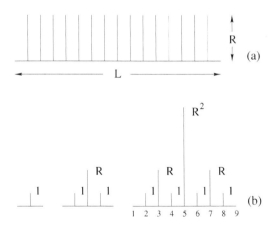

Fig. 5.4. (a) A regular comb with sidebranches of length R that are spaced at unit intervals on a backbone of length L. (b) First three iterations of the bifurcating hierarchical comb. At each stage, the backbone doubles in length, whereas the length of each sidebranch increases by a factor R. New sidebranches of unit length are also created in the middle of each new backbone segment of length 2.

can sometimes get stuck either in a dead end or in a region of negligible flow. It is shown that the competition between free flow and getting stuck leads to a transition in the first-passage properties of the system. An appealing aspect of the comb structures is that many first-passage properties can be obtained either exactly or asymptotically. This presentation follows that given initially in Kahng & Redner (1989), but some of these results are also obtained in the other cited references on comb structures.

In a regular comb, the sidebranches are all identical. The resulting first-passage properties of a sufficiently long regular comb are one dimensional in character, but with a smaller diffusion coefficient than that of the linear chain because of the sidebranching. However, if the sidebranches are all infinite, the mean time between hopping events on the backbone is infinite and the longitudinal transport is best described as a continuous-time random walk. If the sidebranch lengths are hierarchically organized, then first-passage properties are governed sensitively by the relative extent of the sidebranching. As a function of this relative sidebranching, first-passage properties change abruptly from exhibiting regular scaling to exhibiting multiscaling. We discuss these general examples in the following subsections.

5.4.2. Homogeneous Comb

For isotropic hopping between nearest-neighbor sites on the comb, a particle may spend considerable time within a sidebranch before hopping between sites on the backbone. This is equivalent to a one-dimensional hopping process on the backbone in which the particle "waits" at each site, with a waiting-time distribution that corresponds to the sidebranch hopping. The sidebranch length governs this distribution of waiting times and ultimately the first-passage properties of the comb. We first determine this waiting-time distribution for a single sidebranch and then build on these results to solve the first-passage probability of the homogeneous comb.

5.4.2.1. One Sidebranch

Consider a linear chain of sites $\{i\}$, with a single attached sidebranch of length R (Fig. 5.5). A particle hops isotropically between neighboring sites along both the backbone and the sidebranch at discrete time steps. We solve the dynamics of this system by successively eliminating the master equations for the sidebranch sites and replacing them with a renormalized master equation for the junction site.

Let $P_i(t)$ denote the probability that a particle is at site i at time t and define the generating function $P_i(z) = \sum_{t=0}^{\infty} P_i(t) z^t$. In terms of the generating

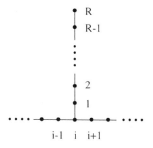

Fig. 5.5. A linear chain to which a sidebranch of length R is attached.

function, the master equations for the sidebranch sites are (Fig. 5.5)

$$P_i(z) = a P_{i-1}(z) + a P_{i+1}(z) + a P_1(z),$$
$$P_1(z) = \frac{2a}{3} P_i(z) + a P_2(z),$$
$$P_2(z) = a P_1(z) + a P_3(z),$$
$$\vdots$$
$$P_{R-1}(z) = a P_{R-2}(z) + 2a P_R(z),$$
$$P_R(z) = a P_{R-1}(z),$$

where i denotes the backbone site at the end of the sidebranch and now $a = z/2$. We first eliminate $P_R(z)$ so that the master equation for $P_{R-1}(z)$ becomes

$$(1 - 2a^2) P_{R-1}(z) = a P_{R-2}(z).$$

Next we eliminate P_{R-1} to yield the master equation for $P_{R-2}(z)$:

$$\left(1 - \frac{a^2}{1 - 2a^2} \right) P_{R-2}(z) = a P_{R-3}(z).$$

Continuing this process to the beginning of the sidebranch, we find that the effective master equation for the junction site is

$$w_R(z) P_i(z) = a P_{i-1}(z) + a P_{i+1}(z), \tag{5.4.1}$$

where $w_R(z)$ is the waiting-time polynomial that encodes the effect of the sidebranch. This polynomial is defined by the continued fraction $w_R(z) = 1 - 2u_R(z)/3$, with $u_1 = 2a^2$, and $u_n = a^2/(1 - u_{n-1})$ for $n > 1$.

We compute this waiting-time polynomial by following step by step the approach presented in Subsection 2.4.2.4; this gives

$$u_R = \frac{z}{2} \left[\frac{\cos(R-1)\phi}{\cos R\phi} \right], \tag{5.4.2}$$

with $\phi = \tan^{-1}\sqrt{z^2 - 1}$. For an infinite sidebranch, the recursion relation for u_R reduces to the algebraic relation $u_\infty = z^2/4\,(1 - u_\infty)$, with the solution

$$u_\infty = \frac{z^2}{2}(1 - \sqrt{1 - z^2}).$$ (5.4.3)

This latter form of the waiting-time polynomial is essentially the same as the generating function for the first-passage probability on a linear chain quoted in approximation (1.5.6). Thus, as a function of time, the probability that a random walk reemerges at time t from an infinite sidebranch asymptotically decays as $t^{-3/2}$. From our discussion in Subsection 2.2.2, the probability for a walk to reemerge at time t for a finite sidebranch of length R decays as $t^{-3/2}$ for $t < R^2$ and as e^{-t/R^2} thereafter. From this distribution, the mean waiting time in a single branch is simply

$$\begin{aligned} \langle t \rangle &\approx \int_0^\infty t \times t^{-3/2} F e^{-t/R^2}\,dt \\ &\sim \int^{R^2} t^{-1/2}\,dt \\ &\sim R. \end{aligned}$$ (5.4.4)

This mean waiting time can also be obtained by a simple physical argument. From our discussion of the finite interval (Subsection 2.2.2), we know that if a discrete random walk makes a single unit-length step into a long sidebranch, then the probability of reaching the far end is $1/R$ and the time needed to make this excursion is of the order of R^2. The contribution of this extreme excursion to the waiting time of a sidebranch is therefore of the order of $(1/R) \times R^2 = R$.

5.4.2.2. Periodic Sidebranching

From the single-sidebranch results, it is now easy to determine the qualitative features of the mean exit time on a periodic comb. A comb of length N with finite-length sidebranches of length R emanating from each backbone site (Fig. 5.4) can be viewed as an effective linear chain, with typical time between hops that is also of the order of R. This overall time rescaling leads to a mean time to traverse the comb that is proportional to RN^2. More precisely, nearest-neighbor hopping on the periodic comb is equivalent to an effective *continuous-time random walk* (CTRW) along the backbone, in which the time between successive steps is just the time for a particle to reemerge from a sidebranch [see, e.g., Bouchaud & Georges (1990), and Weiss (1994)].

We now compute the first-passage probability of this regular comb. As we have seen in Subsection 5.4.2.1, the effect of a sidebranch is to multiply the generating function for the occupation probability at a backbone site by a waiting-time polynomial. Thus the master equations for the comb may be written as

$$P_1(z) = 1 + bP_2(z),$$
$$w(z)P_2(z) = 3bP_1(z) + bP_3(z),$$
$$w(z)P_3(z) = bP_2(z) + bP_4(z),$$
$$\vdots$$
$$w(z)P_{N-1}(z) = bP_{N-2}(z) + bP_N(z),$$
$$w(z)P_N(z) = bP_{N-1}(z),$$
$$P_{N+1}(z) = bP_N(z), \tag{5.4.5}$$

where $b = z/3$ and $w(z)$ is the waiting-time polynomial for each sidebranch. For simplicity the subscript R on $w(z)$ is not written.

These master equations have the same form as those given in Subsection 2.4.2.2 for the linear chain [Eqs. (2.4.10)] and can be solved by the same methods. Now we define

$$f_2 = \frac{b^2/w^2}{1 - 3b^2/w}, \qquad f_n = \frac{b^2/w^2}{1 - f_{n-1}} \tag{5.4.6}$$

for $n \geq 3$. Then, following exactly the same decimation procedure as that in the linear chain, we find the first-passage probability

$$F_{N+1} = 3\frac{w^{N-1}}{b^{N-2}} f_2 f_3 \cdots f_N. \tag{5.4.7}$$

To complete this solution, we require f_n. Again using the approach given in Subsection 2.4.2.2, we write $f_n = g_n/h_n$, and then Eqs. (5.4.6) give the following recursion formulas for g_n and h_n:

$$g_n = \frac{b^2}{w^2} h_{n-1},$$
$$h_n = h_{n-1} - g_{n-1} = h_{n-1} - \frac{b^2}{w^2} h_{n-2}.$$

The solution for h_n has the form $h_n \propto \lambda^n$, with $\lambda_\pm = (1 \pm \sqrt{1 - 4b^2/w^2})/2$. The general solution is $h_n = B_+\lambda_+^{n-1} + B_-\lambda_-^{n-1}$, where, for later convenience, we choose h_n to be proportional to λ^{n-1} rather than to λ^n.

We now fix B_+ and B_- by the initial conditions. Starting with

$$f_2 = \frac{b^2/w^2}{1 - 3b^2/w} = \frac{b^2}{w^2}\frac{h_1}{h_2},$$

we have

$$\frac{h_2}{h_1} = 1 - \frac{3b^2}{w}.$$

Because it is only the ratio of h_2 to h_1 that ultimately determines f_n, we may specify one of h_1 or h_2 arbitrarily and then the other member of this pair is determined. For simplicity, we choose $h_1 = 1$. This fixes h_2, from which we find

$$B_\pm = \frac{1}{2}\left(1 \pm \frac{1 - 6b^2/w}{\sqrt{1 - 4b^2/w^2}}\right).$$

Assembling these results, the first-passage probability on a regular comb is

$$\begin{aligned}
F_{N+1}(z) &= \frac{3w^{N-1}}{b^{N-2}} f_2 f_3 \cdots f_N \\
&= \frac{3w^{N-1}}{b^{N-2}} \left(\frac{b^2}{w^2}\right)^{N-1} \frac{h_1}{h_2} \frac{h_2}{h_3} \cdots \frac{h_{N-1}}{h_N} \\
&= \frac{3b^N}{w^{N-1} h_N}.
\end{aligned} \tag{5.4.8}$$

Although this implicit expression appears formidable, it is relatively easy to extract asymptotic information. Consider the simplest such property, namely, the mean first-passage time. We can determine this either from $\langle t \rangle = F'_{N+1}(z)|_{z=1}$ or, equivalently, by expanding F in a series about $z = 1$ and extracting the linear term in z. Here we do the latter. We define $s = (1 - z)$ and expand Eq. (5.4.8) to first order in s. In the limit $s \to 0$, we may replace b with $1/3$ and $\phi = \tan^{-1}\sqrt{z^2 - 1}$ by $i\sqrt{2s}$ to obtain, to first order in s,

$$\begin{aligned}
w &= 1 - \frac{z}{3}\frac{\cos(R - 1)\phi}{\cos R\phi}, \\
&\approx \frac{2}{3}\left[1 + s\left(R + \frac{1}{2}\right)\right].
\end{aligned}$$

Similarly, h_N has the lowest-order expansion:

$$h_N \approx \frac{1}{2^{N-1}}\left\{1 + s\left(R + \frac{3}{2}\right)[(N - 1) + (N - 1)^2]\right\}.$$

Assembling these results, we can easily write the expansion of F to lowest order in s, from which the mean first-passage time is

$$\langle t \rangle \propto RN^2. \tag{5.4.9}$$

As expected, this first-passage time scales quadratically with the length of the backbone, whereas the effect of the waiting in each sidebranch is to introduce an effective time unit for motion on the backbone that is proportional to R.

5.4.2.3. Infinite-Length Sidebranches

Finally, consider the limit in which all the sidebranch lengths are infinite. The mean sidebranch residence time is also infinite, and its distribution has an asymptotic $t^{-3/2}$ tail. As in the case of finite-length sidebranches, we study the induced one-dimensional CTRW, whose distribution of times between steps, $\psi(t)$, has a $t^{-3/2}$ power-law tail, to determine first-passage properties. The asymptotic properties of such a long-tailed CTRW are well known and extensively discussed, for example, in the review article by Bouchaud and Georges (1990) and in Weiss' book (1994). For completeness, the main first-passage properties for the comb geometry are briefly presented. The two most natural questions are the following:

- What is the probability that the particle first returns to its starting point at time t?
- What is the probability that the particle first reaches $x = L$ at time t?

In general, the probability for a particle to first reach a point x on the comb backbone at time t, $F(x, t)$, may be written as

$$F(x, t) = \sum_N F(x, N)\, P(N, t), \tag{5.4.10}$$

where $F(x, N)$ is the one-dimensional first-passage probability as a function of the number steps N along the backbone and $P(N, t)$ is the probability that N steps of a CTRW occur in a time t. To determine this latter probability, we exploit the simplifying fact that, for a long-tailed CTRW, extreme events control many of statistical properties. From this, we can determine the asymptotic behavior of $P(N, t)$ with essentially no calculation.

For a general CTRW with $\psi(t) \sim t^{-(1+\mu)}$, we first apply an *extreme-statistics* argument [see, e.g., Galambos (1987)] to determine the time T_N required for N steps to occur. Now in N steps, the longest waiting time is determined by the criterion

$$\int_{t_{\max}}^{\infty} \psi(t)\, dt = 1/N.$$

This merely states that out of N steps there will be one whose waiting time is in the range $[t_{\max}, \infty]$. From this, we immediately find $t_{\max} \sim N^{1/\mu}$. In a finite ensemble of N steps, the apparent waiting-time distribution therefore equals

$\psi(t)$ for $t < t_{\max}$, but vanishes for $t > t_{\max}$. For this truncated distribution, the mean waiting time for a single step is then simply

$$\langle t \rangle_N \sim \int^{t_{\max}} t \, \psi(t) \, dt \sim N^{-1+1/\mu}. \qquad (5.4.11)$$

Finally, we estimate the time required for N steps as $T_N \sim N \langle t \rangle_N \sim N^{1/\mu}$. This applies for μ in the range $(0, 1]$. Parenthetically, from this result the mean-square displacement of a random walk along the backbone is simply $\langle x_N^2 \rangle \propto N \propto t^\mu$ which, for a comb with infinite-length sidebranches, varies as $t^{1/2}$. Thus transport along the backbone is extremely slow compared with a normal random walk.

More generally, the probability that N steps occur in a time t scales as $Q(t, N) \to N^{-1/\mu} L_{\mu, 1}(t/N^{1/\mu})$ as $N \to \infty$, where $L_{\mu, \beta}$ is the Lévy stable distribution of order μ and asymmetry parameter β [Gnedenko & Kolmogorov (1954) and Bouchaud & Georges (1990)]. For the CTRW, the asymmetry parameter takes the extremal value $\beta = 1$, as the time of each step is necessarily positive. A basic aspect of this Lévy distribution is that for large N, $L_{\mu, \beta}(Z) \sim Z^{-(1+\mu)}$; that is, the distribution of the sum of N variables has the same asymptotic tail as the distribution of each variable itself. From this latter fact, it is now straightforward to transform from the distribution of elapsed time at fixed N to the distribution for the number of steps at fixed time, $P(N, t)$. Because the former scales as $N^{-1/\mu}$, the latter distribution scales as $t^{-\mu}$.

We now use this last result to determine the time dependence of the probability of first returning to the origin. We have

$$F(0, t) = \sum_N F(0, N) P(N, t)$$

$$\sim \int^{t^\mu} N^{-3/2} t^{-\mu} dN$$

$$\sim t^{-3\mu/2}, \quad 0 < \mu < 1. \qquad (5.4.12)$$

Here we replaced the sum with an integral and the infinite upper limit with t^μ, which is the longest time for N steps to occur. Because the induced CTRW has exponent $\mu = 1/2$ for the comb system, we conclude that the probability of first returning to the origin scales as $F(0, t) \sim t^{-3/4}$.

Similarly for first passage to the end of the comb $x = L$, we know that as a function of N the mean first-passage time is proportional to L^2 and the first-passage probability asymptotically scales as e^{-N/L^2}. Using this in Eq. (5.4.12), we now find

$$F(L, t) \sim \frac{1}{t^\mu} e^{-t^\mu/L^2}, \qquad (5.4.13)$$

with again $\mu = 1/2$. In contrast to a linear chain, the first-passage probability to traverse a finite-length comb now contains an intermediate-time power-law decay because of the influence of the infinitely long sidebranches.

5.4.3. Hierarchical Comb

We now study the hierarchical comb, in which, at each iteration, the length of the backbone doubles, the lengths of each branch increase by a factor of R, and sidebranches of length 1 are "born" at each new backbone site created by this length doubling [Fig. 5.4(b)]. This type of system was introduced as an idealization of hierarchical relaxation that typifies glassy systems [Huberman & Kerszberg (1985), Weiss & Havlin (1986), and Teitel, Kutasov, & Domany (1987)]. The hierarchical comb naturally incorporates such a wide distribution of time scales while still being exactly soluble [Kahng & Redner (1989)].

The existence of the two distinct length scales – the backbone length (2^N) and the longest sidebranch length (R^N) – leads to a phase transition in the nature of diffusion on the comb. When $R < 2$ the role of the sidebranches is asymptotically negligible and the first-passage probability exhibits one-dimensional behavior. For $R > 2$, wanderings in sidebranches predominate and first-passage behavior is characterized by multiple time scales. In the marginal case of $R = 2$, there is a logarithmic correction to the one-dimensional results that portends the transition to multiscale behavior.

5.4.3.1. First-Passage Time

We determine the mean first-passage time to traverse the backbone of the comb by solving the discrete analogs of the adjoint equations for the first-passage time in a continuum system (Subsection 1.6.3.2). Our solution is based on systematically eliminating the equations associated with high-order sites, analogous to the renormalization of the hierarchical tree.

To start this calculation, we need a judicious site-labeling scheme. We denote sites along the backbone by $a, b, c, \ldots$, and we denote the nth site on a sidebranch attached to backbone site i as (i, n), with $n = 1, 2, 3, \ldots$ (Fig. 5.6). A random walk starts at site a and is absorbed at the output site. Let t_j be the mean first-passage time to reach the output starting from site j. Finally, we define T_x and T_y as the time between nearest-neighbor hops in the x and the y directions, respectively. Although these nearest-neighbor hopping times both equal one, they rescale differently under renormalization, and it is useful to define them in the more general fashion.

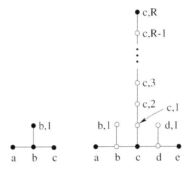

Fig. 5.6. First two iterations of the hierarchical comb and the associated site labeling. First-order sites are indicated by filled circles and second-order sites by open circles.

On the first-order comb, the mean first-passage times satisfy the recursion relations, or adjoint equations,

$$t_a = T_x + t_b,$$
$$t_b = \frac{1}{3}T_x + \frac{1}{3}(T_x + t_a) + \frac{1}{3}(T_y + t_{b,1}),$$
$$t_{b,1} = T_y + t_b. \tag{5.4.14}$$

Solving, we find that the mean first-passage for a random walk that begins at the input site a is

$$t_a \equiv \langle t \rangle = 2(2T_x + T_y) = 6. \tag{5.4.15}$$

Now we consider the second-order comb with R sites on the longest side-branch. The corresponding equations for the first-passage times starting at any site are

$$t_a = T_x + t_b,$$
$$t_b = \frac{1}{3}(T_x + t_a) + \frac{1}{3}(T_y + t_{b,1}) + \frac{1}{3}(T_x + t_c),$$
$$t_{b,1} = T_y + t_b,$$
$$t_c = \frac{1}{3}(T_x + t_b) + \frac{1}{3}(T_y + t_{c,1}) + \frac{1}{3}(T_x + t_d),$$
$$t_{c,1} = \frac{1}{2}(T_y + t_{c,2}) + \frac{1}{2}(T_y + t_c),$$
$$t_{c,n} = \frac{1}{2}(t_{c,n-1} + t_{c,n+1}) + T_y, \quad \text{for } n = 2, \ldots, R-1,$$
$$t_{c,R} = T_y + t_{c,R-1},$$
$$t_d = \frac{1}{3}(T_x + t_c) + \frac{1}{3}(T_y + t_{d,1}) + \frac{1}{3}T_x,$$
$$t_{d,1} = T_y + t_d. \tag{5.4.16}$$

Although these equations appear ugly, they are exactly renormalizable, and this allows us to find the first-passage time on an arbitrary-order comb. We first eliminate the equations associated with the second-order sites (open circles in Fig. 5.6) to obtain the renormalized recursion relations for the first-passage times of the first-order sites:

$$t_a^{(1)} = T_x^{(1)} + t_b^{(1)},$$

$$t_b^{(1)} = \frac{1}{3}T_x^{(1)} + \frac{1}{3}\left[T_x^{(1)} + t_a^{(1)}\right] + \frac{1}{3}\left[T_y^{(1)} + t_{b,1}^{(1)}\right],$$

$$t_{b,1}^{(1)} = T_y^{(1)} + t_b^{(1)}, \tag{5.4.17}$$

with

$$T_x^{(1)} = 2(2T_x + T_y), \quad T_y^{(1)} = 2RT_y.$$

Here we have relabeled the remaining sites (filled circles in the right half of Fig. 5.6) as

$$a \to a^{(1)} \quad c \to b^{(1)} \quad c, R \to b, 1^{(1)}$$

to emphasize that this renormalization has recast the recursion relations for the first-passage time in the same form as that of the first-order comb. Thus the hierarchical comb is exactly renormalizable. Because Eqs. (5.4.17) have the same form as that of Eqs. (5.4.14), we immediately infer that the mean first-passage time for the second-order comb is [compare with Eq. (5.4.15)]

$$t_a^{(1)} \equiv \langle t \rangle_1 = 2\left[2T_x^{(1)} + T_y^{(1)}\right].$$

We now apply this same renormalization repeatedly to an Nth-order comb, with the final result

$$\langle t \rangle_N = 2\left[2T_x^{(N)} + T_y^{(N)}\right], \tag{5.4.18}$$

with

$$T_x^{(N)} = 2\left[2T_x^{(N-1)} + T_y^{(N-1)}\right], \quad T_y^{(N)} = 2RT_y^{(N-1)}.$$

The first-passage time may be obtained directly from this recursion, but it is instructive to solve for this time by generating function methods. Introducing $g_x(z) = \sum_{N=0}^{\infty} T_x^{(N)} z^N$ and $g_y(z) = \sum_{N=0}^{\infty} T_y^{(N)} z^N$ into these equations, we find

$$g_x(z) = \frac{T_x}{1 - 4z} + \frac{2zT_y}{(1 - 2Rz)(1 - 4z)} \to \frac{1}{1 - 4z} + \frac{2z}{(1 - 2Rz)(1 - 4z)},$$

$$g_y(z) = \frac{T_y}{1 - 2Rz} \to \frac{1}{1 - 2Rz}. \tag{5.4.19}$$

For $R < 2$, the closest singularity to the origin in $g_x(z)$ is a simple pole at $z = 1/4$, and this gives, for the mean time to traverse the comb, $\langle t \rangle_N \sim 4^N = (2^N)^2 = L^2$. In this limit, the effect of the sidebranches is asymptotically irrelevant and the mean first-passage time has the same scaling behavior as that of a linear chain. Conversely, for $R > 2$, the closest singularity of $g_x(z)$ is a simple pole at $z = 1/2R$, leading to $\langle t \rangle_N \sim (2R)^N$. Here the mean first-passage time is controlled by the longest sidebranch rather than by the backbone. In the marginal case of $R = 2$, the confluence of the two singularities gives a second-order pole at $z = 1/4$, so that $\langle t \rangle_N = (N/2) \times 4^N$. Because $L = 2^N$, this leads to a logarithmic correction to the mean first-passage time, $\langle t \rangle_N \sim L^2 \ln L/(2 \ln 2)$.

5.4.3.2. First-Passage Probability

Finally, a formal solution is given for the first-passage probability on the hierarchical comb. As in the case of the homogeneous comb, we eliminate the equations associated with all sidebranch sites, leading to effective master equations for the backbone sites. The new feature is that the waiting-time polynomial for each backbone site is distinct, as it depends on the length of the eliminated sidebranch. In close analogy with Eqs. (5.4.5), the master equations for the hierarchical comb are

$$
\begin{aligned}
w_0(z)P_1(z) &= 1 + b\,P_2(z), \\
w_1(z)P_2(z) &= z\,P_1(z) + b\,P_3, \\
w_2(z)P_3(z) &= b\,P_2(z) + b\,P_4, \\
w_1(z)P_4(z) &= b\,P_3(z) + b\,P_5, \\
w_3(z)P_5(z) &= b\,P_4(z) + b\,P_6, \\
&\;\;\vdots
\end{aligned}
\tag{5.4.20}
$$

with $b = z/3$. The different subscripts on w in these equations reflect the hierarchical sidebranch lengths, with $w_N(z)$ defined as the waiting-time polynomial of a sidebranch of length R^N.

We now systematically eliminate the sites associated with the shortest sidebranches. That is, we first eliminate the equations associated with sites $2, 4, 6, \ldots$ (Fig. 5.4). After some algebra, we obtain the renormalized master equations for the odd-index sites that have the same structure as that of

Eqs. (5.4.20), but with the rescaled waiting-time polynomials,

$$w_0^{(1)}(z) = \frac{3}{z}\left[w_1(z)w_0(z) - \frac{z^2}{3}\right]$$

$$w_N^{(1)}(z) = \frac{3}{z}\left[w_1(z)w_{N+1}(z) - \frac{2z^2}{9}\right], \quad N \geq 1, \qquad (5.4.21)$$

and the initial condition factor of 1 in the equation for $P_1(z)$ is replaced with $3w_1/z$. By this decimation, we thus obtain the master equations for a comb of one-half the original length, but with rescaled coefficients. We then successively apply this decimation to the shortest branches at each stage, each time reducing the backbone length by a factor of 2 and ultimately arriving at a renormalized zeroth-order structure. Finally, we obtain the formal solution for the first-passage probability on an Nth-order structure,

$$F^{(N)}(z) = \frac{z^2}{\left[3w_0^{(N)}(z)w_1^{(N)}(z) - z^2\right]}, \qquad (5.4.22)$$

where the superscripts on the w's indicate N iterations of the renormalization in Eqs. (5.4.21).

The essential aspect of this solution is that it is dominated by the longest sidebranch when $R > 2$. If we focus on the contribution of only this largest sidebranch, it is relatively straightforward, but tedious, to determine the asymptotic behavior of the first-passage probability. As we expect, however, the mean first-passage time is proportional to R^N, whereas the characteristic decay time in the long-time exponential tail of the first-passage probability is proportional to R^{2N}. Although the renormalization approach becomes computationally formidable for the hierarchical tree, it is worth emphasizing that the method is exact and that the interesting asymptotic behavior can be extracted with a reasonable amount of calculation.

5.5. Hydrodynamic Transport

We now study the first-passage properties in media in which the flow is hydro-dynamically driven by an external pressure drop. This leads to a heterogeneous flow field because of the local disorder of the medium. The archetypical example of this type of transport process is hydrodynamic dispersion [Saffman (1959, 1960)]. Here, a steadily flowing fluid in a porous medium carries suspended Brownian particles that follow the streamlines. Because of both the multiplicity of different flow paths through the system and molecular diffusion, there is a distribution of times for particles to pass through this system, and our basic goal, as usual, is to determine the first-passage properties of these tracer particles.

As we shall discuss, the first-passage probability is determined both by the probability that a particle selects a particular path and the transit time associated with each path. In the spirit of preceding sections, we study this dispersion process on simple but prototypical systems from which we can calculate the first-passage probability analytically and where we anticipate that the behavior is representative of realistic media with hydrodynamically driven flows.

A crucial element in hydrodynamically driven systems is the interplay between convection and molecular diffusion. This basic feature will be brought into sharp focus in our discussion in Subsection 5.5.2 of a single-junction system with fast flow in one branch and slow flow in the other. We are particularly interested in determining what happens to the first-passage time when the slow branch is so slow that molecular diffusion becomes the dominant transport mechanism. The basic result is at first sight surprising, but ultimately quite simple to understand: The first-passage time becomes *independent* of the flow rate in the slow bond. This is a simple example of the equal-time theorem mentioned in the introduction to this chapter.

5.5.1. Single-Sidebranch Network

As a preliminary, we return to the role of a single sidebranch on the first-passage characteristics of a linear system (Fig. 5.7). Because there is only a single junction point, it is simpler to treat this network in the continuum limit rather than by the large set of master equations for a random walk on a discrete version of this structure (as was done in Subsection 5.4.2.1).

5.5.1.1. Solution to the Diffusion Equation

To find the first-passage probability, we solve the diffusion equation in each backbone segment $(0, x)$ and (x, L), as well as in the sidebranch, subject

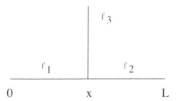

Fig. 5.7. A linear system of length $L = \ell_1 + \ell_2$, with a sidebranch of length ℓ_3 attached at $x = \ell_1$. There is a current input at 0 and an output at L. The end of the sidebranch is reflecting.

to appropriate boundary conditions at the input, output, junction point, and the end of the sidebranch. In terms of Laplace transforms, these boundary conditions are

$$
\begin{array}{ll}
j_1(x, s) = 1 & \text{at } x = 0, \\
j_3(y, s) = 0 & \text{at } y = \ell_3, \\
c_2(x, s) = 0 & \text{at } x = L, \\
c_1(x, s) = c_2(x, s) = c_3(x, s) & \text{at junction}, \\
j_1(x, s) - j_2(x, s) - j_3(y, s) = 0 & \text{at junction}.
\end{array}
\tag{5.5.1}
$$

The first condition imposes a unit impulse of flux at the input at $t = 0$, and the second ensures that there is no loss of concentration at the end of the sidebranch. The absorbing boundary condition at the output means that once a particle reaches the output it leaves the system. Additionally, the concentration is continuous and the net current vanishes at the junction.

The solution to the diffusion equation in the ith bond ($i = 1, 2, 3$) has the general form $c_i(x, s) = A_i e^{x\sqrt{s/D}} + B_i e^{-x\sqrt{s/D}}$, where A_i and B_i are determined by the boundary conditions. Following the reasoning given in Subsection 2.2.2, we find that the absorbing boundary condition at $x = L$ reduces c_2 to

$$
c_2(x, s) = A_2 \sinh\sqrt{\frac{s}{D}}(L - x).
\tag{5.5.2a}
$$

Similarly, the reflecting boundary at the end of the sidebranch implies that c_3 has the form

$$
c_3(y, s) = A_3 \cosh\sqrt{\frac{s}{D}}(\ell_3 - y).
\tag{5.5.2b}
$$

Then, the condition $j_1(x = 0, s) = 1$ reduces c_1 to

$$
c_1(x, s) = A_1 \cosh\sqrt{\frac{s}{D}}x + \frac{1}{\sqrt{sD}}e^{-x\sqrt{s/D}}.
\tag{5.5.2c}
$$

Finally, continuity of the concentration and conservation of flux at the junction provide three more equations to determine the remaining coefficients A_i and thereby the concentration in each bond. For the unit-flux initial condition, the first-passage probability is just the outgoing flux at $x = L$, $j_2(x, s)$. From Eq. (5.5.2a), this flux is merely $A_2\sqrt{sD}$. By solving for the A_i, we find that the Laplace transform of the first-passage probability is

$$
j_2(L, s) = \left(\cosh\sqrt{\frac{s}{D}}L + \cosh\sqrt{\frac{s}{D}}\ell_1 \sinh\sqrt{\frac{s}{D}}\ell_2 \tanh\sqrt{\frac{s}{D}}\ell_3\right)^{-1}.
\tag{5.5.3}
$$

This result has several important physical implications. First, note that if the sidebranch is absent ($\ell_3 = 0$), the first-passage probability immediately reduces to that of a linear chain, $j_2(L, s) = \operatorname{sech}(L\sqrt{s/D})$, as quoted in Subsection 2.2.2. Second, by expanding $j_2(x, s)$ in a power series in s, we obtain

$$
j_2(L, s) = 1 - \frac{s}{D}\left(\frac{L^2}{2} + \ell_2\ell_3\right)
$$
$$
+ \left(\frac{s}{D}\right)^2 \left[\frac{5}{24}L^4 + \ell_2\ell_3\left(\frac{\ell_1^2}{2} + \frac{5}{6}\ell_2^2 + \frac{1}{3}\ell_3^2 + 2\ell_1\ell_2\right)\right] + \dots.
$$

Thus the mean and the mean-square first-passage times for the single-sidebranch network are

$$
\langle t \rangle = \frac{1}{D}\left(\frac{1}{2}L^2 + \ell_2\ell_3\right), \tag{5.5.4a}
$$

$$
\langle t^2 \rangle = \frac{1}{D^2}\left[\frac{5}{12}L^4 + \ell_2\ell_3\left(\ell_1^2 + \frac{5}{3}\ell_2^2 + \frac{2}{3}\ell_3^2 + 4\ell_1\ell_2\right)\right]. \tag{5.5.4b}
$$

As in the case of the comb structure, note that $\langle t \rangle$ is a *linear* function of the sidebranch length ℓ_3.

We can also compute higher moments of the first-passage time $\langle t^k \rangle$ from successive terms in the power-series expansion of j_2. This procedure quickly becomes unwieldy; however, when ℓ_3 is the longest component of the network, it is easy to see that the dominant contribution to $\langle t^k \rangle$ is proportional to $\ell_2\ell_3^{2k-1}$. This behavior again has a simple physical explanation in terms of an extreme excursion into the sidebranch. From the discussion of first-passage in a finite interval (Subsection 2.2.2), we know that the probability of reaching the end of the sidebranch scales as $1/\ell_3$, whereas the kth power of the time for this excursion is ℓ_3^{2k}. The contribution of this event to the overall mean first-passage time is therefore ℓ_2^{2k-1}. We then apply this same reasoning to the segment of the network from the junction to the exit to conclude that the kth moment must also be proportional to ℓ_2. This then gives the result for $\langle t^k \rangle$.

5.5.1.2. Infinite-Length Sidebranch

A natural question is what happens when the sidebranch length ℓ_3 is infinite? According to the series expansion of the outgoing flux, all moments of the first-passage time are divergent. How is this reflected in the nature of the first-passage probability? This question has a simple yet profound answer.

We start by reconsidering the coupled equations for diffusion on the single-sidebranch network. The new feature imposed by the infinite length is that the sidebranch concentration consists of only a decaying exponential as a function of y, that is, $c_3(y, s) = A_3 e^{-y\sqrt{s/D}}$ [compare with Eq. (5.5.2b)]. This automatically ensures that the flux leaving the sidebranch is zero as $y \to \infty$. The resulting first-passage probability turns out be just Eq. (5.5.3) in the limit $\ell_3 \to \infty$. This is

$$j_2(x, s) = \left(\cosh \sqrt{\frac{s}{D}} L + \cosh \sqrt{\frac{s}{D}} \ell_1 \sinh \sqrt{\frac{s}{D}} \ell_2 \right)^{-1}. \quad (5.5.5)$$

The interesting feature of this expression emerges when we expand j_2 in a power series in s to find the moments of the first-passage time. This expansion is

$$j_2(x, s) = 1 - \sqrt{\frac{s}{D}} \ell_2 - \frac{s}{D} \frac{L^2}{2} + \mathcal{O}(s^{3/2}) + \dots.$$

The presence of the leading correction term of the order of $\sqrt{s}$ implies that the mean first-passage time is infinite! Then from the basic connection between a function and its Laplace transform given in Chap. 1, the first-passage probability has a $t^{-3/2}$ long-time tail. This is the underlying source of the divergent first-passage time.

5.5.1.3. The Role of Bias

It is also instructive to study the single-sidebranch network when there is a constant bias along the backbone. Thus for bonds 1 and 2 we solve the convection–diffusion equation,

$$\frac{\partial c(x, t)}{\partial t} + v \frac{\partial c(x, t)}{\partial x} = D \frac{\partial^2 c(x, t)}{\partial x^2},$$

where v is the bias velocity, whereas for bond 3 we solve the diffusion equation. Now the general forms for the bond concentrations are $c_i(x, s) = A_i e^{\alpha_+ x} + B_i e^{\alpha_- x}$, for $i = 1$ and 2, with $\alpha_\pm = (v \pm \sqrt{v^2 + 4Ds})/2D$, whereas for bond 3, the general solution has a similar form, with v set to zero. Applying the boundary conditions of Eqs. (5.5.1) and solving leads, after some tedious steps, to the first-passage probability

$$j_2(x, s) = - \frac{\sqrt{v^2 + 4Ds}\, e^{-\alpha_- \ell_1} \left(1 + \frac{s}{D\alpha_-}\right) u(\ell_1)}{s u(\ell_1) u(-\ell_2) + \sqrt{s D} \tanh \sqrt{\frac{s}{D}} \ell_3 \, u'(\ell_1) u(-\ell_2) + D u'(\ell_1) v(\ell_2)}, \quad (5.5.6)$$

where

$$u(\ell) \equiv e^{\alpha_+ \ell} - e^{\alpha_- \ell},$$
$$v(\ell) \equiv \alpha_- e^{\alpha_- \ell} - \alpha_+ e^{\alpha_- \ell},$$

and $u'(\ell)$ denotes differentiation with respect to ℓ.

Perhaps the most interesting feature of this system is the mean first-passage time in the limit of large bias velocity. Expanding Eq. (5.5.6) to first order in s gives the extraordinarily simple result

$$j_2(x, s) = 1 - s \left[\frac{\ell_1 + \ell_2 + \ell_3}{v} - \frac{3D}{2v^2} + \mathcal{O}\!\left(e^{-Pe}\right) \right] + \dots,$$

where $Pe = vL/2D$ is the Péclet number. Thus, as $v \to \infty$, the mean first-passage time approaches

$$\langle t \rangle \to \frac{\ell_1 + \ell_2 + \ell_3}{v}. \tag{5.5.7}$$

That is, the mean first-passage time is simply the system volume divided by the flow velocity, independent of any other geometrical features of the system! This strikingly simple result is one example of the general equal-time theorem mentioned in the introduction.

5.5.2. Single-Junction Network

We can give a deeper and more general meaning to the single sidebranch result by considering the closely related single-junction network that again consists of three bonds (Fig. 5.8). This toy system embodies the general behavior of an arbitrary bond network with a pressure drop applied across the network. The interesting situation is that of a large overall pressure drop ΔP, but with the bias in one of the two bonds in parallel, either 2 or 3, vanishingly small. Although the convection time across this slow bond becomes large, the entrance probability correspondingly becomes small in such a way that the mean time to traverse the network becomes independent of the slow bond. This occurs because when the convection time across the slow bond becomes equal to the diffusion time, the entrance probability and the local transit time both

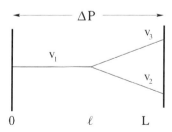

Fig. 5.8. A three-bond single-junction network with an overall pressure drop ΔP that leads to local bias velocities $v_1 = v$, v_2, and v_3. There is a unit-flux input at 0 and an output at L.

"stick" at their values at the diffusive–convective crossover. It is because of this mechanism that the first-passage time across the single-junction network continues to satisfy the equal-time theorem even when there is a slow bond. This physical picture holds for an arbitrary network and only requires that there be at least one fast path through the network where the Péclet number is large. The single-junction network allows us to understand this physical picture analytically.

We now solve the convection–diffusion equations for the single-junction network by following steps similar to those in the single-sidebranch network. We assume bias velocities $v_1 = v, v_2,$ and $v_3,$ with $v_2 + v_3 = v$ for flux conservation. The convection–diffusion equations for each bond are supplemented by the boundary conditions

$$j_1(x, s) = 1 \qquad\qquad \text{at } x = 0,$$
$$c_2(x, s) = c_3(y, s) = 0 \qquad \text{at } x = L,$$
$$c_1(x, s) = c_2(x, s) = c_3(x, s) \qquad \text{at junction,}$$
$$j_1(x, s) - j_2(x, s) - j_3(y, s) = 0 \qquad \text{at junction.}$$

Although the solution is, in principle, mechanical and straightforward, the details were performed with Mathematica, and we merely outline the important steps and the interesting results.

From the general solution to the convection–diffusion equation, the absorbing boundary conditions at $x = L$ imply that the concentrations in bonds 2 and 3 have the general form $c_i(x, s) = A_i[e^{\alpha_i(x-L)} - e^{\beta_i(x-L)}]$ for $i = 1, 2,$ where $\alpha_i, \beta_i = (v_i \pm \sqrt{v_i^2 + 4Ds})/2D$ and $\ell \le x \le L.$ Similarly, the condition of unit flux input implies that c_1 has the form

$$c_1(x, s) = A_1(\alpha_1 e^{\alpha_1 x} - \beta_1 e^{\beta_1 x}) + \frac{1}{D\alpha_1} e^{\beta_1 x}$$

for $0 \le x \le \ell.$ Finally, we apply the remaining boundary conditions of continuity of the flux and the concentration at $x = \ell$ to fix the remaining constants.

The first basic result concerns the splitting probability, namely, the probability that a particle reaches the output via bond 2 or bond 3. By definition, the time-integrated output flux via bond i is simply $j_i(L, s = 0).$ The $s = 0$ limit of the flux can still be easily obtained by hand calculation, and the result is

$$j_2(L, 0) = \frac{v_2(e^{-Pe_3} - 1)}{v_2(e^{-Pe_3} - 1) + v_3(e^{-Pe_2} - 1)},$$
$$j_3(L, 0) = \frac{v_3(e^{-Pe_2} - 1)}{v_2(e^{-Pe_3} - 1) + v_3(e^{-Pe_2} - 1)}, \qquad (5.5.8)$$

where the local Péclet numbers are defined as $Pe_i = v_i(L - \ell)/D$, for $i = 2, 3$. When the bias in both downstream bonds is large, these fluxes reduce to the obvious result $j_2 = v_2/(v_2 + v_3)$ and $j_3 = v_3/(v_2 + v_3)$, whereas if the bias in both bonds vanishes, then $j_2 = j_3 = 1/2$.

The most interesting situation is the "mixed" case in which, say, $v_3 \to 0$ while v_2 remains large. Then the exit probabilities via each bond are

$$j_2(L, 0) = \frac{v_2 v_3(L - \ell)/D}{v_2 v_3(L - \ell)/D + v_3},$$

$$j_3(L, 0) = \frac{v_3}{v_2 v_3(L - \ell)/D + v_3}, \tag{5.5.9}$$

so that their ratio is

$$\frac{j_3}{j_2} \to \frac{D}{v_2(L - \ell)} = \frac{(L - \ell)/v_2}{(L - \ell)^2/D} \equiv \frac{\tau_{3,D}^{-1}}{\tau_{2,c}^{-1}}. \tag{5.5.10}$$

The physical meaning of these results is simple yet far reaching. At the junction, the ratio of entrance probabilities into the two bonds is just the corresponding ratio of fluxes. If parameters are varied so that one bond becomes slower (for example, by making it more perpendicular to the flow direction), the entrance probability decreases linearly with the local bond velocity. However, when this bond becomes sufficiently slow that the diffusive rate τ_D^{-1} of entering this bond becomes comparable with the convective entrance rate τ_c^{-1}, then the entrance probability sticks at the limiting value determined by the diffusive flux. Thus the entrance probability into a slow bond cannot become arbitrarily small.

The other basic feature is the behavior of the mean first-passage time across the network. This can be obtained from the first correction term in the expansion of the output flux at $x = L$ in a power series in s. Although the calculation is straightforward, the explicit results for the mean first-passage time are too long to display explicitly. However, this system again illustrates the equal-time theorem. As long as the Péclet number is large throughout at least one path through the network, then $\langle t \rangle$ approaches the value predicted by the equal-time theorem. This result for the mean first-passage time has been generalized to a network with an arbitrary topology [Koplik et al. (1988)].

5.5.3. The Hierarchical Blob

As our final illustration of first-passage on networks, we consider purely convective transport on the hierarchical blob illustrated in Fig. 5.9 [Redner, Koplik, & Wilkinson (1987)]. This model is again minimal in the sense that it

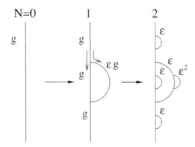

Fig. 5.9. Self-similar hierarchical blob at levels $N = 0$, 1, and 2. At each iteration, a bond of permeability g is replaced with four new bonds with permeabilities g and ϵg, as shown. The small arrows on the $N = 1$ system indicate the two possible flow paths through the system.

involves the smallest number of bonds that preserves a coordination number of 3 at all interior sites and leads to multiple paths with differing first-passage times. We seek to understand the first-passage properties that arise from the interplay between the primary backbone and the many, more tortuous paths.

We construct this network iteratively by replacing a single bond, with fluid permeability g, with a first-order blob that consists of three bonds in series, each of permeability g, and a fourth bond of permeability ϵg, with $\epsilon < 1$, in parallel with the middle series bond. Because the fluid flux in a bond equals its permeability times the imposed pressure drop, $g \Delta P$, the time to traverse this bond may be taken as $1/(g \Delta P)$. We also adopt the *perfect-mixing* assumption, in which a particle chooses a given outgoing bond with a probability proportional to the fractional flux in this bond. Thus at a junction with outgoing fluxes ϕ_1 and ϕ_2, bond i is chosen with probability $\phi_i/(\phi_1 + \phi_2)$. These rules are equivalent to viewing the system as a resistor network with the correspondences

$$\text{pressure drop} \longleftrightarrow \text{voltage drop},$$
$$\text{bond permeability} \longleftrightarrow \text{bond conductance},$$

and the rules for composing bonds and determining local fluxes are just the same as those of the corresponding resistor network.

For example, at the first junction in the $N = 1$ blob of Fig. 5.9, we define the entrance probabilities into the straight and the curved bonds as $p \equiv 1/(1 + \epsilon)$ and $q = 1 - p \equiv \epsilon/(1 + \epsilon)$, respectively. Then the corresponding transit times are $1 + \epsilon$ and $(1 + \epsilon)/\epsilon$. The first-passage properties of the system are controlled by this asymmetry parameter ϵ, as it characterizes the relative role of the slow regions of the system. We wish to understand the influence of this parameter on first-passage properties.

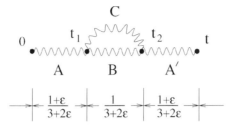

Fig. 5.10. Decomposition of an Nth-order blob into four $(N - 1)$th-order blobs A, B, A', and C. For a unit pressure drop across the Nth-order blob, the corresponding pressure drops between each lower-order blob is indicated. A particle that starts at $t = 0$ reaches the first junction at time t_1 and the second at t_2 before exiting at time t.

5.5.3.1. First-Passage Time

We first determine the average time for a particle to traverse an Nth-order blob, $\langle T_N \rangle$, when a unit pressure difference is imposed between the end points. We start by decomposing the Nth-order blob into the four $(N - 1)$th-order blobs, as indicated in Fig. 5.10. For a unit pressure difference across the Nth-order blob, we use the resistor network correspondence to derive that the pressure drops across each of the lower-order blobs A, B, and A' are $(1 + \epsilon)/(3 + 2\epsilon)$, $1(3 + 2\epsilon)$, and $(1 + \epsilon)/(3 + 2\epsilon)$, respectively, as indicated in the figure. Therefore if a unit pressure drop is applied across the Nth-order blob, the transit times across A, B, A', and C are $a \equiv (3 + 2\epsilon)/(1 + \epsilon)$, $b \equiv 3 + 2\epsilon$, a, and $c \equiv (3 + 2\epsilon)/\epsilon$, respectively.

Because the particle takes the straight path with probability p and the curved path with probability q, $\langle T_N \rangle$ and $\langle T_{N-1} \rangle$ are simply related by

$$\langle T_N \rangle = p(2a + b)\langle T_{N-1} \rangle + q(2a + c)\langle T_{N-1} \rangle$$
$$\equiv a_1 \langle T_{N-1} \rangle, \qquad (5.5.11)$$

with $a_1 = 4a$. Because $\langle T_0 \rangle = 1$, we obtain

$$\langle T_N \rangle = a_1^N = \langle T_1 \rangle^N. \qquad (5.5.12)$$

We can recast this last result more simply by noting that, in $a_1^N = (4a)^N$, the factor 4^N is the number of bonds in the network, whereas the factor a^N is just the inverse of the permeability of the Nth-order blob. The former may be identified with the system volume Ω. In a similar vein, the permeability of the blob coincides with the total current I when a unit pressure drop is applied across the system. This gives the equal-time-theorem result

$$\langle T_N \rangle = \Omega / I. \qquad (5.5.13)$$

As discussed previously in Subsection 5.5.2, this simple result arises because there is an *exact* cancellation between the entrance probability into any bond and the transit time across this bond, so that a particle spends an equal amount of time in each bond, on average, *independently* of the network geometry and the external conditions.

5.5.3.2. First-Passage Probability

We now calculate the first-passage probability by exploiting the network self-similarity. Let $F_N(\Delta P; t)$ be the probability for a particle to traverse an Nth-order blob at time t under an applied pressure drop ΔP. Referring to Fig. 5.10, we find that $F_N(\Delta P; t)$ satisfies the convolution relation

$$F_N(\Delta P; t)$$
$$= \int_0^{t_2} dt_1 \int_0^t dt_2 \left[p F_{N-1}\left(\frac{\Delta P}{a}; t_1\right) F_{N-1}\left(\frac{\Delta P}{b}; t_2 - t_1\right) F_{N-1}\left(\frac{\Delta P}{a}; t - t_2\right) \right.$$
$$\left. + q F_{N-1}\left(\frac{\Delta P}{a}; t_1\right) F_{N-1}\left(\frac{\Delta P}{c}; t_2 - t_1\right) F_{N-1}\left(\frac{\Delta P}{a}; t - t_2\right) \right]. \quad (5.5.14)$$

The first line accounts for the contribution of the straight path, and the second line is the contribution that is due to the curved path between the two middle nodes. We introduce the Laplace transform, $F_N(\Delta P; s) = \int F_N(\Delta P; t) e^{-st} dt$, and use the rescaling $F(\Delta P/a, t) = F(\Delta P, at)$ to reduce Eq. (5.5.14) to the functional relation

$$F_N(s) = [F_{N-1}(as)]^2 \left[p F_{N-1}(bs) + q F_{N-1}(cs) \right], \quad (5.5.15)$$

where the argument $\Delta P (= 1)$ has been dropped because it is identical in all factors of F.

We may extract the moments of the first-passage time by using the fact that the Laplace transform is the moment generating function, that is, $F_N(s) = \langle e^{-st} \rangle = 1 - s\langle T_N \rangle + \frac{s^2}{2}\langle T_N^2 \rangle - \ldots$. Thus, expanding both sides of Eq. (5.5.15) as power series in s and comparing like powers, we find the moment recursion relations

$$\frac{\langle T_N^k \rangle}{k!} = \sum_{\substack{\alpha, \beta, \gamma \\ \alpha+\beta+\gamma=k}}^k \langle T_{N-1}^\alpha \rangle \langle T_{N-1}^\beta \rangle \langle T_{N-1}^\gamma \rangle \frac{a^{\alpha+\beta}}{\alpha! \beta! \gamma!} [p b^\gamma + q c^\gamma]. \quad (5.5.16)$$

That is, the kth moment on the Nth-order blob involves moments of up to order k on the $(N - 1)$th-order blob. This construction of high-order moments from lower-order moments represents the formal solution for all the moments of the first-passage time.

To appreciate the consequences of this decomposition, let's consider the second moment. Eq. (5.5.16) then gives

$$\langle T_N^2 \rangle = a_2 \langle T_{N-1}^2 \rangle + b_2 \langle T_{N-1} \rangle^2 \qquad (5.5.17)$$

with

$$a_2 = \left(\frac{1 + 4\epsilon + \epsilon^2}{\epsilon} \right) \left(\frac{3 + 2\epsilon}{1 + \epsilon} \right)^2, \quad b_2 = 10 \left(\frac{3 + 2\epsilon}{1 + \epsilon} \right)^2.$$

We iterate Eq. (5.5.17) to ultimately arrive at

$$\langle T_N^2 \rangle = (1 - c_2) a_2^N + c_2 \left(a_1^2 \right)^N, \qquad (5.5.18)$$

with $c_2 = b_2 / (a_1^2 - a_2) = 10\epsilon / (12\epsilon - \epsilon^2 - 1)$.

There are two fundamentally different behaviors, depending on which term on the right-hand side of Eq. (5.5.18) dominates. When $a_2 > a_1^2$, which occurs when $\epsilon < 0.08392 \cdots \equiv \epsilon_2$, the first term dominates, so that, as $N \to \infty$, $\langle T_N^2 \rangle \sim (1 - c_2) a_2^N$. Because $a_2 > a_1^2$, this means that $\langle T_N^2 \rangle \gg \langle T_N \rangle^2$. In fact, by similar (but progressively more tedious) analysis, arbitrary-order moments behave similarly and exhibit *multiscaling*, in which

$$\frac{\langle T_N^k \rangle^{1/k}}{\left\langle T_N^{(k-1)} \right\rangle^{1/(k-1)}} \to \infty \qquad (5.5.19)$$

as $N \to \infty$.

This behavior arises from the effect of the small fraction of particles that enter the slowest bond in the network. The probability of entering this slowest bond is simply ϵ^N, and, from the discussion at the beginning of Subsection 5.5.3.1, the time to traverse this bond scales as $(3/\epsilon)^N$ as $\epsilon \to 0$. This gives a contribution to the kth moment that varies as

$$\epsilon^N \left(\frac{3}{\epsilon} \right)^{Nk}.$$

Thus for $\epsilon < \epsilon_2$, an infinity of time scales characterizes the distribution of first-passage times.

On the other hand, when $a_2 < a_1^2$, or $\epsilon > \epsilon_2$, the second term on the right-hand side of Eq. (5.5.18) dominates and

$$\langle T_N^2 \rangle \sim c_2 \left(a_1^2 \right)^N = c_2 \langle T_N \rangle$$

as $N \to \infty$. In this regime, $\langle T_N^2 \rangle$ scales as $\langle T_N \rangle^2$ and the first two moments of the first-passage time (but *only* the first two moments) are governed by a *single* characteristic time as $N \to \infty$.

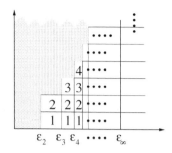

Fig. 5.11. Illustration of the scaling of the first-passage time moments as functions of ϵ. The multiscaling zone is shaded. For $\epsilon_{k-1} < \epsilon < \epsilon_k$, moments $\langle T_N^j \rangle$ with $j \geq k$ scale differently than $\langle T_N \rangle^j$ whereas all lower-order moments obey scaling. Each segmented column denotes the number of moments that obey scaling. For $\epsilon > \epsilon_\infty = 1/3$, all moments obey single-parameter scaling.

Continuing with this analysis of Eq. (5.5.16) for higher-order moments, we find that there is an infinite sequence of transitions at $\epsilon = \epsilon_k$, with $\epsilon_2 = 0.08392\ldots$, $\epsilon_3 = 0.14727\ldots$, $\epsilon_4 = 0.188003\ldots$, etc., with $\lim_{k\to\infty} \epsilon_k = 1/3$ (Fig. 5.11). For $\epsilon_{k-1} < \epsilon \leq \epsilon_k$, moments of order $j \leq k$ exhibit scaling in which

$$\langle T_N^j \rangle \sim c_j \left(a_1^j \right)^N = c_j \langle T_N \rangle^j .$$

Conversely, moments of order $j > k$ exhibit the multiscaling behavior

$$\langle T_N^j \rangle \sim a_j^N .$$

These transitions terminate when ϵ reaches $1/3$. In the scaling regime $\epsilon > \frac{1}{3}$, there is a *single* time scale that characterizes all moments of the first-passage probability, namely,

$$\langle T_N^k \rangle \sim c_k \langle T_N \rangle^k \tag{5.5.20}$$

holds for all k, where c_k is a numerical coefficient that is a slowly varying and nonsingular function of k.

In the regime $\epsilon > \epsilon_\infty$, we can exploit scaling to determine all moments of the first-passage time recursively. Using scaling relation (5.5.20) in the recursion for the first-passage time moments, Eq. (5.5.16), and defining $D_k = c_k / k\,!$, we find the closed set of equations

$$D_k = \sum_{\substack{\alpha,\beta,\gamma \\ \alpha+\beta+\gamma=k}}^{k} D_\alpha D_\beta D_\gamma a^{\alpha+\beta} [p b^\gamma + q c^\gamma]$$

or

$$D_k = \frac{\sum_{\alpha,\beta,\gamma<k} D_\alpha D_\beta D_\gamma \left[(1+\epsilon)^{\gamma-1} + \left(\frac{1+\epsilon}{\epsilon}\right)^{\gamma-1}\right]}{\left[4^k - 2 - (1+\epsilon)^{k-1} - \left(\frac{1+\epsilon}{\epsilon}\right)^{k-1}\right]}. \qquad (5.5.21)$$

The condition for the denominator to remain positive for all k coincides with the condition $a_1^k > a_k$, which, in turn, gives $\langle T_N^k \rangle$ scaling as $\langle T_N \rangle^k$. Although unwieldy, this formula provides the coefficients $c_k = D_k\, k!$ and thus all the moments of the first-passage time.

The single time-scale relation (5.5.20) also implies that the first-passage probability itself obeys scaling. To see this, we substitute relation (5.5.20) into the definition of the Laplace transform. This yields, for large N,

$$F_N(s) = \langle e^{-st} \rangle = 1 - s\langle T_N \rangle + \frac{s^2}{2}\langle T_N^2 \rangle - \cdots$$

$$\sim 1 - s\langle T_N \rangle + c_2 \frac{s^2}{2}\langle T_N \rangle^2 - \cdots$$

$$\equiv f(s\langle T_N \rangle). \qquad (5.5.22)$$

That is, $F_N(s)$ is a function of only the scaled variable $s\langle T_N \rangle = s(4a)^N$, which, in turn, implies that $F_N(t)$ has the scaling form $F_N(t) = f(t/\langle T_N \rangle)$. We now exploit this scaling to determine the asymptotic properties of the first-passage probability itself. Using Eq. (5.5.22) in (5.5.15), we find that the scaling function $f(s)$ obeys the functional relation

$$f(s) = f^2\left(\frac{s}{4}\right)\left[pf\left(\frac{bs}{4a}\right) + qf\left(\frac{cs}{4a}\right)\right]$$

$$= f^2\left(\frac{s}{4}\right)\left[\frac{1}{1+\epsilon}f\left(\frac{1+\epsilon}{4}s\right) + \frac{\epsilon}{1+\epsilon}f\left(\frac{1+\epsilon}{4\epsilon}s\right)\right]. \qquad (5.5.23)$$

To determine the asymptotic solution of this equation, let us recall some basic facts about the structure of the Laplace transform $f(s)$. By definition, $f(s)$ is a monotonically decreasing function of s with $f(s = 0) = 1$ by normalization of the first-passage probability. From Eq. (5.5.23), the value of f at any s is a nonlinear combination of values of f at neighboring locations. Consequently $f(s)$ must be finite for any finite s and diverge only as $s \to -\infty$.

We now use this dependence of $f(s)$ on s to determine the asymptotics of the time-dependent first-passage probability. For $s \to \infty$, the first term on the right-hand side of Eq. (5.5.23) is dominant. Keeping only this term gives

$$f(s) \sim \frac{1}{1+\epsilon} f^2 \left(\frac{s}{4}\right) f \left(\frac{1+\epsilon}{4} s\right),$$

whose asymptotic solution is $f(s) \sim \exp(-s^\alpha)$, with α determined by the root of $4^\alpha = 2 + (1 + \epsilon)^\alpha$. This gives $\alpha = 1$ at $\epsilon = 1$, and α decreasing monotonically as ϵ increases. The Laplace inverse gives the controlling factor in the short-time behavior for the first-passage probability as

$$f(t) \sim \exp\left[-t^{-\alpha/(1-\alpha)}\right].$$

Physically, this essentially singular short-time behavior is dominated by the contribution of the "fast" path (see Fig. 5.10) in the equation for $f(s)$.

Conversely, for $s \to -\infty$, the dominant contribution to $f(s)$ in Eq. (5.5.23) leads to

$$f(s) \sim \frac{\epsilon}{1+\epsilon} f^2 \left(\frac{s}{4}\right) f \left(\frac{1+\epsilon}{4\epsilon} s\right). \tag{5.5.24}$$

This has the asymptotic solution $f(s) \sim \exp(|s|^\beta)$, with β determined by the root of $4^\beta = 2 + (\frac{1+\epsilon}{\epsilon})^\beta$. The Laplace inverse now gives the controlling factor in the long-time limit of the first-passage probability

$$f(t) \sim \exp\left[-t^{-\beta/(\beta-1)}\right]. \tag{5.5.25}$$

This result arises from the contribution of the "slow" path. The exponent $\beta/(\beta - 1)$ is always greater than one (compressed exponential decay) and approaches one as $\epsilon \to \frac{1}{3}$ (pure exponential decay) and approaches infinity as $\epsilon \to 1$ from below.

6

Systems with Spherical Symmetry

6.1. Introduction

This chapter is devoted to first-passage properties in spherically symmetric systems. We shall see how the contrast between persistence, for spatial dimension $d \leq 2$, and transience, for $d > 2$, leads to very different first-passage characteristics. We will solve first-passage properties both by the direct time-dependent solution of the diffusion equation and by the much simpler and more elegant electrostatic analogy of Section 1.6.

The case of two dimensions is particularly interesting, as the inclusion of a radial potential drift $v(r) \propto 1/r$ is effectively the same as changing the spatial dimension. Thus general first-passage properties for isotropic diffusion in d dimensions are closely related to those of diffusion in two dimensions with a superimposed radial potential bias. This leads to nonuniversal behavior for the two-dimensional system.

As an important application of our study of first-passage to an isolated sphere, we will obtain the classic Smoluchowski expression for the chemical reaction rate, a result that underlies much of chemical kinetics. Because of the importance of this result, we will derive it by time-dependent approaches as well as by the quasi-static approximation introduced in Section 3.6. The latter approach also provides an easy way to understand detailed properties of the spatial distribution of reactants around a spherical trap.

6.2. First Passage between Concentric Spheres

We begin by computing the splitting (or exit) probabilities and the corresponding mean hitting times to the inner and outer the boundaries of the annular region $R_- \leq r \leq R_+$ as functions of the starting radius r (Fig. 6.1). This represents the natural extension of first passage in an absorbing interval (Chap. 2). We then study the same properties in the region exterior to a sphere and finally consider the role of radial flow in two dimensions.

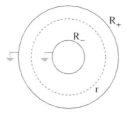

Fig. 6.1. The annular geometry. A normalized shell of probability starts at r and diffuses until it hits either the outer or the inner boundary with respective probabilities $\mathcal{E}_{\pm}(r)$.

6.2.1. Splitting Probabilities

By the Laplacian formalism of Section 1.6, the splitting probabilities to each boundary satisfy the Laplace equation

$$\nabla^2 \mathcal{E}_{\pm}(r) = 0, \qquad (6.2.1)$$

subject to the boundary conditions $\mathcal{E}_-(R_-) = 1$, $\mathcal{E}_-(R_+) = 0$, and $\mathcal{E}_+(R_-) = 0$, $\mathcal{E}_+(R_+) = 1$. Here the dependent variable r is the starting position of the particle. By spherical symmetry, only the radial part of the Laplacian

$$\frac{1}{r^{d-1}} \frac{\partial}{\partial r} \left(r^{d-1} \frac{\partial}{\partial r} \right)$$

is nonzero. Thus the solution to Eq. (6.2.1) is $\mathcal{E} = A + B/r^{d-2}$ for $d \neq 2$ and $\mathcal{E} = A + B \ln r$ for $d = 2$. Matching to the boundary conditions gives

$$\mathcal{E}_+(r) = \begin{cases} \dfrac{1 - (R_-/r)^{d-2}}{1 - (R_-/R_+)^{d-2}}, & d \neq 2 \\[2mm] \dfrac{\ln(R_+/r)}{\ln(R_+/R_-)}, & d = 2 \end{cases}, \qquad (6.2.2)$$

and $\mathcal{E}_-(r) = 1 - \mathcal{E}_+(r)$.

There are many instructive ramifications of Eq. (6.2.2). Perhaps the most striking is the correspondence to elementary electrostatics in the limiting case in which the outer sphere is infinitely far away. Here, the probability of hitting the inner sphere when starting at r is simply

$$\mathcal{E}_-(r) = \left(\frac{R_-}{r} \right)^{d-2}$$

for $d > 2$. This is exactly the electrostatic potential at r when the inner sphere is fixed at unit potential! Conversely (and much more trivially), if the inner sphere is shrunk to zero, then the probability of hitting the outer sphere is one.

This is equivalent to the elementary statement that the electric field within any conductor-surrounded cavity is zero.

Another interesting feature is the relative affinity of a diffusing particle to each boundary. This may be characterized by the initial radius r_{equal} such that the exit probabilities to either boundary are equal. Setting $\mathcal{E}_+ = 1/2$ in Eq. (6.2.2), we find that this radius is given by

$$
r_{equal} =
\begin{cases}
\dfrac{1}{2}(R_-^{2-d} + R_+^{2-d})^{1/(2-d)}, & d \neq 2 \\[2mm]
\sqrt{R_- R_+}, & d = 2
\end{cases}
. \qquad (6.2.3a)
$$

For diffusion exterior to a small sphere at the origin in a large spherical domain (R_- fixed and R_+ diverging), the initial radius for equal exit probabilities approaches the limiting values

$$
r_{equal} \rightarrow
\begin{cases}
2^{1/(d-2)} R_- & \text{for } d > 2 \\[2mm]
R_+/2^{1/(2-d)} & \text{for } d < 2 \\[2mm]
\sqrt{R_- R_+} & \text{for } d = 2
\end{cases}
. \qquad (6.2.3b)
$$

Thus equal splitting is achieved by a diffusing particle starting near the inner boundary for $d > 2$, near the outer boundary for $d < 2$, and for $d = 2$ the particle must start at the geometric mean of the two radii.

These results have a natural interpretation in terms of the transience and the recurrence of diffusion. For $d > 2$, a particle that is initially exterior to a small sphere is transient and likely wanders to infinity. The particle must therefore start close to the inner sphere for an appreciable probability to be absorbed by this sphere. Complementary reasoning applies for $d < 2$. This dimension dependence of r_{equal} shows that the spatial dimension acts as an effective bias that is inward for $d < 2$ and outward for $d > 2$. We will make this connection more quantitative at the end of this section.

We may also obtain the average unrestricted hitting time to either boundary, $t(r)$. This time obeys the Poisson equation $D\nabla^2 t(r) = -1$ [Eq. (1.6.21)], subject to the boundary conditions $t(R_-) = t(R_+) = 0$. This corresponds to a vanishing exit time if the particle starts at one of the boundaries. The solution to this differential equation is straightforward and gives

$$
t(r) =
\begin{cases}
\dfrac{1}{2Dd}\left[(R_+^2 - R_-^2)\dfrac{r^{2-d} - R_-^{2-d}}{R_+^{2-d} - R_-^{2-d}} - (r^2 - R_-^2)\right], & d \neq 2 \\[4mm]
\dfrac{1}{4D}\left[(R_+^2 - R_-^2)\dfrac{\ln(r/R_-)}{\ln(R_+/R_-)} - (r^2 - R_-^2)\right], & d = 2
\end{cases}
. \qquad (6.2.4)
$$

As in the case of the splitting probabilities, it is illuminating to examine the interplay between the starting radius and the spatial dimension in this result. Consider the starting radius r_{max} that maximizes the mean exit time as well as the value of this maximal exit time. When $t(r)$ is maximized with respect to r, this optimal starting radius is

$$
r_{max} = \begin{cases} \left(\dfrac{2-d}{2} \dfrac{R_+^2 - R_-^2}{R_+^{2-d} - R_-^{2-d}} \right)^{1/d}, & d \neq 2 \\[3mm] \left(\dfrac{R_+^2 - R_-^2}{2 \log(R_+/R_-)} \right)^{1/2}, & d = 2 \end{cases} \tag{6.2.5a}
$$

In the limit $R_+/R_- \to \infty$, this result simplifies to

$$
r_{max} = \begin{cases} \left(\dfrac{2-d}{2} \right)^{1/d} R_+, & d < 2 \\[3mm] \dfrac{R_+}{\sqrt{2 \ln(R_+/R_-)}}, & d = 2 \\[3mm] \left(\dfrac{d-2}{2} \right)^{1/d} \left(\dfrac{R_+}{R_-} \right)^{2/d} R_+, & d < 2 \end{cases} \tag{6.2.5b}
$$

As d increases, r_{max} moves to the inner boundary. This is analogous to the behavior of the starting radius for equal splitting probability [Eq. (6.2.2)] – as the dimension increases it becomes less likely for a diffusing particle to hit the inner sphere and the exit time is maximized by starting close to the inner sphere. This maximal exit time, which we obtain by evaluating $t(r)$ at r_{max}, is

$$
t_{max} = \frac{1}{2D} \left\{ \frac{1}{2-d} \left[(2-d) \frac{R_+^2 - R_-^2}{R_+^{2-d} - R_-^{2-d}} \right]^{2/d} + \frac{R_+^2 R_-^2}{d} \frac{R_+^{-d} - R_-^{-d}}{R_+^{2-d} - R_-^{2-d}} \right\} . \tag{6.2.6}
$$

Figure 6.2 shows the dimension dependence of r_{equal}, r_{max}, and t_{max}. Note that an increasing spatial dimension is analogous to an increasing outward bias. Thus the dimension dependences qualitatively mirror the bias dependence of these quantities for diffusion in the one-dimensional interval given in Fig. 2.5.

6.2.2. *First Passage to a Sphere in Radial Potential Flow*

When a diffusing particle also experiences a radial potential bias, $\vec{v}(\vec{r}) = v_0 \hat{r}/r^{d-1}$, in d spatial dimensions, there is an essential interplay between the spatial dimension and the flow that leads to different first-passage behavior for $d < 2$ and $d > 2$. For $d \neq 2$, the drift and the diffusion terms in the convection–diffusion equation have different r dependences – the former is

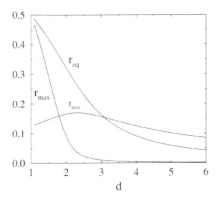

Fig. 6.2. Dimension dependences of r_{equal}, r_{max}, and t_{max}. The two optimal radii are normalized as $(r - R_-)/(R_+ - R_-)$ and the maximal exit time is normalized by $(R_+ - R_-)^2$. The sphere radii are $R_- = 1$ and $R_+ = 100$.

v_0/r^{d-1}, and the latter is D/r. These two terms are equal at a length scale $\lambda = (v_0/D)^{1/(d-2)}$. When $d > 2$, drift dominates for $r < \lambda$ and diffusion dominates for $r > \lambda$. Thus for $d > 2$, diffusion is sufficient to carry a particle to large distances and outward radial drift is irrelevant. Conversely, for $d < 2$, outward drift keeps the particle from returning to the origin infinitely often. Conversely, inward drift is important for $d > 2$ but asymptotically irrelevant for $d \leq 2$.

These features are illustrated by $\mathcal{E}_-(r)$, the probability of eventually hitting the inner sphere of radius R_- when the particle starts at radius $r > R_-$. This exit probability obeys the generalized Laplace equation [Eq. (1.6.18)]

$$D\mathcal{E}_-''(r) + \left(D\frac{d-1}{r} + \frac{v_0}{r^{d-1}} \right) \mathcal{E}_-'(r) = 0, \qquad (6.2.7)$$

subject to the boundary conditions $\mathcal{E}(R_-) = 1$ and $\mathcal{E}_-(\infty) = 0$ (except for $d < 2$ and $v_0 \leq 0$). The solution is

$$\mathcal{E}_-(r) = \begin{cases} \dfrac{\exp{(br^{2-d})} - 1}{\exp{(bR_-^{2-d})} - 1} & \text{for } d > 2 \\[2mm] \exp{[b(r^{2-d} - R_-^{2-d})]} & \text{for } d < 2 \text{ and } v_0 > 0 \,, \\[2mm] 1 & \text{for } d < 2 \text{ and } v_0 \leq 0 \end{cases} \qquad (6.2.8)$$

where $b = [v_0/D(d-2)]$.

The qualitative dependence of this exit probability on starting radius is shown in Fig. 6.3. For $d > 2$, the interesting situation is inward flow, in which it counterbalances the transience of diffusion. Here $\mathcal{E}_-(r)$ remains of the order of 1 for $r < \lambda$ – where the drift is stronger than diffusion – and then drops sharply to zero for larger r – where diffusion and ultimate transience wins

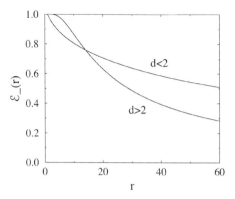

Fig. 6.3. Dependence of the probability of eventually hitting a sphere of radius $R_- = 1$ on initial radius r for radial potential flow in d dimensions. Shown are the cases of $d > 2$ with inward flow (specifically $d = 3$ and $v_0 = -20$) and $d < 2$ with outward flow (specifically $d = 1.5$ and $v_0 = 0.05$).

out over the drift. Conversely, for $d < 2$, the interesting situation is outward flow, in which diffusion and the flow are in opposition. Now the recurrence of diffusion implies that a particle that starts close to the sphere still has an appreciable chance of being trapped, even with an outward bias.

6.2.3. Connection between Diffusion in General Dimension and Radial Drift in Two Dimensions

The foregoing examples suggest that the spatial dimension can be viewed as an effective radial drift whose direction changes sign at two dimensions. We now make this notion precise by relating isotropic diffusion in d dimensions and diffusion in two dimensions with a superimposed radial potential flow. This connection provides a convenient route for solving one system in terms of its counterpart.

Consider the diffusion equation for a spherically symmetric system:

$$\frac{\partial c(\vec{r}, t)}{\partial t} = D\nabla^2 c(\vec{r}, t). \qquad (6.2.9)$$

This can be rewritten as

$$\frac{\partial c(r, t)}{\partial t} = D\left[\frac{\partial^2 c(r, t)}{\partial r^2} + \frac{(d - 1)}{r}\frac{\partial c(r, t)}{\partial r}\right]. \qquad (6.2.10a)$$

On the other hand, for radial potential flow in two dimensions, $\vec{v}(r) = v_0\hat{r}/r$, the probability distribution of a diffusing particle that is passively carried by this flow obeys the convection–diffusion equation, which we rewrite suggestively as

$$\frac{\partial c(r,t)}{\partial t} = D\left[\frac{\partial^2 c(r,t)}{\partial r^2} + \frac{(1-v_0/D)}{r}\frac{\partial c(r,t)}{\partial r}\right]. \qquad (6.2.10b)$$

Comparing these two equations, we see that the "centrifugal" term in the d-dimensional diffusion equation can be viewed as an effect radial bias of magnitude $v_0 = D(2-d)$ in a two-dimensional system.

There is a subtlety with this connection, as decreasing the spatial dimension d seems equivalent to increasing v_0. This is in contrast to the fact that decreasing d makes a system more recurrent and therefore should be equivalent to *decreasing* the radial velocity. The point is that the relevant quantity is the probability density to be within r and $r + dr$ (integrated over all angles), which is proportional to $\rho(r,t) = r^{d-1}c(r,t)$, and *not* the concentration itself. This integrated density satisfies an equation of motion in which the sense of the bias is outward for $d > 2$ and inward for $d < 2$, as we should expect. However, the formal connection between the solutions to Eqs. (6.2.10a) and (6.2.10b) is based on $v_0/D = 2 - d$.

6.3. First Passage to a Sphere

6.3.1. Image Solution

We now consider the related problem of determining *where* on the surface of an absorbing sphere a diffusing particle hits (Fig. 6.4). In Section 1.6, we showed that this probability is the same as the electric field at the impact point $\vec{r}$ when a point charge of magnitude $q = 1/(\Omega_d D)$ is initially at r_0 and the surface of the sphere is held at zero potential.

We may solve this problem easily by the image method [Jackson (1999), Section 2.2]. For $d > 2$ and with a point charge q located at $\vec{r}_0$ exterior to the

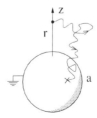

Fig. 6.4. A diffusing particle is released at radius r on the z axis and hits the point (a, θ, ϕ) on the sphere surface.

sphere, the electrostatic potential at $\vec{r}$ is

$$\Phi(\vec{r}) = \frac{q}{|\vec{r} - \vec{r}_0|^{d-2}} + \frac{q'}{|\vec{r} - \vec{r}_0'|^{d-2}}, \tag{6.3.1a}$$

where q' and $\vec{r}_0'$ are the magnitude and the location of the image charge, respectively. The potential at any point on the surface of the sphere will be zero if

$$q' = -q \left(\frac{a}{r_0}\right)^{d-2}, \quad r_0' = \frac{a^2}{r_0}.$$

Similarly, for $d = 2$ the electrostatic potential generically has the form

$$\Phi(\vec{r}) = q \ln|\vec{r} - \vec{r}_0| + q' \ln|\vec{r} - \vec{r}_0'| + A, \tag{6.3.1b}$$

where the constant A is nonzero. For this potential to vanish everywhere on the surface of the grounded circle, we must choose $q' = -q$ and $r_0' = a^2/r_0$, and $A = q \ln(a/r_0)$.

Let us use these results to find the probability that the particle hits a given point on the sphere. Without loss of generality, we assume that the particle is initially on the z axis. The hitting probability then depends on only the polar angle of the impact point and we may write the electrostatic potential for $d > 2$ as

$$\Phi(\vec{r}) = q \left[\frac{1}{\left(r^2 + r_0^2 - 2rr_0\cos\theta\right)^{(d-2)/2}} - \frac{(a/r_0)^{d-2}}{\left(\frac{a^4}{r_0^2} + r^2 - 2\frac{a^2 r}{r_0}\cos\theta\right)^{(d-2)/2}} \right]. \tag{6.3.2}$$

Then, taking the radial component of the electric field for any point on the sphere surface, using $q = 1/(\Omega_d D)$, and multiplying by $-D$ to compute the flux, we find that the hitting probability to any point (a, θ) on the sphere surface is

$$\mathcal{E}(\theta) = \frac{(d-2)}{\Omega_d} \frac{1}{ar_0^{d-2}} \frac{\left(1 - \frac{a^2}{r_0^2}\right)}{\left(1 - \frac{2a}{r_0}\cos\theta + \frac{a^2}{r_0^2}\right)^{d/2}}, \quad d > 2. \tag{6.3.3a}$$

For $d = 2$, the corresponding hitting probability is

$$\mathcal{E}(\theta) = \frac{1}{2\pi a} \frac{\left(1 - \frac{a^2}{r_0^2}\right)}{\left(1 - \frac{2a}{r_0}\cos\theta + \frac{a^2}{r_0^2}\right)}, \quad d = 2. \tag{6.3.3b}$$

The integral of the hitting probability over the sphere surface gives one for $d = 2$ and $(a/r)^{d-2}$ for $d > 2$, to reproduce our previous results about exit probabilities. More interesting is the angular dependence of hitting probability. For example, the ratio of the probabilities of hitting the north pole (most likely) and the south pole (least likely) is

$$\frac{\text{FPP}_{\text{max}}}{\text{FPP}_{\text{min}}} = \left(\frac{1 + a/r_0}{1 - a/r_0}\right)^d.$$

Thus in $d = 3$, for example, if $r_0 = 2a$, the particle is 27 times more likely to hit the north pole rather than the south pole of the sphere.

The same image approach also gives the hitting probability to any point on the sphere when a particle starts in the interior. By inversion symmetry, the electrostatic potential interior to a grounded sphere that is due to a point charge q that is located at $|\vec{r}_0| < a$ on the z axis is the sum of the potential of this point charge and that of a larger image charge $q' = -q(a/r_0)^{d-2}$ that is located at $r_0' = a^2/r_0$. With this fact, the hitting probability is again given by Eqs. (6.3.3a) and (6.3.3b) for $d > 2$ and $d = 2$, respectively.

6.3.2. Efficient Simulation of Diffusion-Limited Aggregation

Our results about the hitting probability provide an elegant and efficient way to simulate diffusion-limited aggregation (DLA) [for a general reference about this process see, e.g., Family & Landau (1984)]. Our discussion also illustrates an important maxim about simulations of diffusion. It is almost always pointless to simulate diffusion by the step-by-step motion of discrete random walks. Especially for processes with absorbing boundaries, such simulations can be performed more elegantly by exploiting first-passage ideas.

DLA is a nonequilibrium growth process that is defined by the following simple rules:

1. Start with a single "sticky" seed particle at the origin.
2. From afar, launch a diffusing particle. On touching the aggregate, the particle is incorporated and also becomes sticky.
3. Repeat step (2), until an aggregate of a desired size is grown.

This simulation leads to a ramified structure, as crudely illustrated in Fig. 6.5. Although this algorithm is extremely simple, it is also grossly inefficient. In the early days of DLA simulations, this led to sometimes misguided efforts at improvement. These issues provide a natural segue to the use of first-passage ideas to construct an efficient simulation.

The first problem with the above algorithm is the notion of launching a particle from afar. How far is far? Clearly, we may launch a particle equiprobably from any point on the surface of smallest sphere of radius R_{circ} that circumscribes the aggregate. A more serious issue is what to do if the particle

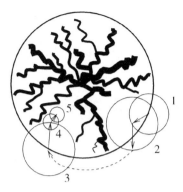

Fig. 6.5. Illustration of a "good" algorithm to grow a DLA. Shown is a trajectory of a diffusing particle that is about to be incorporated into the aggregate. When the particle strays outside the circumscribing circle it is returned to this circle by implementing Eq. (6.3.3b). When the particle is within a "fjord" of the DLA, it is computationally efficient to move the particle isotropically to the circumference of the largest circle that is centered at the current particle position and that does not contact the DLA.

wanders far outside the circumscribing sphere. Because the average time for a diffusing particle to reach an absorbing sphere is infinite for $d = 2$ and, even worse, a particle has a finite probability of not hitting the sphere for $d > 2$, a naive simulation should get bogged down in following a particle that strays infinitely far from the aggregate. In early simulations, this pathology was avoided by terminating the random walk if it reached an arbitrarily determined outer radius and reinjecting the particle at R_{circ}.

We can avoid this uncontrolled procedure by using the hitting probability of Eq. (6.3.3a) to move a particle that strays outside the circumscribing sphere back to the sphere surface *in a single step*. This feature is very nicely discussed in the article by Sander, Sander, and Ziff (1994). Additionally, we can speed up the motion of a particle that is inside the circumscribing circle by moving the particle equiprobably to the surface of the largest inscribed circle centered at the particle that does not touch the aggregate. Thus the particle makes large-distance hops when it is in the mouth of a wide "fjord" of the aggregate and much smaller hops once deep inside the fjords (Fig. 6.5). When these simple tricks are implemented, the best DLA simulations now appear to be memory limited, rather than CPU-time limited. Some of these features were originally developed by Ball and Brady (1985), Ball et al. (1985), and Meakin (1985, 1986).

In fact, when the particle is close to the aggregate, it is more efficient to use an off-center inscribed circle with the maximum possible radius such that the particle is included in this circle and the aggregate is not touched. We can use the image method to determine where on the surface the particle hits this inscribed circle. This ensures that the average step length of the particle step length is always as large as possible and would lead to an even more efficient growth algorithm.

6.4. Time-Dependent First-Passage Properties

6.4.1. Overview

We now determine the full time dependence of first-passage probability between two concentric spheres by using Green's function methods [Koplik, Redner, & Hinch (1994)]. Although this is a somewhat pedantic exercise, it is included for completeness. We treat both isotropic diffusion in d dimensions between two concentric spheres of inner and outer radii R_- and R_+, respectively, and also diffusion with superimposed radial drift in two dimensions.

There are three generic situations that are roughly analogous to the modes of absorption, transmission, and reflection as discussed in Chapter 2:

  (i) "Absorbing" – both boundaries absorbing.
 (ii) "Inner" – absorption at $r = R_-$ and reflection at $r = R_+$.
(iii) "Outer" – reflection at $r = R_-$ and absorption at $r = R_+$.

For the absorbing system, we restrict ourselves to isotropic diffusion. For the latter two cases it is also interesting and physically relevant to consider nonzero radial drift. A physical example of an inner system would be a spherical fluid drain in a spherically symmetric system.

6.4.2. Both Boundaries Absorbing

As in the corresponding one-dimensional interval, we compute the hitting probabilities and the mean first-passage times to each boundary. Because these first-passage characteristics do not involve angular variables, the system is effectively one dimensional. It is therefore convenient to take the initial condition to be a spherical shell of concentration that is initially at $r = r_0$, $c(r, t = 0) = \delta(r - r_0)/\Omega_d r_0^{d-1}$. Here Ω_d is the surface area of a d-dimensional unit sphere for normalization. To solve for the Green's function, we first introduce the Laplace transform and also reexpress the radius in terms of the dimensionless radial coordinate $x = r\sqrt{s/D}$. With these steps, the diffusion equation becomes

$$c''(x, s) + \frac{d-1}{x}\, c'(x, s) - c(x, s) = -\frac{s^{(d-2)/2}}{D^{d/2}}\frac{\delta(x - x_0)}{\Omega_d x_0^{d-1}}, \quad (6.4.1)$$

where the prime now denotes differentiation with respect to x.

For each subdomain $x > x_0$ and $x < x_0$, this is a Bessel equation in which the solution is a superposition of the combinations $x^\nu I_\nu(x)$ and $x^\nu K_\nu(x)$, with $\nu = 1 - d/2$. Here I_ν and K_ν are the modified Bessel functions of the first and the second kind, respectively. We now apply the same reasoning as that in the absorbing one-dimensional interval to restrict the form of the Green's function. First, because both the interior ($x_- \leq x < x_0$) and the exterior ($x_0 < x \leq x_+$) Green's functions vanish at their respective boundaries, these constituent Green's functions must have the form [compare with Eqs. (2.27) and (2.28)]

$$c_<(x, s) = Ax^\nu I_\nu(x) + Bx^\nu K_\nu(x)$$
$$\rightarrow Ax^\nu [I_\nu(x)K_\nu(x_-) - I_\nu(x_-)K_\nu(x)],$$
$$c_>(x, s) = Bx^\nu [I_\nu(x)K_\nu(x_+) - I_\nu(x_+)K_\nu(x)].$$

Next, because of the continuity of the Green's function at $x = x_0$, this function must be expressible in a symmetric form under the interchange $x_+ \leftrightarrow x_-$. Consequently, the Green's function reduces to

$$c(x, s) = Ax^\nu \mathcal{C}_\nu(x_>, x_+)\mathcal{C}_\nu(x_<, x_-), \tag{6.4.2}$$

where $x_> = \max(x, x_0)$ and $x_< = \min(x, x_0)$, and where we have introduced the space-saving notation $\mathcal{C}_\nu(a, b) = I_\nu(a)K_\nu(b) - I_\nu(b)K_\nu(a)$ for the linear combination of the two types of Bessel functions that automatically equals zero at each boundary. The product form in Eq. (6.4.2) ensures continuity of the Green's function at $x = x_0$ and is a direct extension of Eq. (2.28) to the annular geometry.

Finally, to fix A, we integrate Eq. (6.4.1) across the discontinuity at x_0 to give the joining condition

$$c_>'|_{x_0^+} - c_<'|_{x_0^-} = -\frac{s^{(d-2)/2}}{D^{d/2}}\frac{1}{\Omega_d x_0^{d-1}}.$$

These derivatives may be found from the recursion relations $I_\nu' = (\nu/x)I_\nu + I_{\nu-1}$ and $K_\nu' = (\nu/x)K_\nu - K_{\nu-1}$ [Abramowitz & Stegun (1972)]. Evaluating the derivatives at x_0, using the Wronskian relation

$$I_{\nu-1}(x_0)K_\nu(x_0) + I_\nu(x_0)K_{\nu-1}(x_0) = \frac{1}{x_0}, \tag{6.4.3}$$

and after some straightforward but tedious algebra, we obtain the Green's function

$$c(x, s) = \frac{s^{(d-2)/2}}{D^{d/2}\Omega_d}(x\,x_0)^\nu \frac{\mathcal{C}_\nu(x_<, x_-)\,\mathcal{C}_\nu(x_>, x_+)}{\mathcal{C}_\nu(x_-, x_+)}. \tag{6.4.4}$$

We can now obtain first-passage properties by integrating the flux to each boundary over the surface of the sphere. The Laplace transforms of the total current to the inner and the outer boundaries are

$$
\begin{aligned}
J_{\pm}(x_0) &\equiv \int_{S_{\pm}} D \frac{\partial c}{\partial r}\bigg|_{r=R_{\pm}} \\
&= \Omega_d R_{\pm}^{d-1} \left(\frac{D}{s}\right)^{(d-1)/2} \left(\frac{s}{D}\right)^{1/2} D \frac{\partial c}{\partial x}\bigg|_{x=x_{\perp}} \\
&= \pm \left(\frac{x_{\pm}}{x_0}\right)^{\nu} \frac{C_{\nu}(x_0, x_{\mp})}{C_{\nu}(x_-, x_+)},
\end{aligned}
\tag{6.4.5}
$$

where $\pm$ refer to the flux to the outer and the inner boundaries, respectively. In the limit $s \to 0$, these currents reduce to the eventual hitting probabilities by means of the outer and the inner boundaries, $\lim_{s \to 0} J_{\pm}(x_0) = \mathcal{E}_{\pm}(r_0)$, respectively. This $s \to 0$ limit coincides with the eventual hitting probability already derived in Eq. (6.2.2) by the electrostatic analogy.

6.4.3. One Reflecting and One Absorbing Boundary

For mixed boundary conditions, the solution for the inner and the outer problems are trivially related by the interchange of R_+ and R_-, and we focus on the former case of reflection at the outer boundary and absorption at the inner boundary. The solvable situations are isotropic diffusion in d dimensions, or, by the equivalence of Section 6.2, two dimensions with radial drift.

6.4.3.1. Isotropic Diffusion in Arbitrary Spatial Dimension

By following the standard methods given in Subsection 6.4.2, we find that the Green's function for absorption at the inner boundary and reflection at the outer boundary is

$$
c(x, s) = \frac{1}{s} \left(\frac{s}{D}\right)^{d/2} \frac{(x\, x_0)^{\nu}}{\Omega_d} \frac{C_{\nu}(x_<, x_-)\, \mathcal{D}_{\nu, -}(x_>, x_+)}{\mathcal{D}_{\nu, -}(x_-, x_+)},
\tag{6.4.6}
$$

with $\mathcal{D}_{\nu, \pm}(a, b) \equiv I_{\nu}(a)K_{\nu \pm 1}(b) + K_{\nu}(a)I_{\nu \pm 1}(b)$, another space-saving notation. The total current to the absorbing inner sphere, that is, the Laplace transform of the first-passage probability, is $J_-(x_0) = \int_{S_-} D \frac{\partial c}{\partial r}\big|_{r=R_-}$. We find

$$
J_-(x_0) = \left(\frac{x_0}{x_-}\right)^{\nu} \frac{\mathcal{D}_{\nu, -}(x_0, x_+)}{\mathcal{D}_{\nu, -}(x_-, x_+)}.
\tag{6.4.7}
$$

The first-passage probability in the complementary outer problem – reflection at $r = R_-$ and absorption at $r = R_+$ – may be obtained by the transformation $R_- \leftrightarrow R_+$.

The interesting first-passage characteristic in the long-time limit is $R_+ \to \infty$, i.e., the first passage exterior to an absorbing sphere in an unbounded domain. In this limiting case, the first-passage probability of Eq. (6.4.7) simplifies to

$$J_-(x_0) \to \left(\frac{x_0}{x_-}\right)^\nu \frac{K_\nu(x_0)}{K_\nu(x_-)}. \tag{6.4.8}$$

From the small-s behavior of this function (with all co-ordinates multiplied by $\sqrt{s/D}$ to restore the full s dependence), we may then obtain the time-dependent asymptotics.

We now study the various basic cases in detail:

(i) $\nu > 0$ ($d < 2$). The expansion of $J_-(x_0)$ gives

$$J_-(x_0) = 1 - \left(\frac{s}{4D}\right)^\nu \frac{\Gamma(1-\nu)}{\Gamma(1+\nu)}(a^{2\nu} - x_0^{2\nu}) - \frac{s}{4D}\frac{\Gamma(1-\nu)}{\Gamma(2-\nu)}(a^2 - x_0^2) + \dots.$$

Because the coefficient of the zeroth power of s equals 1, the probability of eventually reaching the sphere is 1. This is just the recurrence of diffusion for $d < 2$. From the leading correction term in this expansion, the considerations of Section 1.4 imply that first-passage probability has the long-time tail $t^{-(1+\nu)}$. This gives an infinite first-passage time to the sphere. Thus even though return to the sphere is certain for all $d < 2$, the mean time to return is infinite, since $0 < \nu < 1$.

(ii) $\nu = 0$ ($d = 2$). The expansion of $J_-(x_0)$ is

$$J_-(x_0) = 1 - 2\frac{\ln(x_0/a)}{\ln s} + \dots.$$

Once again, the probability of eventually hitting the circle equals 1, and the first correction varying as $1/\ln s$ corresponds to the first-passage probability having an asymptotic $1/(t \ln^2 t)$ time dependence.

(iii) $\nu < 0$ ($d > 2$). Here we define $\mu = -\nu > 0$, and the expansion of $J_-(x_0)$ has the small-s behavior

$$J_-(x_0) = \left(\frac{a}{x_0}\right)^{2\mu}\left[1 - \left(\frac{s}{4D}\right)^\mu \frac{\Gamma(1-\mu)}{\Gamma(1+\mu)}(a^{2\mu} - x_0^{2\mu})\right.$$
$$\left. - \frac{s}{4D}\frac{\Gamma(1-\mu)}{\Gamma(2-\mu)}(a^2 - x_0^2) + \dots\right].$$

From the $s \to 0$ limit of this expression, the probability of eventually hitting the sphere is $(a/x_0)^{d-2}$, as expected by electrostatics. The correction terms give the time dependence of the hitting probability for the subset of particles that eventually do reach the sphere. The leading correction term varying as

s^μ means that the hitting probability at time t varies as $t^{-(1+\mu)}$. When $d = 4$, the leading correction term in $J_-(x_0)$ is linear in s, so that the hitting time for the returnees is finite. This situation of a finite hitting time but of a return probability which is less than one is termed strong transience [Jain & Orey (1968)]. Physically, this phenomenon reflects the fact that a particle must reach the absorber in a finite time for large spatial dimension if it is to be absorbed at all. This occurs for $d > 4$ in a purely diffusive system.

6.4.3.2. Radially Biased Diffusion in Two Dimensions

In dimensionless coordinates, the Laplace transform of the two-dimensional convection – diffusion equation with the "shell" initial condition again has the Bessel form,

$$c''(x, s) - \frac{Pe - 1}{x} c'(x, s) - c(x, s) = -\frac{1}{D} \frac{\delta(x - x_0)}{2\pi x_0},$$

with the Péclet number now defined as $Pe = v_0/D$. We follow the same calculational steps as in the case of d dimensions and obtain the Green's function

$$c(x, s) = \frac{1}{2\pi D} \left(\frac{x}{x_0}\right)^\nu \frac{\mathcal{C}_\nu(x_<, x_-)\mathcal{D}_{\nu,+}(x_>, x_+)}{\mathcal{D}_{\nu,+}(x_-, x_+)}, \qquad (6.4.9)$$

where now $\nu = Pe/2$. The total flux to the absorbing circle is $J|_{x=x_-} = (2\pi x_-)Dc'|_{r=x_-}$. The convective contribution to the flux, vc, can be ignored because the concentration vanishes at $x = x_-$. Performing this derivative and again using the Wronskian relation, Eq. (6.4.3), gives

$$J_-^{(2)}(x_0) = \left(\frac{x_-}{x_0}\right)^\nu \frac{\mathcal{D}_{\nu,+}(x_0, x_+)}{\mathcal{D}_{\nu,+}(x_-, x_+)}. \qquad (6.4.10a)$$

We may then obtain the first-passage probability in the complementary outer problem – reflection at $r = R_-$ and absorption at $r = R_+$ – by interchanging $R_- \leftrightarrow R_+$ in the above expression.

In the interesting case of first passage exterior to an absorbing circle in an unbounded domain, the first-passage probability of Eq. (6.4.10a) reduces to

$$J_-^{(2)}(x_0) = \left(\frac{x_-}{x_0}\right)^\nu \frac{K_\nu(x_0)}{K_\nu(x_-)}. \qquad (6.4.10b)$$

The near coincidence between this and the first-passage probability given in Eq. (6.4.8) for isotropic diffusion in arbitrary dimensions is no accident, but rather, is a consequence of the general relation between these two systems in which the Péclet number is equivalent to $2 - d$.

There are two interesting cases – outward and inward flow. Outward flow corresponds to $v > 0$ or, by the aforementioned connection, to isotropic diffusion in $d > 2$. Therefore the probability of eventually hitting the circle equals $(a/x_0)^{Pe}$. Inward flow corresponds to $d < 2$, so that the eventual hitting probability equals 1. The mean first-passage time is infinite for small negative Péclet number, but becomes finite once the Péclet number reaches -1.

For the outer problem, there are also two different asymptotic behaviors that depend on the magnitude of the bias. For $v > -1/2$, the first-passage probability decays exponentially in time and the mean first-passage time is finite. On the other hand for $v < -1/2$ there is a power-law long-time tail in the first-passage probability with a velocity-dependent exponent, $t^{-(\mu+1)}$. Thus a sufficiently strong inward drift compensates for the finite-size cutoff of the system and gives a divergent mean first-passage time and an attendant power-law tail in the first-passage probability.

6.5. Reaction-Rate Theory

6.5.1. Background

We now turn to the relation between first passage to a sphere and the von Smoluchowski rate theory for the kinetics of diffusion-controlled reactions [von Smoluchowski (1917); for a more contemporary discussion, see, e.g., Rice (1985)]. In such processes, two molecules react immediately when they approach within a reaction radius and the overall process is limited by their encounter rate. A generic example is diffusion-controlled single-species annihilation, $A + A \rightarrow 0$. The evolution of the reactant density, $\rho_A(\vec{r}, t)$, may be described by the diffusion–reaction equation

$$\frac{\partial \rho_A(\vec{r}, t)}{\partial t} = D\nabla^2 \rho_A(\vec{r}, t) - k\rho_A(\vec{r}, t)^2, \qquad (6.5.1)$$

where the second term accounts for the annihilation rate in a homogeneous system. The reaction rate k quantifies the efficiency of an encounter between reactants. The goal of the Smoluchowski theory is to compute this rate.

This rate may be reframed as the following idealized first-passage process. Fix one particle (a sphere of radius a) as the "target." In the rest frame of the target, the other reactants diffuse and are absorbed if they hit the target. The density is taken to be small, so that reactions among the background mobile particles can be ignored. We then identify the flux to the target as the reaction rate k.

To compute this flux, we first determine the concentration around an absorbing target particle by solving the diffusion equation

$$\frac{\partial c(\vec{r}, t)}{\partial t} = D\nabla^2 c(\vec{r}, t), \qquad (6.5.2)$$

subject to the initial condition $c(\vec{r}, t = 0) = 1$ for $r > a$ and the boundary conditions $c(r = a, t) = 0$ and $c(r \to \infty, t) = 1$. The reaction rate is then the integral of the flux over the sphere surface:

$$k(t) = -D \int_S \left| \frac{\partial c(\vec{r}, t)}{\partial r} \right|_{r=a} d\Omega. \qquad (6.5.3)$$

We should anticipate two regimes of behavior. For $d > 2$, a diffusing particle is transient and the relatively slow loss of reactants by trapping at the target is compensated for by their resupply from infinity. Thus a steady state is reached and the reaction rate k is finite. Conversely for $d \le 2$, diffusing particles are sure to hit the target and a depletion zone grows continuously about the trap. Correspondingly, the flux and thus the reaction rate decay in time. Although a reaction rate does not strictly exist within the Smoluchowski theory for $d \le 2$, it is still meaningful to think of the theory as providing a time-dependent reaction rate.

6.5.2. Time-Dependent Solution for General d

We now solve Eq. (6.5.2) with the specified initial and boundary conditions. In principle, we can obtain the solution by convolving the Green's function for diffusion in the presence of an absorbing sphere [expression (6.4.8)] with the constant initial density. It is much simpler, however, to solve the inhomogeneous problem directly. The Laplace transform of Eq. (6.5.2) is

$$sc - 1 = D\nabla^2 c,$$

where the factor -1 accounts for the constant initial condition $c(r, t = 0) = 1$. A particular solution is simply $c = 1/s$, and to complete the solution we first need to solve the homogeneous equation

$$c'' + \frac{d-1}{r} c' - \frac{s}{D} c = 0,$$

where the prime denotes differentiation with respect to r.

The elemental solutions to this equation are again $r^\nu I_\nu(r\sqrt{s/D})$ and $r^\nu K_\nu(r\sqrt{s/D})$ with $\nu = 1 - d/2$. Because the solution must be finite as $r \to \infty$, only the K_ν term contributes. The general solution to the diffusion equation therefore is

$$c(r, s) = \frac{1}{s} + Ar^{\nu} K_{\nu}(r\sqrt{s/D}).$$

Imposing the absorbing boundary condition at $r = a$ then gives

$$c(r, s) = \frac{1}{s}\left[1 - \left(\frac{r}{a}\right)^{\nu} \frac{K_{\nu}(r\sqrt{s/D})}{K_{\nu}(a\sqrt{s/D})}\right]. \tag{6.5.4}$$

In determining the flux, we again separately treat the cases of (i) $\nu < 0$ ($d > 2$) and (ii) $\nu \geq 0$ ($d \leq 2$).

6.5.2.1. The Case $\nu < 0$ or $d > 2$

For $\nu < 0$, we define $\mu = -\nu = d/2 - 1 > 0$. In this regime, a steady-state concentration profile exists that we obtain by taking the $s \to 0$ limit of Eq. (6.5.4). This gives the electrostatic solution

$$c(r, s \to 0) = \frac{1}{s}\left[1 - \left(\frac{a}{r}\right)^{d-2}\right].$$

To determine the time dependence of the flux, we use $K_{-\nu} = K_{\nu}$ and the recursion formula $K_{\nu}(x)' = -K_{\nu-1}(x) - \nu K_{\nu}(x)/x$ in Eq. (6.5.4) to give the flux to any point on the absorbing sphere:

$$\phi \equiv -D\frac{\partial c}{\partial r}\Big|_{r=a}$$
$$= \frac{2D\nu}{as} + \sqrt{\frac{D}{s}} \frac{K_{\mu-1}(a\sqrt{s/D})}{K_{\mu}(a\sqrt{s/D})}.$$

Because the leading behavior is proportional to $1/s$, the flux approaches a constant at long times. This is expected because the concentration itself approaches a steady state. There are now two interesting subcases that depend on whether $\mu - 1 \gtrless 0$. For $\mu - 1 > 0$, which corresponds to $d > 4$, the small-argument asymptotics of K_{μ} gives

$$\phi = -\frac{2D\mu}{as} + \sqrt{\frac{D}{s}} \frac{\Gamma(\mu - 1)}{\Gamma(\mu)} \left(\frac{a\sqrt{s/D}}{2}\right) + \cdots$$
$$= -\frac{2D\mu}{as} + \frac{a}{2}\frac{\Gamma(\mu - 1)}{\Gamma(\mu)} + \cdots.$$

Here the second term corresponds to a $1/t$ leading correction to the steady-state flux. We may obtain the total current, or reaction rate k, by integrating the flux over the surface of the sphere. This then gives

$$k = (d - 2)\Omega_d D a^{d-2} + \frac{\text{const}}{t} + \ldots, \quad d > 4. \tag{6.5.5a}$$

On the other hand, for $\mu - 1 < 0$ (and $\mu > 0$), corresponding to $2 < d < 4$, we write $K_{\mu-1} = K_{1-\mu}$ to facilitate taking the $s \to 0$ limit. Then we find

$$\phi = -\frac{2D\mu}{as} + \sqrt{\frac{D}{s}} \frac{\Gamma(1 - \mu)}{\Gamma(\mu)} \left(\frac{a\sqrt{s/D}}{2} \right)^{2\mu-1} + \ldots$$

$$= -\frac{2D\mu}{as} + \text{const} \times s^{(d-4)/2} + \ldots.$$

Now the second term corresponds to a correction that varies asymptotically as $t^{1-d/2}$. Thus

$$k = (d - 2)\Omega_d D a^{d-2} + \text{const} \times t^{1-d/2} \quad 2 < d < 4. \tag{6.5.5b}$$

In the special and physically most relevant case of $d = 3$ ($\mu = 1/2$) the ratio $K_{\mu-1}/K_\mu$ cancels, and this gives the exact result for the flux:

$$\phi = \frac{D}{as} + \sqrt{\frac{D}{s}},$$

from which we immediately obtain the classical Smoluchowski formula for the reaction rate:

$$k = 4\pi a D \left(1 + \frac{a}{\sqrt{\pi D t}} \right). \tag{6.5.5c}$$

Note that this rate is proportional to the diffusion coefficient D times a^{d-2}. Their product then has units of (volume/time), as follows from diffusion-reaction equation (6.5.1). More importantly, the reaction rate is pathological for $d < 2$ because the rate *increases* as the size of the trap *decreases*. This peculiarity gives a strong hint that there is a fundamental difference between the kinetics of diffusion-controlled reactions in $d > 2$ and $d \leq 2$.

6.5.2.2. The Case $v \geq 0$ or $d \leq 2$

Here, there is a fortuitous cancellation of two terms in evaluating the derivative of Eq. (6.5.4), and the flux is now given by

$$\phi = -\frac{D}{s} \frac{K_{1-v}(a\sqrt{s/D})}{K_v(a\sqrt{s/D})}, \quad d < 2.$$

In the $s \to 0$ limit, this expression reduces to

$$\phi \to \left(\frac{2}{a}\right)^{d-1} \left(\frac{D}{s}\right)^{d/2}.$$

Thus the time dependence of the total current is

$$k \propto \frac{(Dt)^{d/2}}{t} \propto \frac{1}{t^{1-d/2}}, \tag{6.5.6}$$

where the first expression emphasizes the correct units of k.

For $v = 0$ or $d = 2$, the flux is

$$\phi = -\frac{D}{s} \frac{K_1(a\sqrt{s/D})}{K_0(a\sqrt{s/D})}$$

$$\sim \frac{4}{a} \frac{1}{s \ln s}, \quad s \to 0.$$

As $t \to \infty$, this corresponds to the total current having the time dependence

$$k \sim \frac{4\pi D}{\ln t}, \quad d = 2. \tag{6.5.7}$$

Note that the reaction rates for $d \leq 2$ are independent of the radius of the target sphere! This is yet another manifestation of the recurrence of diffusion for $d \leq 2$ – even an infinitely small trap is sure to capture all exterior particles, so that the flux, and hence the reaction rate, eventually vanishes. This is the mechanism that ultimately leads to the slow kinetics of diffusion-controlled reactions for spatial dimension $d \leq 2$.

6.5.3. Elementary Time-Dependent Solution for $d = 3$

An intriguing aspect of the Smoluchowski rate theory is that in the physical case of three dimensions the rate can be expressed in terms of elementary functions. This simplification hinges on the fact that the radial part of the Laplacian operator in three dimensions (and three dimensions only) is related to the one-dimensional Laplacian by

$$\nabla_{3d}^2 \left(\frac{u}{r}\right) = \frac{1}{r} \nabla_{1d}^2 u. \tag{6.5.8}$$

With this trick, we can express the solution of a spherically symmetric three-dimensional system in terms of a corresponding one-dimensional system. Thus the solution to the diffusion equation for the concentration $c(r, t)$ in three dimensions,

$$\frac{\partial c}{\partial t} = D\nabla^2_{3d}c,$$

can be written in terms of the solution to the one-dimensional diffusion equation for $u(r, t) = rc(r, t)$:

$$\frac{\partial u}{\partial t} = D\nabla^2_{1d}u.$$

We now determine the reaction rate by this equivalence. As a necessary preliminary, we first solve the one-dimensional problem

$$\frac{\partial c}{\partial t} = D\frac{\partial^2 c}{\partial x^2},$$

subject to $c(x, t = 0) = 1$, $c(0, t) = 0$, and $c(x \to \infty, t) = 1$. The Laplace transform of this equation is $sc(x, s) - 1 = c''(x, s)$. The particular solution is $1/s$, and the solution to the homogeneous equation is $A\,e^{-x\sqrt{s/D}}$. The growing exponential does not appear because of the boundary condition at $x = \infty$. The boundary condition at $x = 0$ then gives

$$c(x, s) = \frac{1}{s}\left(1 - e^{-x\sqrt{s/D}}\right), \tag{6.5.9}$$

from which

$$c(x, t) = \text{erf}\left(\frac{x}{\sqrt{4Dt}}\right). \tag{6.5.10}$$

Finally, the current to the origin is

$$k = -D\frac{\partial c}{\partial x}\bigg|_{x=0} = \sqrt{\frac{D}{\pi t}} \qquad d = 1. \tag{6.5.11}$$

To infer the three-dimensional solution, we now define $u = rc$. In terms of u, the initial condition is now $u = r$ (corresponding to a constant initial concentration in three dimension) and the absorbing boundary condition must be imposed at $r = a$, rather than at $r = 0$, to correspond to a sphere with a finite radius in three dimensions. With these features, the appropriate solution for u in one dimension is

$$u(r, s) = \frac{1}{s}\left[r - ae^{-(r-a)\sqrt{s/D}}\right],$$

from which the concentration in three dimensions equals

$$c(r, t) = \left\{\left(1 - \frac{a}{r}\right) + \frac{a}{r}\text{erf}\left[\frac{(r-a)}{\sqrt{4Dt}}\right]\right\}. \tag{6.5.12}$$

The radial derivative at $r = a$ is simply

$$-\frac{\partial c}{\partial r}\Big|_{r=a} = \frac{1}{a} + \frac{1}{\sqrt{\pi Dt}},$$

and the current $k = -4\pi a^2 D(\partial c/\partial r)$ reproduces Eq. (6.5.5c).

6.5.4. Quasi-Static Approach

We now derive the flux to an absorbing sphere by the quasi-static approximation that was first introduced in Section 3.6. In general, we shall see that this method is extraordinarily simple compared with the exact calculation for the flux. Moreover, the quasi-static solution reproduces the small-r limit of the concentration with the *exact* amplitude for $d \geq 2$, and thus the exact flux. In short, the quasi-static approximation is an extremely simple yet powerful tool for dealing with diffusion in the presence of absorbing boundaries.

By the quasi-static hypothesis, the concentration within the near zone, $r < r^*$, is given by the solution to the Laplace equation, subject to the boundary conditions $c(a, t) = 0$ and $c(r^*, t) = 1$. Here the boundary between the near and the far zones is defined by $r^* = a + \sqrt{Dt}$. As $t \to \infty$ we may therefore approximate r^* by $\sqrt{Dt}$. Conversely, in the far zone ($r > r^*$) the solution just equals the initial concentration of one. We then match these two component solutions at r^* to obtain the concentration for all r.

There are again different behaviors for $d < 2$, $d = 2$, and $d > 2$ and we treat each of these cases in turn.

6.5.4.1. Dimensions $d < 2$

In the near zone, the solution to the Laplace equation has the general form $A + Br^{-(d-2)}$, and by matching to the boundary conditions we easily obtain

$$c(r, t) \approx \frac{(r/a)^{2-d} - 1}{(\sqrt{Dt}/a)^{2-d} - 1}, \qquad r < r^*, \tag{6.5.13}$$

whereas in the far zone, $r > r^*$ $c(r, t) = 1$. At long times, the near-zone concentration is proportional to $(r - a)/(Dt)^{1-d/2}$, which coincides with the exact small-distance time dependence. We may now compute the total incoming flux by integrating over the surface of the sphere. This gives

$$\Phi \propto \frac{(2 - d)\Omega_d D}{(Dt)^{1-d/2}}, \tag{6.5.14}$$

thus reproducing the time dependence in relation (6.5.6).

We may also exploit the small-distance concentration profile to determine the time-dependent first-passage properties with no additional calculation. From Chap. 3, we learned that the time-dependent concentration $c(r, t)$ that arises from a spatially constant initial concentration coincides with the survival probability at time t of a single diffusing particle that is initially at r. Thus from approximation (6.5.13), the survival probability for a particle that starts a distance $(r - a)$ from an absorbing sphere decays at long times as $(r - a)/(Dt)^{1-d/2}$. From this, the first-passage probability asymptotically decays as $(r - a)/t^{2-d/2}$.

6.5.4.2. Two Dimensions

We now solve

$$\frac{D}{r}\frac{\partial}{\partial r}r\frac{\partial c(r, t)}{\partial r} = 0 \qquad (6.5.15)$$

in the near zone $a \leq r \leq \sqrt{Dt}$. This solution is simply $c(r, t) = A + B\ln r$, and matching to the boundary conditions $c(r = a, t) = 0$ and $c(r = \sqrt{Dt}, t) = 1$ then gives the asymptotic solution

$$c(r, t) \sim \frac{\ln(r/a)}{\ln(\sqrt{Dt}/a)}, \qquad a \leq r \leq \sqrt{Dt}. \qquad (6.5.16)$$

This reproduces the correct asymptotic flux to the circle [approximation (6.5.7)]. Finally, by exploiting the equivalence between the concentration and the survival probability mentioned above, the survival probability for a particle that starts at r decays as $\ln(r/a)/\ln t$ at long times, from which the first-passage probability decays as $\ln(r/a)/(t \ln^2 t)$. It is striking that this very simple quasi-static approach can reproduce the delicate logarithmic behavior of the flux in two dimensions.

6.5.4.3. Dimensions $d > 2$

Here the requisite solution to the Laplace equation for $a \leq r \leq \sqrt{Dt}$ is

$$c(r, t) \sim \frac{1 - \left(\frac{a}{r}\right)^{d-2}}{1 - \left(\frac{a}{\sqrt{Dt}}\right)^{d-2}}, \qquad a \leq r \leq \sqrt{Dt}. \qquad (6.5.17)$$

In the long-time limit, the concentration approaches the exact static Laplacian form, $c(r) \to 1 - (a/r)^{d-2}$, and thus also reproduces the exact reaction rate. Note also that the static limit of the concentration is just the asymptotic survival probability of a particle that starts at r.

6.5.5. Closest Particle to an Absorbing Sphere

As a last illustration of the quasi-static approximation, we study the following question: What is the typical distance from the edge of the trap to the closest particle? More generally, what are the statistical properties of this closest particle? Although this problem can be analyzed by exact methods [Weiss, Kopelman, & Halvin (1989), and Havlin et al. (1990)], we will apply the quasi-static approximation, together with some basic facts about extreme value statistics, to determine the properties of the closest particle to an absorbing sphere in d dimensions in a simple manner [Redner & ben-Avraham (1990)]. The basic result is that the particle closest to the sphere is typically at a distance that grows as $t^{1/4}$ in $d = 1$ and as $(\ln t)^{1/2}$ for $d = 2$. In $d > 2$ this nearest distance becomes constant as $t \to \infty$.

Consider first the one-dimensional system with a trap located at $x = 0$ (Fig. 6.6). A crude estimate for the typical distance of the closest particle, x_{min}, is provided by the criterion

$$\int_0^{x_{min}} c(x, t)\, dx = 1. \tag{6.5.18}$$

This states that there is of the order of a single particle in the range $[0, x_{min}]$. Using the exact short-distance form $c(x, t) \sim c_0 x/\sqrt{\pi D t}$ in Eq. (6.5.18) gives

$$x_{min} = 2^{3/4} \pi^{1/4} \left(\frac{Dt}{c_0^2} \right)^{1/4}. \tag{6.5.19}$$

If we were to use the quasi-static approximation $c(x, t)_{qs} \sim c_0 x/\sqrt{Dt}$, then we would get almost the same result; only the factor $\pi^{1/4}$ would be missing. The basic feature is that although the diffusive depletion layer is of the order of $(Dt)^{1/2}$, the closest particle is at a much smaller distance $(Dt/c_0^2)^{1/4}$.

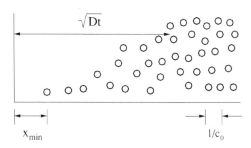

Fig. 6.6. Schematic picture of the concentration near a trap at $x = 0$ in one dimension, showing the definition of x_{min} and the relative magnitudes of the other basic length scales.

We can also compute the full probability distribution for x_{min} by using basic extreme-statistics ideas [Galambos (1987)]. A very useful simplifying trick is to recognize that the distribution of x_{min} involves distances that are in the range of x_{min} itself. Thus we need consider only the linear short-distance tail of the concentration profile in the range $[0, y]$, with y much larger than the closest distance $(Dt/c_0^2)^{1/4}$ and much less than the diffusion length $(Dt)^{1/2}$. Our final result is independent of y, as it must.

Let $N(y, t) = \int_0^y c(x, t) \, dx$ be the number of particles in $[0, y]$ and let

$$P_<(x, t) = \frac{\int_0^x c(x', t) \, dx'}{\int_0^y c(x', t) \, dx'}$$

be the probability that a particle lies in the range $[0, x]$. Now if there are $N(y, t)$ particles in the range $[0, y]$, the probability that the closest particle is located at x is

$$P(x, t) = c(x, t)[1 - P_<(x, t)]^{N(y,t)-1}.$$

This equation merely states that among the $N(y, t)$ particles, the closest is distributed according to $c(x, t)$, whereas the remaining $N - 1$ particles are each distributed according to $1 - P_<$. We thus find

$$P(x, t) \sim c(x, t) e^{-[N(y,t)-1]P_<(x,t)},$$
$$\sim c(x, t) \exp\left[-\int_0^x c(x', t) \, dx'\right],$$
$$= \frac{c_0 x}{\sqrt{\pi Dt}} \exp\left(-\frac{c_0 x^2}{\sqrt{4\pi Dt}}\right). \tag{6.5.20}$$

Note that the dependence on y disappears as long as $y \gg x$. This (normalized) distribution gives the average position for the nearest particle $\langle x \rangle = \int_0^\infty x P(x, t) \, dx = (\pi^3 Dt/4c_0^2)^{1/4}$ and the width of the distribution is also of the order of $(Dt/c_0^2)^{1/4}$.

For $d > 1$, the trap must have a nonzero radius $a > 0$ so that a diffusing particle can actually hit the trap. For $d = 2$, we again use the quasi-static concentration profile of approximation (6.5.16) to obtain the properties of the closest particle. Following the same approach as in one dimension, the typical distance of the closest particle is determined by $r_{min} = \int_a^{r_{min}} 2\pi r \, c(r, t) \, dr = 1$. This leads to a transcendental equation for r_{min} whose asymptotic solution is

$$r_{min} \sim \sqrt{\frac{\ln(Dt/a^2)}{2\pi c_0}}. \tag{6.5.21}$$

Following step-by-step the derivation of approximation (6.5.20), we find that the corresponding distribution for the scaled distance $\rho = r/a$ of the nearest particle to the trap is

$$P(\rho, t) \sim \frac{2\pi c_0 a}{\ln(\sqrt{Dt}/a)} \rho \ln \rho \exp \left[-\frac{2\pi c_0 a}{\ln(\sqrt{Dt}/a)} \left(\frac{\rho^2}{2} \ln \rho - \frac{\rho^2 - 1}{4} \right) \right].$$

$$(6.5.22)$$

It is remarkable that such a complicated result, which agrees with the asymptotic formula [Havlin et al. (1990)], can be obtained so simply.

For $d = 3$, $c(r, t)$ approaches a steady state as $t \to \infty$ and r_{min} similarly approaches a constant. We now determine r_{min} from the criterion $r_{min} = \int_a^{r_{min}} 4\pi r^2 c(r, t) \, dr = 1$. This leads to the limiting behaviors

$$\rho \sim \begin{cases} \dfrac{1}{\mathcal{N}^3} & \mathcal{N} \ll 1 \\[3mm] 1 + \left(\dfrac{2}{3\mathcal{N}} \right)^{1/2} & \mathcal{N} \gg 1 \end{cases}, \qquad (6.5.23)$$

where $\mathcal{N} = 4\pi a^3 c_0 / 3$. As expected, the minimum distance reduces to the interparticle spacing for low densities and approaches the sphere radius for high densities.

7

Wedge Domains

7.1. Why Study the Wedge?

We now investigate the first-passage properties of diffusion in wedge domains with absorption when the particle hits the boundary (Fig. 7.1). There are several motivations for studying this system. One is that first passage in the two-dimensional wedge can be mapped onto the kinetics of various one-dimensional diffusion-controlled reaction processes. By this correspondence, we can obtain useful physical insights about these reactions as well as their exact kinetic behavior. These connections will be discussed in detail in the next chapter.

A second motivation is more theoretical and stems from the salient fact the survival probability and related first-passage properties in the wedge decay as power laws in time with characteristic exponents that depend continuously on the wedge opening angle. This arises, in part, because a wedge with an infinite radial extent does not have a unique characteristic length or time scale. One of our goals is to compute this exponent and to develop intuition for its dependence on the wedge angle. These results naturally lead to the "pie wedge" of finite radial extent that exhibits an unexpected discontinuous transition in the behavior of the mean exit time as the wedge angle passes through $\pi/2$. This feature also manifests itself as a pathology in the flow of a viscous fluid in a wedge-shaped pipe [Moffat & Duffy (1979)].

Finally, the first-passage properties to a sharp tip (wedge angle $> \pi$) is an essential ingredient in diffusion-controlled growth processes, such as dendrites, crystals, and diffusion-limited aggregates. The stability of such perturbations depends on whether the first-passage probability is largest at the tip or away from it. Thus understanding first passage in the wedge geometry provides useful intuition for these stochastic growth phenomena, an approach that was nicely developed by Turkevich and Scher (1985) to yield new insights about the structure of diffusion-limited aggregates.

We shall obtain first-passage properties in the wedge geometry both by direct solution of the diffusion equation and also, for two dimensions, in a

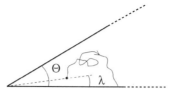

Fig. 7.1. A two-dimensional wedge of opening angle Θ with absorbing boundaries. For a diffusing particle initially at angle λ, the horizontal boundary is eventually hit with probability λ/Θ and the other boundary is hit with probability $1 - \lambda/\Theta$.

more aesthetically pleasing fashion by conformal transformation techniques, in conjunction with the electrostatic formulation. The former approach provides the full time-dependent first-passage properties, whereas conformal methods are more appropriate for time-integrated quantities, such as hitting probabilities and mean first-passage times. Also presented is a heuristic extension of the electrostatic approach that allows us to infer, with little additional computational effort, time-dependent first-passage properties in the wedge from corresponding time-integrated properties.

7.2. Two-Dimensional Wedge

7.2.1. Solution to the Diffusion Equation

We first solve the diffusion equation in the wedge to determine the survival probability of a diffusing particle. Although the exact Green's function for this system is well known, we adopt the strategy of choosing an initial condition that allows us to eliminate angular variables and deal with an effective radial problem. This is suitable if we are interested in only asymptotic first-passage properties.

The diffusion equation for the two-dimensional wedge geometry is most naturally written in plane-polar coordinates,

$$\frac{\partial c}{\partial t} = D \left(\frac{\partial^2 c}{\partial r^2} + \frac{1}{r} \frac{\partial c}{\partial r} + \frac{1}{r^2} \frac{\partial^2 c}{\partial \theta^2} \right), \qquad (7.2.1)$$

where $c = c(r, \theta, t)$ is the particle concentration at (r, θ) at time t, D is the diffusion coefficient, and the boundary conditions are $c = 0$ at $\theta = 0, \Theta$. To reduce this to an effective one-dimensional radial problem, we first note that the exact Green's function can be written as an eigenfunction expansion in which the angular dependence is a sum of sine waves such that an integral number of half-wavelengths fit within $(0, \Theta)$ to satisfy the absorbing boundary conditions [Carslaw & Jaeger (1959)]. These sine waves are $\sin(n\pi\theta/\Theta)$, with n an integer, and they represent the Fourier series expansion of the Green's function.

In this series, each sine wave is multiplied by a conjugate decaying function of time, in which the decay rate increases with n. In the long-time limit, only the lowest term in this expansion dominates the survival probability. Consequently, we may obtain the correct long-time behavior by choosing an initial condition whose angular dependence is a half-sine-wave in the wedge. This ensures that the time-dependent problem will contain only this single term in the Fourier series. This is completely analogous to the long-time approximation we applied for diffusion in spherical geometries (Chap. 6).

We thus define $c(r, \theta, t = 0)\,\pi \sin(\pi\theta/\Theta)\delta(r - r_0)/2\Theta r_0$. With this initial distribution and after the Laplace transform is applied, diffusion equation (7.2.1) becomes

$$sc(r, \theta, s) - \frac{\pi}{2\Theta r_0}\delta(r - r_0)\sin(\pi\theta/\Theta) = D\left(\frac{\partial^2 c}{\partial r^2} + \frac{1}{r}\frac{\partial c}{\partial r} + \frac{1}{r^2}\frac{\partial^2 c}{\partial \theta^2}\right),$$

where $c = c(r, \theta, s)$. Substituting the ansatz $c(r, \theta, s) = R(r, s)\sin(\pi\theta/\Theta)$ into the above differential equation, the angular dependence may now be separated and reduces the system to an effective one-dimensional radial problem. Now, introducing the dimensionless coordinate $x = r\sqrt{s/D}$, we find the modified Bessel equation for the remaining radial coordinate:

$$R''(x, s) + \frac{1}{x}R'(x, s) - \left(1 + \frac{v^2}{x^2}\right)R(x, s) = -\frac{v}{2Dx_0}\delta(x - x_0), \quad (7.2.2)$$

where $v = \pi/\Theta$ and the prime now denotes differentiation with respect to x.

The general solution for $x \neq x_0$ is a superposition of modified Bessel functions of order v. Because the domain is unbounded, the interior Green's function ($x < x_0$) involves only I_v, as K_v diverges as $x \to 0$, whereas the exterior Green's function ($x > x_0$) involves only K_v, as I_v diverges as $x \to \infty$. By imposing continuity at $x = x_0$, we find that the Green's function has the symmetric form $R(x, s) = AI_v(x_<)K_v(x_>)$, with the constant A determined by the joining condition that arises by integration of Eq. (7.2.2) over an infinitesimal radial range that includes r_0. This gives

$$R'_>|_{x=x_0} - R'_<|_{x=x_0} = -\frac{v}{2Dx_0},$$

from which $A = v/2D$. Therefore the radial Green's function in the wedge is

$$R(x, s) = \frac{v}{2D}\, I_v(x_<)K_v(x_>), \quad (7.2.3)$$

and its Laplace inverse has the relatively simple closed form [Carslaw &

Jaeger (1959)]

$$R(r, t) = \frac{\nu}{4Dt} e^{-(r^2 + r_0^2)/4Dt} I_\nu \left(\frac{rr_0}{2Dt} \right).$$ (7.2.4)

With this radial Green's function, the asymptotic survival probability is

$$S(t) \sim \int_0^{\Theta} \sin(\nu\theta) \, d\theta \int_0^\infty r \, R(r, t) \, dr.$$ (7.2.5)

We may easily estimate the long-time behavior of this integral by first integrating over θ and then using the asymptotic approximations $e^{-r_0^2/4Dt} \to 1$ and $I_\nu(x) \sim (x/2)^\nu / \Gamma(\nu + 1)$ to give

$$S(t) \sim \frac{2}{\nu \Gamma(\nu + 1)} \int_0^\infty \frac{r}{2Dt} \left(\frac{rr_0}{8Dt} \right)^\nu e^{-r^2/4Dt} \, dr$$

$$= \frac{2}{\nu} \left(\frac{r_0}{\sqrt{16Dt}} \right)^\nu$$

$$\propto \left(\frac{r_0}{\sqrt{Dt}} \right)^{\pi/\Theta}.$$ (7.2.6)

Alternatively, we can estimate this integral by noting that the radial distance over which the concentration is appreciable extends to the order of $\sqrt{Dt}$. This provides the cutoff $r \approx \sqrt{Dt}$ in the radial integral in approximation (7.2.5), within which the Gaussian factors in $R(r, t)$ can be replaced with one. This approximation is in the same spirit as that used in Section 1.4 to find the asymptotic behavior of a generating function by approximating an exponential by a step function cutoff. Using the small-argument expansion of the Bessel function, we then obtain

$$S(t) = \int_0^{\Theta} \sin(\nu\theta) \, d\theta \int_0^\infty r \, R(r, t) \, dr,$$

$$\propto \int_0^{\sqrt{Dt}} \frac{1}{Dt} \left(\frac{rr_0}{Dt} \right)^\nu r \, dr,$$

$$\propto \left(\frac{r_0}{\sqrt{Dt}} \right)^{\pi/\Theta}.$$ (7.2.7)

In summary, the basic result is that the survival probability of a diffusing particle in a wedge of opening angle Θ decays with time as

$$S(t) \sim t^{-\pi/2\Theta}.$$ (7.2.8)

7.2.2. Physical Implications

It is instructive to consider illustrative special cases of this important result, both to appreciate its meaning and also to discuss various heuristic approaches that give additional insights about the wedge-angle dependence of the survival probability exponent. One such example is wedge angles that are an even-integer submultiple of 2π, for example, $\Theta/2\pi = 1/2, 1/4, 1/6$, etc. In these cases, the survival probability can be obtained independently by an obvious extension of the image method for the planar interface $\Theta = \pi$ that was presented in Section 3.2. For example, suppose that $\Theta = \pi/2$, with a particle initially located at (x_0, x_0). The exact probability distribution is then the sum of four Gaussians – two with positive amplitudes centered at (x_0, x_0) and at $(-x_0, -x_0)$ and two with negative amplitudes centered at $(x_0, -x_0)$ and at $(-x_0, x_0)$ – and the survival probability is the integral of this superposition in the first quadrant. Although this superposition is simple, it is not pleasant to compute the resulting concentration over the first quadrant to obtain the survival probability. Instead we extend the simple-minded version of image method from Section 3.2, based on the blip function picture, to the wedge.

In our naive blip-function image method for the semi-infinite interval, we replaced Gaussians with square waves. In the two-dimensional quadrant, we similarly replace each Gaussian with a square-shaped constant function with amplitude $1/(Dt)$ and linear dimension $\sqrt{Dt}$ (Fig. 7.2). The resultant

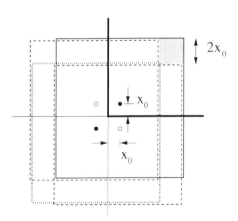

Fig. 7.2. Simplified image method for an absorbing wedge of opening angle $\pi/2$. The probability distribution inside the wedge is generated by a square blip that is due to the initial particle (full line) and three image blips (dashed for negative images, dotted for positive image). Each blip is approximated as a square with constant magnitude $1/Dt$ and linear dimension $\sqrt{Dt}$. Their superposition gives the shaded square of linear dimension $2x_0$; hence the asymptotic survival probability is proportional to x_0^2/Dt.

distribution contains the essential scaling features of the exact Gaussian form. Although a circular shape might be more appropriate for this cartoon representation, the corresponding superposition is more difficult to visualize, whereas the asymptotic behavior of the survival probability is not changed. From Fig. 7.2, the only element that is not canceled by the image contributions is the small square region at the upper-right corner. From this sketch, it is clear that the asymptotic survival probability is proportional to x_0^2/Dt, in agreement with the asymptotic result of approximation (7.2.6). This image argument also shows that the the mean displacement of the surviving particles along the wedge bisector is $\langle r(t) \rangle \sim A(\Theta)\sqrt{Dt}$, with $A(\Theta)$ a slowly varying function of Θ. Thus particles that do survive to time t must retreat from the wedge apex as $\sqrt{Dt}$.

Another amusing feature of approximation (7.2.6) is that it also applies to wedge angles $\Theta > 2\pi$. For Θ that is an integer multiple of 2π, i.e., $\Theta = 2\pi k$, the diffusing particle can be interpreted as entering a distinct Riemann sheet each time the winding number changes by ± 1. The survival probability again decays as a power law in time, but with an exponent that approaches zero as $\Theta \to \infty$.

For $\Theta = 2\pi$, that is, an absorbing needle, the survival probability $S(t) \propto t^{-1/4}$. This limiting case can also be understood by the following crude but intuitive argument [Considine & Redner (1989)]. We mentally decompose the diffusive motion into components parallel (x) and perpendicular (y) to the needle. When $x < 0$, the transverse motion is one-dimensional diffusion in the presence of a trapping point at $y = 0$. This would yield a survival probability that decays as $t^{-1/2}$. However, when $x > 0$, the transverse motion is unconstrained and there is no trapping. The time intervals over which the particle is in the regions $x > 0$ and $x < 0$ each grow as $t^{1/2}$. Accordingly, a plot of the survival probability on a double logarithmic scale should consist of an alternating sequence of line segments of slope $-\frac{1}{2}$, when $x < 0$, and slope 0, when $x > 0$ [Fig. 7.3(a)]. Each line segment extends over the same abscissa range on a logarithmic scale. The overall effect of this line-segment composite is a straight line of slope $-1/4$, corresponding to the survival probability decaying as $t^{-1/4}$.

7.3. Three-Dimensional Cone

For the corresponding three-dimensional problem of diffusion within an absorbing cone, the diffusion equation can again be separated into a Bessel equation for the radial coordinate and a noninteger Legendre equation for the angular coordinate. The solution to the diffusion equation can then be written

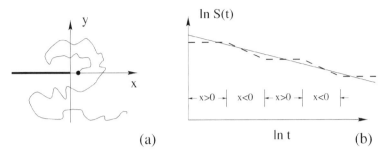

(a)

(b)

Fig. 7.3. (a) Diffusing particle in the presence of an absorbing needle (thick line) and
(b) illustration of the behavior for the survival probability. In (b) the segments labeled
$x > 0$ and $x < 0$, with alternating slopes of $-1/2$ and 0 (dashed), correspond to sections
of the trajectory in (a). The overall survival probability has a slope $-1/4$ (solid).

as an eigenfunction expansion, in which the angular factors are the non-
integral Legendre functions, $P_\nu(\cos\theta)$ [Jackson (1999)]. Here θ is the polar
angle, with $0 < \theta < \Theta/2$, and the discrete set of indices ν are determined by
the boundary conditions. For the asymptotic survival probability, we again
retain only the first term in this eigenfunction expansion, corresponding to
one half-wavelength of the Legendre function inside the cone. This is in the
same spirit as the wedge solution. This leading term has the form [Carslaw &
Jaeger (1959)]

$$P(r,\theta,t) \propto \frac{1}{4\pi Dt\sqrt{rr_0}}\, e^{-(r^2+r_0^2)/4Dt}\, (2\nu+1)I_{\nu+\frac{1}{2}}\left(\frac{rr_0}{2Dt}\right) P_\nu(\cos\theta).$$
(7.3.1)

We now estimate the survival probability by using the same heuristic approach
as that used in two dimensions. Namely, we cut off the radial integral at
$\sqrt{Dt}$, set the Gaussian factors to one for $r < \sqrt{Dt}$, and finally use the small-
argument expansion of the Bessel function. This gives

$$S(t) = \int d\Omega \int_0^\infty \frac{e^{-(r^2+r_0^2)/4Dt}}{4\pi Dt\sqrt{rr_0}}(2\nu+1)I_{\nu+\frac{1}{2}}\left(\frac{rr_0}{2Dt}\right) P_\nu(\cos\theta)\, r^2\, dr$$

$$\approx \int_0^{\sqrt{Dt}} \frac{1}{4\pi Dt\sqrt{rr_0}}\left(\frac{rr_0}{2Dt}\right)^{\nu+\frac{1}{2}} r^2\, dr$$

$$\propto \left(\frac{r_0}{\sqrt{Dt}}\right)^{\nu_0},$$
(7.3.2)

where the integral $\int d\Omega$ is over the solid angle of the cone.

To obtain numerical results, we require the lowest index of the non-
integral Legendre function that vanishes at the cone boundaries, (Fig. 7.4).

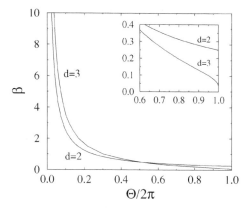

Fig. 7.4. Comparison of the survival probability exponent β for a diffusing particle in an absorbing wedge (two dimensions) and in an absorbing cone (three dimensions). The exponents have similar dependences on wedge angle, except as $\Theta \to 2\pi$, where $\beta_{3d} \to 0$ while β_{2d} remains finite. The inset shows details for $0.6 \leq \Theta/2\pi \leq 1.0$.

The asymptotic forms of this index are [Watson (1952)]

$$
\nu \approx
\begin{cases}
\dfrac{4.91}{\Theta} - \dfrac{1}{2}, & \Theta \to 0 \\[2ex]
\dfrac{1}{2 \ln \frac{4}{2\pi - \Theta}}, & \Theta \to 2\pi
\end{cases}
$$

in the limit of small and large opening angles, respectively. For small opening angle, the survival probability exponents for both the two-dimensional wedge and the three-dimensional cone both vanish as the inverse of the opening angle (Fig. 7.4). However, as $\Theta \to 2\pi$ (absorbing needle), the exponent of $S(t)$ in two dimensions remains finite, while the exponent in three dimensions goes to zero. This means that $S(t)$ decays slower than a power law in the presence of an absorbing needle in three dimensions. This is an especially intriguing manifestation of the difference between recurrence and transience of diffusion in two and three dimensions.

7.4. Conformal Transformations and Electrostatic Methods

The connection between first-passage properties and electrostatics takes on added significance in two dimensions, where conformal transformation methods provide an elegant and powerful method to solve electrostatic problems. In this section, we will exploit conformal transformations, together with the electrostatic formulation, to obtain time-integrated first-passage properties for diffusion within the wedge. This formulation also provides interesting results about *where* on the boundary a diffusing particle gets absorbed. For example, a diffusing particle is likely to be absorbed very close to the tip of an absorbing needle precisely because the electric field is relatively strong near the tip. Conversely, it is highly improbable that a diffusing particle will penetrate and get absorbed deep within a narrow crack.

It is also important to appreciate that conformal transformation methods can provide the first-passage characteristics in *any* geometry that is amenable to conformal analysis. Our treatment of the wedge geometry represents only a small subset of what can be accomplished by such methods.

7.4.1. Point Source and Line Sink

To set the stage for the wedge geometry, consider the first-passage probability for a diffusing particle in two dimensions to an absorbing infinite line. We already obtained this result in Section 3.2 by applying the image method. For a particle that starts at (x_0, y_0) and an absorbing line $y = 0$, this approach led to the Cauchy distribution for the probability that a diffusing particle is ultimately trapped at x:

$$\mathcal{E}(x; x_0, y_0) = \frac{1}{\pi} \frac{y_0}{(x - x_0)^2 + y_0^2}. \qquad (7.4.1)$$

As mentioned in Section 3.2, the x^{-2} large-x decay of this distribution (Fig. 7.5) implies that the mean position of the probability density trapped

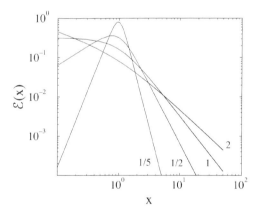

Fig. 7.5. Probability for a diffusing particle to eventually hit a point a distance x from the wedge apex for different wedge angles Θ. Shown are the cases $\Theta/\pi = 1/5$, $1/2$, 1 (absorbing line) and 2 (absorbing needle or half-line). The initial particle position is a unit distance from the wedge apex along its bisector. For large x, $\mathcal{E}(x)$ decays as $x^{-1-\pi/\Theta}$, whereas for small x, $\mathcal{E}(x)$ varies as $x^{\pi/\Theta-1}$.

along the absorbing line, $\langle x \rangle = \int_0^\infty x \mathcal{E}(x; x_0, y_0) \, dx$, is infinite. In fact, all moments $\langle x^n \rangle$ with $n \geq 1$ are infinite, whereas moments with $n < 1$ are finite.

This problem may also be solved more elegantly by the electrostatic formulation of Section 1.6. In this approach, the time-integrated concentration $\mathcal{C}(x, y) = \int_0^\infty c(x, y, t) \, dt$ obeys the Laplace equation

$$\nabla^2 \mathcal{C}(z) = -\frac{1}{2\pi D} \delta(z - z_0),$$

where $z = (x, y)$ is the complex coordinate and the factor $1/(2\pi D)$ ensures the correct normalization. Using the image method for two-dimensional electrostatics, we find that the complex potential is

$$\mathcal{C}(z) = \frac{1}{2\pi D} \ln \frac{z - z_0}{z - z_0^*} = \frac{1}{2\pi D} \ln \frac{x - x_0 + i(y - y_0)}{x - x_0 + i(y + y_0)}, \qquad (7.4.2)$$

where the asterisk denotes complex conjugation. Finally, the total flux that is absorbed at x coincides with the electric field at this point. This is

$$
\begin{aligned}
\mathcal{E}(x; x_0, y_0) &= -\frac{1}{2\pi D} \frac{\partial \mathcal{C}(x, y)}{\partial y} \bigg|_{y=0} \\
&= \frac{1}{2\pi} \left[\frac{1}{y_0 + i(x - x_0)} + \frac{1}{y_0 - i(x - x_0)} \right] \\
&= \frac{1}{\pi} \frac{y_0}{(x - x_0)^2 + y_0^2},
\end{aligned}
\qquad (7.4.3)
$$

which reproduces the result of the direct calculation, Eq. (7.4.1). This same approach may be extended in an obvious way to the semi-infinite slab geometry to reproduce Eq. (3.2.24).

There are several useful implications of this result. For example, the probability that a particle that starts at (x_0, y_0) eventually hits a point somewhere on the positive x axis is just the integral of the local hitting probability over this domain. This is given by

$$
\begin{aligned}
\mathcal{E}_+(x_0, y_0) &= \int_0^\infty \mathcal{E}(x; x_0, y_0) \, dx \\
&= \frac{y_0}{\pi} \int_0^\infty \frac{dx}{(x - x_0)^2 + y_0^2} \\
&\equiv 1 - \frac{1}{\pi} \tan^{-1}(y_0/x_0).
\end{aligned}
\qquad (7.4.4)
$$

More generally, Eq. (7.4.4) gives the probability for a particle, that starts at (x_0, y_0), to hit an arbitrary interval that subtends the angle $\theta_2 - \theta_1$ (Fig. 7.6). From Eq. (7.4.4), the probability of hitting the interval $[x_1, \infty]$ is $1 - (\theta_1/\pi)$,

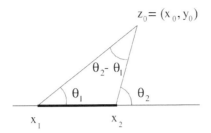

Fig. 7.6. Geometry for the probability of a hitting segment that subtends the angle $\theta_2 - \theta_1$. In making a correspondence to Eq. (7.4.4), the point x_1 is considered the origin, whereas for Eq. (7.4.5), x_1 and x_2 are arbitrary.

whereas the probability of hitting the interval $[x_2, \infty]$ is $1 - (\theta_2/\pi)$. Consequently, the probability of hitting the finite interval $[x_1, x_2]$ is

$$\mathcal{E}_{\text{interval}} = \frac{\theta_2 - \theta_1}{\pi}. \tag{7.4.5}$$

In fact, this result has an even simpler origin that arises from the fact that the conformal transformation

$$w = \frac{z - z_0}{z - z_0^*}$$

maps the upper half-plane onto the unit circle and maps the point z_0 to the origin of this circle. In this circularly symmetric geometry, it is obvious that for a random walk that starts at the center, the probability of hitting a finite arc length of the circle with subtending angle ψ is just $\psi/2\pi$. The effect of the inverse conformal transformation to the upper half-plane is merely to halve the subtending angle. This immediately reproduces the hitting probability of a finite interval given above.

7.4.2. General Wedge Angles

We now use a conformal transformation to extend our hitting probability results for the half-plane to the corresponding hitting probabilities in the wedge. Consider the transformation $w = f(z) = z^{\pi/\Theta}$ that maps the interior of the wedge of opening angle Θ to the upper half-plane. In complex coordinates, the electrostatic potential in the wedge is

$$\mathcal{C}(z) = \frac{1}{2\pi} \ln \frac{z^{\pi/\Theta} - z_0^{\pi/\Theta}}{z^{\pi/\Theta} - (z_0^*)^{\pi/\Theta}}. \tag{7.4.6}$$

From this expression, we can extract complete information about time-integrated first-passage properties in the wedge. For example, the probability of being absorbed at a distance x from the wedge apex, when a particle begins at a unit distance from the apex along the wedge bisector, is just the electric field at this point:

$$\mathcal{E}(x; x_0, y_0) = \frac{1}{\Theta} \frac{x^{\pi/\Theta - 1}}{1 + x^{2\pi/\Theta}}. \tag{7.4.7}$$

For large x, this hitting probability varies as $x^{-(1+\pi/\Theta)}$, as shown in Fig. 7.5. Thus for $\Theta < \pi$, the probability of being ultimately being trapped a distance x from the wedge apex decays more rapidly than x^2 and the mean distance from the wedge apex for the distribution of trapped particles is finite. Another interpretation of the case $\Theta < \pi$ is that if a diffusing particle starts somewhere in an absorbing crack the particle is very unlikely to penetrate any deeper. This is simply a consequence that the electric field is very small deep inside a narrow crack when the boundary is a grounded conductor. Conversely, for $\Theta > \pi$, the hitting probability diverges as $x^{\pi/\Theta - 1}$ as $x \to 0$. This means that a diffusing particle is most likely to hit near the tip of a thin needle. This is the source of the generic tip instability of diffusion-controlled growth processes in two dimensions. Another feature of the hitting probability for an obtuse wedge ($\Theta > \pi$) is that the hitting probability at large distances x from the tip decays more slowly than x^2. Consequently the mean distance from the wedge apex for the distribution of trapped particles is infinite.

Another very simple but useful application of conformal mapping is to the splitting probability. In the wedge geometry, we are interested in the probability that a particle ultimately gets trapped on the horizontal generator or on the inclined generator of the wedge as a function of its starting position. From the conformal transformation $w = z^{\pi/\Theta}$, the probability $\mathcal{E}_+^{\text{wedge}}$ that a diffusing particle, which starts at $z_0 = Ae^{i\lambda}$ in the wedge, hits the positive x axis is the same as the probability that a particle, which starts at $w_0 = A^{\pi/\Theta} e^{i\lambda\pi/\Theta}$ in the upper half-plane, hits the positive x axis. This equivalence immediately gives

$$\mathcal{E}_+^{\text{wedge}} = 1 - \lambda/\Theta. \tag{7.4.8}$$

As is shown in the next chapter, this probability directly translates to useful properties about eventual reaction partners for interacting diffusing particles on the line.

7.5. First-Passage Times

We now investigate the time until a diffusing particle hits the boundary of an absorbing wedge or cone. For the infinite wedge, that is, the radial coordinate is unbounded, this exit time may be finite or infinite, depending on the wedge opening angle. We can understand this transition in a deep way by studying the exit time in a pie wedge, in which r is restricted to $(0, R)$ with an absorbing boundary condition at $r = R$. In this system, the exit time is always finite. However, for wedge angles $\Theta < \pi/2$ the exit time remains finite as $R \to \infty$, whereas for $\Theta \geq \pi/2$, the exit time diverges as $R \to \infty$. This is a harbinger of the transition in the exit time for the infinite wedge.

It is also worth mentioning that the adjoint equation for the exit time in the pie wedge is identical in form to the equation for the velocity field of viscous fluid flow in an infinite pipe with the same pie-wedge cross section [Moffat & Duffy (1979)]. The transition in the behavior of the exit time translates to a corresponding pathology for the flow problem. Namely, for $\Theta < \pi/2$ the flow velocity $r = 0$ remains finite as $R \to \infty$, whereas for $\Theta > \pi/2$, the velocity near $r = 0$ diverges as the distant boundary $R \to \infty$! This type of anomaly is generally described in terms of intermediate asymptotics of the second kind; a variety of beautiful examples of this general phenomenon are given in Barenblatt (1996).

7.5.1. Infinite Two-Dimensional Wedge

Let us begin by determining the mean time for a diffusing particle to hit the sides of an infinite two-dimensional wedge of opening angle Θ. It is convenient to orient the wedge bisector along the $\theta = 0$ axis, with the two wedge generators at $\theta = \pm\Theta/2$. As should now be familiar, the (unrestricted) mean exit time $t(\vec{r})$ to hit anywhere on the boundary, for a particle that starts at $\vec{r}$, obeys the adjoint (Poisson) equation $D\nabla^2 t(\vec{r}) = -1$ [Eq. (1.6.21)]. This is subject to the boundary conditions $t = 0$ for $\theta = \pm\Theta/2$, corresponding to an exit time of zero for a particle that starts at the boundary. In polar coordinates, this adjoint equation is

$$D\left(\frac{\partial^2 t}{\partial r^2} + \frac{1}{r}\frac{\partial t}{\partial r} + \frac{1}{r^2}\frac{\partial^2 t}{\partial \theta^2}\right) = -1. \tag{7.5.1}$$

By dimensional analysis, $t(r, \theta)$ must have the units of r^2/D. This suggests that the exit time can be written in the scaling form,

$$t(r, \theta) = \frac{r^2}{D}f(\theta), \tag{7.5.2}$$

with f a function of θ only. To determine $f(\theta)$ we substitute this ansatz into Eq. (7.5.1) and find that f satisfies $4f + f'' = -1$; here the prime means differentiation with respect to θ. The general solution of this equation is simply

$f(\theta) = A\cos(2\theta) + B\sin(2\theta) - \frac{1}{4}$ and imposing the boundary conditions $t(\pm\Theta/2) = 0$ then gives

$$t(r, \theta) = \frac{r^2}{4D}\left[\frac{\cos(2\theta)}{\cos\Theta} - 1\right]. \qquad (7.5.3)$$

As expected both physically and on dimensional grounds, this exit time scales as r^2/D. For fixed r, this time is obviously maximal for a particle starting along the wedge bisector and smoothly goes to zero as the initial location approaches the wedge.

More intriguingly, $t(r, \theta)$ diverges as $\Theta \to \pi/2$! This feature also follows from that fact that the survival probability $S(t)$ decays as $t^{-\pi/2\Theta}$ [Eq. (7.2.7)]. As a result, the mean exit time, $t = \int_0^\infty S(t')\,dt'$, is divergent for $\Theta \geq \pi/2$. To understand the meaning of this divergence, we are naturally led to consider the exit time in a pie wedge with finite outer radius R. The properties of this finite-system exit time help us understand the exit time in the infinite wedge.

7.5.2. Pie Wedge in Two Dimensions

For a pie wedge with opening angle $\Theta < \pi/2$ and outer radius R, the dependence of the exit time on R must be compatible with a finite limit for the exit time when $R \to \infty$. On the other hand, for $\Theta \geq \pi/2$, the R dependence must be compatible with a divergent exit time as $R \to \infty$. The formal way to understand this phenomenon is to solve the Poisson equation in the pie wedge. That is, we solve $D\nabla^2 t(r, \theta) = -1$ subject to $t = 0$ on the boundaries. The general solution to this equation is

$$t(r, \theta) = \frac{r^2}{4D}\left(\frac{\cos 2\theta}{\cos\Theta} - 1\right) + \sum_{n=0}^{\infty} A_n r^{\lambda_n}\cos(\lambda_n\theta), \qquad (7.5.4)$$

where the infinite series represents the general solution to the homogeneous problem. Here $\lambda_n\Theta/2 = (2n+1)\pi/2$ for $n = 0, 1, 2, \ldots$, to satisfy the boundary condition on the sides of the wedge. The boundary condition $t = 0$ for $r = R$ then leads to

$$A_n = \frac{(-1)^{n+1}4R^{2-\lambda_n}}{D\Theta\lambda_n(\lambda_n^2 - 4)}.$$

The leading behavior of the solution in Eq. (7.5.4) is ultimately determined by the lowest angular eigenvalue λ_0. If $\lambda_0 > 2$ then the particular solution dominates and we recover the infinite wedge result, Eq. (7.5.3). This range of λ corresponds to a wedge angle $\Theta < \pi/2$. However, for $\lambda_0 \leq 2$, corresponding to wedge angle $\Theta \geq \pi/2$, then the eigenfunction expansion dominates the solution for large R. In this case, the leading behavior of the solution for small r is

$$t(r, \theta) \sim r^{\pi/\Theta} R^{2-\pi/\Theta}. \tag{7.5.5}$$

In the context of the exit-time problem, this means that a particle that starts near the wedge apex takes an infinite time to exit the wedge as $R \to \infty$. In the corresponding viscous-fluid-flow problem in an infinite pipe with a pie-wedge cross section of opening angle Θ, this translates to a pathology in the flow field near the wedge apex [Moffat & Duffy (1979)]. For opening angle $\Theta < \pi/2$, the velocity near the apex converges as the radius R of outer boundary diverges, whereas for $\Theta \geq \pi/2$, the velocity near the apex diverges as $R \to \infty$!

These anomalous properties for the pie wedge with $\Theta \geq \pi/2$ can also be obtained by a simple physical argument that relies on the power-law decay of the survival probability. In an infinite wedge with $\Theta < \pi/2$, the survival probability decays faster than t^{-1}, so that the mean exit time $t = \int_0^\infty S(t') \, dt'$ converges. In this case, the role of the circumferential boundary is irrelevant. On the other hand, for $\Theta > \pi/2$, $S(t)$ decays more slowly than t^{-1} and the integral for an infinite system is formally divergent. To compute the exit time in a finite pie wedge, however, we should include a finite-size exponential cutoff in $S(t)$ for $t > R^2/D$. Thus the exit time should scale as

$$t(r, \theta; \Theta > \pi/2) = \int_0^\infty S(t') \, dt'$$
$$\sim \int^{R^2/D} (t')^{-\pi/2\Theta} \, dt'$$
$$\propto R^{2-\pi/\Theta}. \tag{7.5.6}$$

Here, we have replaced the exponential cutoff in $S(t)$ with a sharp cutoff at $t = R^2/D$; this construction still gives the correct asymptotic behavior. Now, by dimensional analysis, t must be a quadratic function of length. Because the only other length in the problem is the starting location r, we infer that $t \propto r^{\pi/\Theta} R^{2-\pi/\Theta}$, in agreement with approximation (7.5.5).

With this crude approach, we can also determine the behavior of the exit time in the borderline case $\Theta = \pi/2$. Here the integral in Eq. (7.5.6) diverges logarithmically, and it is necessary to include both a short-time and a long-time cutoff to get a dimensionally correct result. Clearly, the lower cutoff should be of the order of r^2/D, where r is the starting radial coordinate of the particle. Thus

$$t(r, \theta; \Theta = \pi/2) = \int_0^\infty S(t') \, dt'$$

$$\sim \int_{r^2/D}^{R^2/D} (t')^{-1} \, dt'$$

$$\propto \ln(R/r). \tag{7.5.7}$$

Then, by dimensional analysis, we conclude that for the case $\Theta = \pi/2$ the mean exit time varies as $t \propto r^2 \ln(R/r)$.

7.5.3. Infinite Three-Dimensional Cone

We now study the exit time from a three-dimensional cone. The dimensional argument of Eq. (7.5.2) applies for any spatial dimension and reduces the system to an effective two-dimensional angular problem that is trivially soluble. In the cone, the exit time obeys the Poisson equation in spherical coordinates,

$$D\left[\frac{\partial^2 t}{\partial r^2} + \frac{2}{r}\frac{\partial t}{\partial r} + \frac{1}{r^2 \sin\theta}\frac{\partial}{\partial \theta}\left(\sin\theta \frac{\partial t}{\partial \theta}\right)\right] = -1, \tag{7.5.8}$$

again subject to the absorbing boundary conditions that $t = 0$ for $\theta = \Theta/2$. With the ansatz of Eq. (7.5.2), the angular function now obeys

$$6f + \frac{1}{\sin\theta}\frac{\partial}{\partial \theta}\left(\sin\theta \frac{\partial f}{\partial \theta}\right) = -1.$$

This is just the inhomogeneous Legendre equation with eigenvalue $\ell = 2$. The general solution is simply $f = -\frac{1}{6} + A(3\cos^2\theta - 1)$, that is, a linear combination of a particular solution and the Legendre polynomial of order $\ell = 2$. Imposing the boundary conditions fixes $f(\theta)$, from which the exit time is

$$t(r, \theta) = \frac{r^2}{6D}\left(\frac{3\cos^2\theta - 1}{3\cos^2\Theta/2 - 1} - 1\right). \tag{7.5.9}$$

Qualitatively, the dependence of t on starting location is similar to that for the wedge in two dimensions. One curious feature is the cone angle that marks the transition between a finite and an infinite exit time. This is given by the condition $3\cos^2\Theta/2 - 1 = 0$, from which $\Theta \approx 109.47°$. The exit time is finite for cones narrower than this angle and is divergent for wider cones.

7.5.4. Conditional Exit Time for the Infinite Wedge

One last problem that is easy to treat with the Laplacian formalism of Section 1.6 is the *conditional* exit time, namely, the time for a particle to exit on either the wedge generator $\theta = +\Theta/2$ or on $\theta = -\Theta/2$. For convenience, we define here the wedge bisector as polar angle $\theta = 0$. This problem has a natural meaning only in two dimensions. Define $t_{\pm}(r, \theta)$ as the mean time for a particle that starts at (r, θ) to hit the generator at $\pm\Theta/2$. From the Laplacian formalism, these two conditional exit times satisfy

$$D\nabla^2[\mathcal{E}_{\pm}(\theta)t(r, \theta)] = -\mathcal{E}_{\pm}(\theta),$$

where, from Eq. (7.4.8), the exit probabilities are $\mathcal{E}_{\pm}(\theta) = \frac{1}{2} \pm (\theta/\Theta)$. The products $\mathcal{E}_{\pm}(\theta)t_{\pm}(r, \theta)$ are also zero on both boundaries; on one, $t_{\pm}$ itself vanishes, whereas on the other, $\mathcal{E}_{\pm}$ vanishes. Following the same steps as those leading to Eq. (7.5.3), we find that the conditional exit times are

$$t_{\pm}(r, \theta) = \frac{r^2}{4D}\left[\frac{\cos(2\theta)}{2\mathcal{E}_{\pm}(\theta)\cos\Theta} \pm \frac{\sin(2\theta)}{2\mathcal{E}_{\pm}(\theta)\sin\Theta} - 1\right]. \quad (7.5.10)$$

7.6. Extension to Time-Dependent First-Passage Properties

Although the electrostatic formulation ostensibly gives only time-integrated first-passage properties, we can adapt it in an appealing way to also give time-dependent features. By this trick, we can obtain the asymptotic time dependence of the diffusion equation merely by solving the time-independent Laplace equation and applying basic physical intuition about the relation between Laplacian and diffusive solutions. The wedge geometry provides a nice illustration of this approach.

This adaptation of the time-integrated approach is based on the following reinterpretation of the equivalence between electrostatics and diffusion. We have already learned that an electrostatic system with a point charge and specified boundary conditions is identical to a diffusive system in the same geometry and boundary conditions, in which a continuous source of particles is fed in at the location of the charge. Suppose now that the particle source is "turned on" at $t = 0$. Then, in a near zone that extends out to a distance of the order of $\sqrt{Dt}$ from the source, the concentration has sufficient time to reach its steady-state value. In this zone, the diffusive solution converges to the Laplacian solution. Outside this zone, however, the concentration is very close to zero. More relevant to our discussion is that this almost-Laplacian solution provides the time integral of the survival probability up to time t. We can then deduce the survival probability by differentiating the concentration that is due to this finite-duration source.

As a concrete example, suppose that there is a constant source of diffusing particles at z_0 within the absorbing wedge that is turned on at $t = 0$. Then within the region where the concentration has had time to reach the steady state, $|z_0| < |z| < \sqrt{Dt}$, the density profile is approximately equal to the Laplacian solution, $C(z) \sim |z|^{-\pi/\Theta}$. We can neglect the angular dependence of $C(z)$ in this zone, as it is immaterial for the survival probability. Conversely, for $|z| > \sqrt{Dt}$, the particle concentration is vanishingly small because a particle is unlikely to diffuse such a large distance. From the general discussion in Section 1.6, the near-zone density profile is just the same as the time integral of the diffusive concentration. Thus, by using the equivalence between the spatial integral of this near-zone concentration in the wedge and the time integral of the survival probability, we have

$$\int_0^t S(t')\,dt' \approx \int_0^{\sqrt{Dt}} r^{-\pi/\Theta}\, r\, dr \sim t^{1-\pi/2\Theta}. \tag{7.6.1}$$

Because the total particle density injected into the system is t, the survival probability in the wedge is roughly $\int_0^t S(t')\,dt'/t \sim t^{1-\pi/2\Theta}/t$. This gives $S(t) \sim t^{-\pi/2\Theta}$, in agreement with the known result.

This approach should be viewed as a version of the quasi-static method that hinges on the slow propagation of diffusion. Within the diffusive zone, the concentration profile has time to achieve the steady-state Laplacian form. Beyond the diffusive zone, the concentration is essentially zero. The time dependence induced by the diffusive zone growing as $\sqrt{Dt}$ provides the key to obtaining time-dependent first-passage properties from the static Laplacian solution.

8

Applications to Simple Reactions

8.1. Reactions as First-Passage Processes

In this last chapter, we investigate simple particle reactions whose kinetics can be understood in terms of first-passage phenomena. These are typically diffusion-controlled reactions, in which diffusing particles are immediately converted to a product whenever a pair of them meets. The term diffusion controlled refers to the fact that the reaction itself is fast and the overall kinetics is controlled by the transport mechanism that brings reactive pairs together. Because the reaction occurs when particles *first* meet, first-passage processes provide a useful perspective for understanding the kinetics.

We begin by treating the trapping reaction, in which diffusing particles are permanently captured whenever they meet immobile trap particles. For a finite density of randomly distributed static traps, the asymptotic survival probability $S(t)$ is controlled by rare, but large trap-free regions. We obtain this survival probability exactly in one dimension and by a Lifshitz tail argument in higher dimensions that focuses on these rare configurations [Lifshitz, Gredeskul, & Pastur (1988)]. At long times, we find that $S(t) \sim \exp(-At^{d/d+2})$, where A is a constant and d is the spatial dimension. This peculiar form for the survival probability was the focus of considerable theoretical effort that ultimately elucidated the role of extreme fluctuations on asymptotic behavior [see, e.g., Rosenstock (1969), Balagurov & Vaks (1974), Donsker & Varadhan (1975, 1979), Bixon & Zwanzig (1981), Grassberger & Procaccia (1982a), Kayser & Hubbard (1983), Havlin et al. (1984), and Agmon & Glasser (1986)].

We next discuss diffusion-controlled reactions in one dimension. As a starting point, we treat the reactions of three diffusing particles on the line, where complete results can be obtained by a simple geometric mapping of the three-particle system onto diffusion of a single particle in a wedge with absorbing boundaries [ben-Avraham (1988), Fisher & Gelfand (1988), and Redner & Krapivsky (1999)]. This provides a simple route toward understanding the kinetics of the diffusion-controlled reactions of capture, coalescence, annihilation, and aggregation in one dimension.

In the capture reaction, a diffusing "prey" p dies whenever it meets any one of a group of N diffusing "predators" P, as represented by $p + P \to P$. In the interesting situation of one dimension and predators all to one side of the prey, the prey survival probability decays as $t^{-\beta_N}$, with β_N increasing as $\ln N$ [Bramson & Griffeath (1991), Kesten (1992), and Krapivsky & Redner (1996b, 1999)]. In coalescence, $A + A \to A$, two reactants merge on meeting, with the reaction product identical in character to the incident particles. Here the particle density asymptotically decays as $t^{-1/2}$; we can obtain this result easily by mapping the reaction onto the survival probability of a one-dimensional random walk with an absorber at the origin [Doering & ben-Avraham (1988), ben-Avraham, Burschka, & Doering (1990a), and ben-Avraham (1998); see also the collection of articles in Privman (1997)]. A similar $t^{-1/2}$ decay arises in the annihilation reaction $A + A \to 0$ [Lushnikov (1987), and Torney & McConnell (1983a, 1983b)]. This kinetics can be obtained easily by constructing a correspondence between the particle density in the reaction and the density of domain walls in the Ising–Glauber model [Rácz (1985) and Amar & Family (1989)].

In aggregation, two reactants stick together irreversibly in a mass-conserving reaction [Chandrasekhar (1943), Drake (1972), and Friedlander (1977)], as represented by $A_i + A_j \to A_k$, where A_i now denotes an aggregate with (conserved) mass i. Here the density of clusters of mass k, $c_k(t)$, asymptotically varies as $(k/t^{3/2})e^{-k^2/t}$ [Bramson & Griffeath (1980a, 1980b), Spouge (1988a, 1988b), Takayasu, Nishikawa, & Tasaki (1988), and Thompson (1989)]. Amazingly, this is just the first-passage probability to the origin for a diffusing particle in one dimension! As we shall see, this coincidence is no accident, but arises from an unexpected but close connection between aggregation and diffusion on the semi-infinite interval with an absorbing boundary at the origin.

Finally, we discuss some basic aspects of ballistic annihilation in one dimension, in which particles move at constant velocity and annihilation occurs whenever two particles meet [Elskens & Frisch (1985), and Krapivsky, Redner, & Leyvraz (1995)]. By judicious constructions, the evolution of this deterministic system in various situations can be described by an underlying first-passage process. When there are only two distinct velocities, we can map the kinetics onto the familiar survival probability of a one-dimensional random walk with an absorbing boundary at the origin. We also discuss the generalization to three velocities, where the unusual kinetics can again be understood in terms of a suitably defined first-passage process.

8.2. Kinetics of the Trapping Reaction

The physical process in the trapping reaction is illustrated in Fig. 8.1. Static traps are randomly distributed in space. Whenever a diffusing particle hits a trap it is immediately and permanently trapped. This process underlies a variety of basic chemical processes such as exiton trapping, fluorescence quenching, and spin relaxation processes [see, e.g., Zumofen & Blumen (1981), Movaghar, Sayer, & Würtz (1982), and Seiferheld, Bässler, & Movaghar (1983), and references cited therein]. We are interested in the survival probability $S(t)$, namely, the probability that particle does not hit any trap until time t.

In one dimension, we can obtain the exact survival probability by direct solution of the diffusion equation and then averaging over all configurations of traps and all particle trajectories. This solution relies on the fact that the only topology of trap-free regions is just the one-dimensional interval, so that the average over all these regions is simple to perform. In higher dimensions, it is not possible to perform this average directly. As a much simpler alternative, we will apply a Lifshitz argument to obtain the asymptotic behavior of the survival probability. Part of the reason for presenting this latter approach is its simplicity and wide range of applicability. One sobering aspect, however, is that the asymptotic survival probability does not emerge until the density has decayed to a vanishingly small value. Such a pathology typically arises when a system is controlled by rare events. This serves as an important reality check for the practical relevance of the Lifshitz argument.

Fig. 8.1. A configuration of traps (filled circles) and the trajectory of a diffusing particle. Also shown is the trap-free circle of radius a that is centered about the initial particle position. The probability that the particle remains in this circle is a lower bound for the exact particle survival probability in this configuration of traps.

8.2.1. Exact Solution in One Dimension

A diffusing particle in the presence of traps in one dimension "sees" only the absorbing interval defined by the nearest surrounding traps. We can therefore adapt the solution for the concentration inside an absorbing interval $[0, L]$ (Chap. 2) to determine the survival probability. For a particle initially at $x = x_0$, the concentration at time $t > 0$ is obtained from Eq. (2.2.3), with the coefficients A_n determined by matching the eigenfunction expansion to the initial condition. For a particle initially at x_0, A_n is thus given by the Fourier series inversion

$$c(x, t = 0) = \delta(x - x_0) = \sum_{n=1}^{\infty} A_n \sin\left(\frac{n\pi x}{L}\right),$$

which gives

$$A_n = \frac{2}{L} \sin\left(\frac{n\pi x_0}{L}\right).$$

Therefore the concentration within the interval is

$$c_L(x, t \mid x_0) = \frac{2}{L} \sum_{n=1}^{\infty} \sin\left(\frac{n\pi x}{L}\right) \sin\left(\frac{n\pi x_0}{L}\right) e^{-\left(\frac{n\pi}{L}\right)^2 Dt}. \quad (8.2.1)$$

For a fixed-length interval, we compute the survival probability by averaging over all initial particle positions and also integrating over all x. This gives

$$\overline{S_L(t)} = \frac{1}{L} \int_0^L \int_0^L c_L(x, t \mid x_0) \, dx \, dx_0$$

$$= \frac{8}{\pi^2} \sum_{m=0}^{\infty} \frac{1}{(2m+1)^2} e^{-\frac{(2m+1)^2\pi^2}{L^2} Dt}. \quad (8.2.2)$$

Next, we obtain the configuration-averaged survival probability by averaging this expression over the distribution of lengths of trap-free intervals. The simplest and most natural situation is a random distribution of traps at density ρ, for which the interval-length distribution is $P(L) = \rho e^{-\rho L}$. This gives the formal solution for the average survival probability

$$\langle S(t) \rangle \equiv \langle \overline{S_L(t)} \rangle$$

$$= \frac{8\rho}{\pi^2} \sum_{m=0}^{\infty} \frac{1}{(2m+1)^2} \int_0^{\infty} e^{-\frac{(2m+1)^2\pi^2}{L^2} Dt} e^{-\rho L} \, dL. \quad (8.2.3)$$

As indicated in Fig. 8.2, this integral has very different short- and long-time behaviors. In the former case, intervals of all lengths contribute to the survival

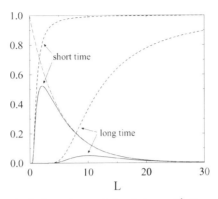

Fig. 8.2. (a) Schematic L dependence of the factors $e^{-\rho L}$ (long-dashed curve) and e^{-Dt/L^2} (short-dashed curves) in Eq. (8.2.3) for short time (steeply rising) and long time (gradually rising). Also shown are the corresponding integrands at short and long times (solid curves).

probability, whereas at long times optimal-length intervals give the main contribution to the survival probability. This latter behavior is not visible until the survival probability has decayed to a vanishingly small and experimentally unattainable value. In fact, the best strategy to observe the long-time behavior (by simulation) is to consider a system with a high concentration of traps [Havlin et al. (1984)]!

8.2.1.1. Long-Time Behavior

In the long-time limit, clearly the first term in the series for $\langle S(t)\rangle$ in Eq. (8.2.3) eventually dominates. If we retain only this term, it is relatively easy to determine the asymptotic behavior of the integral in Eq. (8.2.3). As a function of L, the first exponential factor in this equation rapidly increases 1 to as $L \to \infty$, whereas the second exponential factor decays rapidly with L. Thus the integrand has a peak as a function of L that becomes progressively sharper as $t \to \infty$ (although this is not evident to the eye in Fig. 8.2). We may therefore determine the asymptotic behavior of $\langle S(t)\rangle$ by the Laplace method [Bender & Orszag (1978)].

To apply this method, we first rewrite Eq. (8.2.3) as $\langle S(t)\rangle \sim \int_0^\infty e^{f(L)}\,dL$, and then we fix the location of the maximum by defining the dimensionless length $\ell \equiv L/L^*$ to transform the integral to

$$\langle S(t)\rangle = \frac{8\rho L^*}{\pi^2} \int_0^\infty \exp\left\{-(\rho^2 Dt)^{1/3}\left[(\pi^2/2)^{2/3}\frac{1}{\ell^2} + (2\pi^2)^{1/3}\ell\right]\right\} d\ell$$

$$\equiv \frac{8\rho L^*}{\pi^2} \int_0^\infty \exp\left[-(\rho^2 Dt)^{1/3}\,g(\ell)\right] d\ell. \qquad (8.2.4)$$

The integrand now has an increasingly sharp maximum at a fixed location as $t \to \infty$. We therefore expand $g(\ell)$ to second order about its maximum and perform the resulting Gaussian integral to obtain the leading behavior of $\langle S(t) \rangle$. From the condition that $g'(\ell^*) = 0$, we find $\ell^* = 1$, $g(\ell^*) = 3(\pi/2)^{2/3}$ and $g''(\ell^*) = -3 \times (2\pi^2)^{1/3}$. Therefore

$$
\begin{aligned}
\langle S(t) \rangle &= \frac{8\rho L^*}{\pi^2} \int_0^\infty \exp\left[-(\rho^2 D t)^{1/3} g(\ell)\right] \\
&\sim \frac{8\rho l^*}{\pi^2} \int_0^\infty \exp\left\{-(\rho^2 D t)^{1/3} \left[g(\ell^*) + \frac{1}{2}(\ell - \ell^*)^2 g''(\ell^*)\right]\right\} \\
&\sim \frac{8\rho L^*}{\pi^2} \sqrt{\frac{2\pi}{(\rho^2 D t)^{1/3}|g''(\ell^*)|}} \exp\left[-(\rho^2 D t)^{1/3} g(\ell*)\right] \\
&= \frac{8 \times 2^{2/3}}{3^{1/2}\pi^{7/6}} (\rho^2 D t)^{-1/6} \exp\left(-3(\pi^2\rho^2 D t/4)^{1/3}\right).
\end{aligned}
\tag{8.2.5}
$$

The basic feature of this result is the relatively slow $e^{-t^{1/3}}$ asymptotic decay of $\langle S(t) \rangle$ compared with the exponential decay for the survival probability in a fixed-length interval. This slower decay stems from the contribution of optimal intervals whose length ℓ^* grows as $t^{1/3}$. Although such large intervals are rare, their contribution to the survival probability is asymptotically dominant. In Subsection 8.2.2, we shall see how these extreme intervals are the basis for the Lifshitz tail argument that provides the asymptotic decay of $\langle S(t) \rangle$ in arbitrary spatial dimension. Finally, if one is interested in only the correct controlling factor in the asymptotic survival probability, one can merely evaluate $f(L)$ at its maximum of $L^*(t) = (2\pi^2 D t/\rho)^{1/3}$ and then estimate $\langle S(t) \rangle$ as $e^{f(L^*)} \sim e^{-\text{const.} \times (\rho^2 D t)^{1/3}}$.

8.2.1.2. Short-Time Behavior

It is instructive to study the short-time behavior of $\langle S(t) \rangle$, both because the time dependence is interesting and because this limit indicates that the crossover to the asymptotic behavior for $\langle S(t) \rangle$ is very slow. In fact, the asymptotic decay does not arise until the density has decayed to an extremely small value. Thus although there is considerable theoretical appeal in understanding the long-time decay of the trapping reaction, its practical implications are limited.

There are many ways to estimate the short-time behavior. One crude approach is to note that, at early times, the factor e^{-Dt/L^2} reaches 1 as a function of L (at $L \approx \sqrt{Dt}$) before there is an appreciable decay in the factor $e^{-\rho L}$ (at $L \approx 1/\rho$). Thus, to estimate $\langle S(t) \rangle$, we may cut off the lower limit of the integral at $\sqrt{Dt}$ and replace the factor e^{-Dt/L^2} with 1 (see Fig. 8.2). Using this approximation, the time dependence of the survival probability is

$$\langle S(t) \rangle \approx \int_{\sqrt{Dt}}^{\infty} e^{-\rho L} \, dL$$
$$\approx e^{-\text{const} \times \rho \sqrt{Dt}}. \qquad (8.2.6)$$

This short-time behavior extends until $t \sim 1/(D\rho^2)$, which translates to the diffusion distance being of the order of the mean separation between traps.

A more rigorous approach is to use the fact we should keep all the series terms in Eq. (8.2.3). As shown in Weiss' book (1994), we can evaluate this series easily by defining $\epsilon = \pi^2 Dt/L^2$ and noting that $dS/d\epsilon$ has the form

$$\left\langle \frac{\partial S(t)}{\partial \epsilon} \right\rangle = \frac{8\rho}{\pi^2} \int_0^{\infty} \left[\sum_{m=0}^{\infty} e^{-(2m+1)^2 \epsilon} \right] e^{-\rho L} \, dL.$$

We can estimate the sum by replacing it with an integral, and then we can easily perform the average over L, with the result

$$\langle S(t) \rangle \sim e^{-\rho \sqrt{8Dt/\pi}}. \qquad (8.2.7)$$

Now we may roughly estimate the crossover between the short- and the long-time limits by equating the exponents in Eq. (8.2.5) and in approximation (8.2.7). This gives the numerical estimate $\rho^2 Dt \approx 269$ for the crossover time. Substituting this into the above expression for the short-time survival probability shows that $\langle S(t) \rangle$ must decay to approximately 4×10^{-12} before the long-time behavior represents the main contribution to the survival probability. In fact, because of the similarity of the short- and the long-time functional forms, the crossover is very gradual, and one must wait much longer still before the asymptotic behavior is clearly visible. Although this discussion needs to be interpreted cautiously because of the neglect of the power-law factors in the full expressions for the survival probability, the basic result is that the asymptotic survival probability is of marginal experimental utility. In spite of this deficiency, the question about the asymptotic regime is of fundamental importance, and it helps clarify the role of exceptional configurations in determining the asymptotic survival probability.

8.2.2. Lifshitz Argument for General Spatial Dimension

The Lifshitz approach has emerged as an extremely useful tool to determine asymptotic properties in many disordered and time-varying systems [Lifshitz et al. (1988)]. If we are interested *only* in asymptotics, then it is often the case that a relatively small number of extreme configurations provide the main contribution to the asymptotics. The appeal of the Lifshitz approach is that these extreme configurations are often easy to identify and the problem is typically straightforward to solve on these configurations.

In the context of the trapping reaction, we first identify the large trap-free regions that give the asymptotically dominant contribution to the survival probability. Although such regions are rare, a particle in such a region has an anomalously long lifetime. By optimizing the survival probability with respect to these two competing attributes, we find that the linear dimension of these extreme regions grows as $(Dt)^{1/(d+2)}$ for isotropic diffusion, from which we can easily find the asymptotic survival probability.

8.2.2.1. *Isotropic Diffusion*

It is convenient to consider with a lattice system in which each site is occupied by a trap with probability p and in which a single particle performs a random walk on free sites. The average survival probability $\langle S(t) \rangle$ is obtained by determining the fraction of random-walk trajectories that do not hit any trap up to time t. This fraction must be averaged over all random-walk trajectories *and* over all trap configurations.

An important aspect of these averages is that they may be performed in either order, and it is more convenient to first perform the latter. For a given trajectory, each visited site must not be a trap for the particle to survive, whereas the state of the unvisited sites can be arbitrary. Consequently, a walk that has visited s *distinct* sites survives with probability q^s, with $q = (1 - p)$. Then the average survival probability is

$$\langle S(t) \rangle = z^{-N} \sum_s C(s, t) q^s \equiv \langle q^s \rangle, \qquad (8.2.8)$$

where $C(s, t)$ is the number of random walks that visit s distinct sites at time t and z is the lattice coordination number. Note that the survival probability is an exponential-order moment of the distribution of visited sites. It is this exponential character that leads to the anomalous time dependence of the survival probability.

Clearly the survival probability for each configuration of traps is bounded from below by the contribution that arises from the *largest* spherical trap-free region centered about the initial particle position (Fig. 8.1). This replacement of the configurational average with a simpler set of extremal configurations is the essence of the Lifshitz tail argument. The probability for such a region to occur is simply q^V, where $V = \Omega_d r^d$ is the number of sites in this d-dimensional sphere of radius r. We determine the probability for a particle to remain inside this sphere by solving the diffusion equation with an

absorbing boundary at the sphere surface. This is a standard and readily soluble problem, and the solution is merely outlined.

Because the system is spherically symmetric, we separate the variables as
$c(r, t) = g(r)f(t)$ and then introduce $h(r) = r^\nu g(r)$, with $\nu = [(d/2) - 1]$ to
transform the radial part of the diffusion equation into the Bessel differential
equation

$$h''(x) + \frac{1}{x}h'(x) + \left[1 - \frac{1}{x^2}\left(\frac{d}{2} - 1\right)^2\right]h(x) = 0,$$

where $x = r\sqrt{k/D}$, the prime denotes differentiation with respect to x, and the
boundary condition is $h(a\sqrt{k/D}) = 0$, where a is the radius of the trap-free
sphere. Correspondingly $f(t)$ satisfies $\dot{f} = -kf$. In the long-time limit, the
dominant contribution to the concentration arises from the slowest decaying
mode in which the first zero of the Bessel function $J_{d/2}(r\sqrt{k/D})$ occurs at
the boundary of the sphere. Thus the survival probability within a sphere of
radius a asymptotically decays as

$$S(t) \propto \exp\left(-\frac{\mu_d^2 Dt}{a^2}\right),$$

where μ_d is the location of the first zero of the Bessel function in d dimensions.

To obtain the configuration-averaged survival probability, we average this
survival probability for a fixed-size sphere over the radius distribution of trap-
free spheres. This gives the lower bound for the average survival probability:

$$\langle S(t)\rangle_{\text{LB}} \propto \int_0^\infty \exp\left(-\frac{\mu_d^2 Dt}{r^2} + \Omega_d r^d \ln q\right) r^{d-1}\, dr. \qquad (8.2.9)$$

This integrand becomes sharply peaked as $t \to \infty$, and we can again estimate
the integral by the Laplace method. As in one dimension, we rescale variables
to fix the location of the maximum. Writing the integrand in relation (8.2.9) as
$\exp[-F(r)]$ and differentiating with respect to r, we find that the maximum
of F occurs at

$$r^* = \left(-\frac{2\mu_d^2 Dt}{\Omega_d d \ln q}\right)^{1/(d+2)}.$$

This defines the radius of the trap-free region that gives the dominant contribution to the survival probability at time t. We now rewrite F in terms of the
scaled variable $u = r/r^*$ to give

$$F(u) = -\left(\mu_d^2 Dt\right)^{d/(d+2)}(-\Omega_d \ln q)^{2/(d+2)}\left[\left(\frac{d}{2u}\right)^{2/(d+2)} + \left(\frac{2u}{d}\right)^{d/(d+2)}\right].$$

We now evaluate the integral by expanding $F(u)$ to second order in u and performing the resulting Gaussian. This gives, for the controlling exponential factor in the average survival probability,

$$\langle S(t) \rangle_{\mathrm{LB}} \propto \exp \left[-\mathrm{const} \times (Dt)^{d/(d+2)} (\ln w)^{2/(d+2)} \right]$$
$$\equiv \exp \left[-(t/\tau)^{2/(d+2)} \right]. \qquad (8.2.10)$$

There are two noteworthy points about this last result. First, this type of stretched exponential behavior is not derivable by a classical perturbative expansion, such as an expansion in the density of traps. Second, as in the case of one dimension, the asymptotic decay in relation (8.2.10) again does not set in until the density has decayed to an astronomically small value. We can again obtain a rough estimate for this crossover time by comparing the asymptotic survival probability with the survival probability in the short-time limit. A cheap way to obtain the latter is to expand Eq. (8.2.8) as $\langle q^s \rangle = \langle 1 + s \ln q + (s \ln q)^2/2 + \dots \rangle$, retain only the first two terms, and then reexponentiate. This then gives

$$\langle S(t) \rangle_{\text{short time}} \approx q^{\langle s \rangle} \rightarrow e^{-\rho Dt\, a^{d-2}}, \qquad (8.2.11)$$

where a is the lattice spacing and we have assumed the limit of a small concentration of traps. By comparing the asymptotic form of relation (8.2.10) with short-time approximation (8.2.11), we can infer the crossover time between short-time and asymptotic behavior and then the value of the survival probability at this crossover point. The detailed numerical evaluation of these numbers is tedious and unenlightening; however, the basic result is that the survival probability begins to show its asymptotic behavior only after it has decayed to a microscopically small value. In fact, the crossover to asymptotic behavior occurs earliest when the concentration of traps is large. This is counter to almost all simulation studies of the trapping reaction.

8.2.2.2. Biased Diffusion

If particles undergo biased diffusion, the survival probability decays exponentially with time. This simple behavior can be easily understood by extending the Lifshitz tail argument to account for the bias [Grassberger & Procaccia (1982b), and Kang & Redner (1984c)]. There are two disparate features whose interplay determines the survival probability. The first is the crossover between diffusion and convection. As shown in Chap. 2, this is controlled by the Péclet number $Pe = vL/D$, where L is a characteristic length scale of the system. The second basic feature is the optimal length of the trap-free region in the Lifshitz argument $r^* \propto (-Dt/\ln q)^{1/(d+2)}$.

As time increases, the optimal length of the trap-free region, as well as the Péclet number associated with this length, increases. Once the Péclet number reaches one, the system is effectively in the high-bias limit. Consequently, the asymptotic behavior of the survival probability is always governed by the large-bias limit, for which the decay time is $\tau \approx D/v^2$, independent of the system size. Then the analog of relation (8.2.9) in the long-time and concomitant strong-bias limit is

$$\langle S(t)\rangle_{\text{LB}} \propto \int_0^\infty \exp\left(-\frac{v^2 t}{4D} + \Omega_d r^d \ln q\right) r^{d-1}\, dr. \qquad (8.2.12)$$

Now the average over the radii of trap-free spheres does not influence the essential time dependence that is given by the first exponential factor. Thus the ultimate decay is purely exponential in time:

$$\langle S(t)\rangle \sim \exp(-v^2 t/4D). \qquad (8.2.13)$$

8.3. Reunions and Reactions of Three Diffusing Particles

We now study reunions and reactions of three particles in one dimension. This system is surprisingly rich, but can be understood simply and completely in terms of the first passage of a diffusing particle in a wedge, as discussed in the previous chapter. The three-particle system also provides an important building block toward understanding diffusion-controlled reactions in one dimension.

For certain applications it is useful to view the particles as mutually "vicious," with annihilation occurring whenever any two particles meet [Fisher (1984)]. For other situations it is useful to view one particle as a "prey" that is killed whenever it meets either of the other two mutually independent "predators" [Krapivsky & Redner (1996b), and Redner & Krapivsky (1999)]. Finally, in the context of tracking the relative particle positions it is sometimes useful to view all three particles as independent. With these examples, we investigate the following basic questions:

- What is the survival probability of the prey as a function of time and its initial relative position?
- What is the *first-meeting* probability? More precisely, label the particles as 1, 2, and 3, with initial positions $x_1 < x_2 < x_3$. Either 1 and 2 or 2 and 3 meet first (see Fig. 8.3). What is the probability for these events to occur as a function of the initial particle positions?
- What is the *lead probability*, that is, the probability that the rightmost particle remains in the lead? A related question is, What the probability

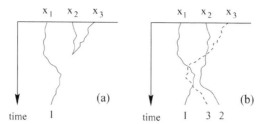

Fig. 8.3. Space–time trajectories of three diffusing particles in one dimension. (a) A configuration that contributes to the probability that 2 and 3 meet first. (b) A configuration that contributes to the probability that particle 1 never leads. The trajectory of 3 is dashed for visual contrast.

that each particle is *never* in the lead? More stringently, we may ask, What is the probability that the particles maintain their order, that is, none of the particles ever meet?

We answer these questions by mapping the three-particle system in one dimension to a "compound" diffusing particle in an appropriately defined two-dimensional wedge [ben-Avraham (1988), and Fisher & Gelfand (1988)]. To accomplish this, we regard the coordinate x_i of each particle on the line as a Cartesian component of a three-dimensional diffusion process. Because the overall length scale is irrelevant, we can then project onto a two-dimensional subspace of relative coordinates. The various constraints enumerated above correspond to this diffusion process being confined to a suitably defined absorbing wedge of the plane, with the process terminating when the sides of the wedge are hit. Because we know the dependence of the survival probability in the wedge as a function of the wedge angle (Chap. 7), we therefore can infer the survival probabilities in the three-particle system.

It is worth emphasizing that this simple approach can still be applied when the diffusion coefficient D_i of each particle is distinct. In this case, we merely redefine the coordinate of the ith particle as $y_i = x_i / \sqrt{D_i}$. In the y_i coordinates the diffusion process is once again isotropic, so that the mapping to the wedge geometry can be used to solve the particle kinetics.

8.3.1. Prey Survival Probability

We first determine the survival probability of a prey as a function of its initial position. We consider two independent cases: (i) "surrounded" prey – a prey between the predators, or (ii) a "chased" prey – prey on one side of both predators.

8.3.1.1. Surrounded Prey

A surrounded prey survives if the coordinates satisfy the inequalities $x_1 < x_2$ – particles 1 and 2 do not meet – *and* $x_2 < x_3$ – particles 2 and 3 do not meet. In the isotropic y_i system these constraints for the compound particle are $y_1\sqrt{D_1} < y_2\sqrt{D_2}$ and $y_2\sqrt{D_2} < y_3\sqrt{D_3}$ [Fig. 8.4(a)]. If the compound particle hits one of the planes, then it is absorbed. In the original one-dimensional system, this corresponds to particle 2 hitting either particle 1 (if $y_2\sqrt{D_2} = y_1\sqrt{D_1}$) or particle 3 (if $y_2\sqrt{D_2} = y_3\sqrt{D_3}$).

The survival probability of the surrounded prey is determined by the opening angle of the accessible wedge for the compound particle, Θ_{mid}. To compute Θ_{mid}, we use the fact that the unit normals to the planes $y_1\sqrt{D_1} = y_2\sqrt{D_2}$ and $y_2\sqrt{D_2} = y_3\sqrt{D_3}$ are

$$\hat{n}_{12} \equiv (-\sqrt{D_1}, \sqrt{D_2}, 0)/\sqrt{D_1 + D_2},$$

$$\hat{n}_{23} \equiv (0, -\sqrt{D_2}, \sqrt{D_3})/\sqrt{D_2 + D_3},$$

respectively. The angle ϕ between these two normals is related to the wedge opening angle by $\Theta_{\text{mid}} = \pi - \phi$ [Fig. 8.4(b)]. Thus

$$\Theta_{\text{mid}} = \cos^{-1}[D_2/\sqrt{(D_1 + D_2)(D_2 + D_3)}].$$

From Eq. (7.2.8), the corresponding survival probability exponent is

$$\beta_{\text{mid}} = \frac{\pi}{2\Theta_{\text{mid}}} = \frac{\pi}{2\cos^{-1}\left[\frac{D_2}{\sqrt{(D_1+D_2)(D_2+D_3)}}\right]}. \qquad (8.3.1)$$

It is instructive to consider some fundamental examples:

- If $D_1 = D_2 = D_3$, the wedge angle is $\Theta_{\text{mid}} = \pi/3$, giving $\beta_{\text{mid}} = 3/2$. This provides basic results for coalescence and aggregation. Survival of a surrounded prey corresponds to a particle that does not undergo *any* reaction in single-species coalescence. Thus the density of unreacted particles in coalescence decays as $t^{-3/2}$. Similarly, in aggregation, particles that do not undergo any reaction are monomers. Therefore the monomer density and indeed the densities of aggregates of any fixed mass all decay as $t^{-3/2}$. A more complete derivation of this result is given in Subsection 8.4.5.
- If $D_2 = 0$ (stationary prey), then $\Theta_{\text{mid}} = \pi/2$. This gives $\beta_{\text{mid}} = 1$, independent of D_1 and D_3. In this limit, the survival probability factorizes into the product of the survival probabilities of the 12 and the 23

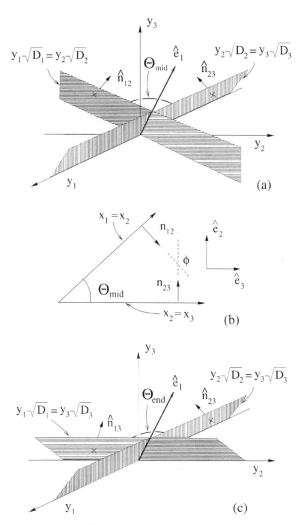

Fig. 8.4. Mapping of three diffusing particles on the line to a single isotropically diffusing compound particle in the three-space $y_i = x_i/\sqrt{D_i}$. (a) For a surrounded prey, its coordinates are subject to the constraints $y_2\sqrt{D_2} > y_1\sqrt{D_1}$ and $y_2\sqrt{D_2} < y_3\sqrt{D_3}$. The compound particle survives if it remains within the indicated wedge-shaped region of opening angle Θ_{mid}. (b) Projection of this wedge onto a plane perpendicular to the $\hat{e}_1 = (1, 1, 1)/\sqrt{3}$ axis. (c) For a chased prey, its coordinates are subject to the constraints $y_1\sqrt{D_1} > y_3\sqrt{D_3}$ and $y_2\sqrt{D_2} < y_3\sqrt{D_3}$, leading to an allowed wedge of opening angle $\Theta_{\text{end}} > \Theta_{\text{mid}}$.

reactions. Because the survival probability for each two-particle system is proportional to $t^{-1/2}$, independent of the particle diffusivities, the survival probability of the middle particle in the combined system decays as $(t^{-1/2})^2 = t^{-1}$. This factorization does *not* apply when $D_2 > 0$, as the motions of the two predators are correlated in the rest frame of the prey.

- Finally, for $D_2 \gg D_1 = D_3 \equiv \delta$, the wedge opening angle vanishes as $\Theta_{mid} \to \sqrt{2\delta/D}$ and the decay exponent therefore diverges as $\beta_{mid} \to \pi\sqrt{D/8\delta}$.

Here, the optimal survival strategy for a surrounded prey is to remain stationary.

8.3.1.2. Chased Prey

For the chased prey – defined to be particle 3 – to survive, both pairs 1 and 3 and 2 and 3 must not meet, whereas meetings of particles 1 and 2 are immaterial. This gives the simultaneous constraints $y_3\sqrt{D_3} > y_1\sqrt{D_1}$ and $y_3\sqrt{D_3} > y_2\sqrt{D_2}$ that define another wedge in the space of the compound particle [Fig. 8.4(c)]. With the same reasoning as that for the surrounded prey, the wedge opening angle for the chased prey is now

$$\Theta_{end} = \pi - \cos^{-1}\frac{D_1}{\sqrt{(D_1 + D_2)(D_1 + D_3)}},$$

and the corresponding survival exponent is

$$\beta_{end} = \frac{\pi}{2\Theta_{end}} = \left[2 - \frac{1}{\pi}\cos^{-1}\frac{D_1}{\sqrt{(D_1 + D_2)(D_1 + D_3)}}\right]^{-1} \quad (8.3.2)$$

We again consider several illustrative cases:

- When $D_1 = D_2 = D_3$, the wedge angle is $\Theta_{end} = 2\pi/3$, corresponding to $\beta_{end} = 3/4$. As we expect naively, for equal diffusivities of all particles, a prey initially at one end is much more likely to survive than a prey between two predators ($\beta_{mid} = 3/2$).
- For $D_3 = 0$ (stationary prey), the problem reduces to two independent one-dimensional survival problems on the semi-infinite line and $\beta_{end} = 1$.
- Finally, for $D_3 \gg D_1, D_2$, the motion of the other two particles becomes irrelevant and the survival probability reduces to that of a particle with an immobile absorbing boundary, for which $\beta_{end} = 1/2$.

Here the optimal survival strategy for a chased prey is to diffuse as rapidly as possible.

8.3.2. Pair Meeting Probabilities

We now determine which of the two particle pairs, 12 or 23, meets first, as a function of their initial positions. To solve this problem, we translate the initial positions x_1, x_2, and x_3 of the particles on the line to the initial position of the compound particle within the wedge. A given particle pair reacting in one dimension corresponds to the compound particle hitting one of the generators of the wedge [Frachebourg, Krapivsky, & Redner (1998)]. This latter meeting probability is trivial to compute in the wedge geometry [Eq. (7.4.8)].

To avoid unilluminating complications, consider for the moment equal particle diffusivities, $D_i = D$, $\forall i$. The compound particle then lies within the wedge of opening angle $\pi/3$ defined by $x_1 < x_2$ and $x_2 < x_3$. The apex of the wedge is generated by the unit vector $\hat{e}_1 \equiv (1, 1, 1)/\sqrt{3}$, and two convenient perpendiculars to this generator are [Fig. 8.4(b)]

$$\hat{e}_2 \equiv (0, -1, 1)/\sqrt{2},$$
$$\hat{e}_3 \equiv (-2, 1, 1)/\sqrt{6}.$$

For the initial state $\vec{x} = (x_1, x_2, x_3)$, the corresponding components in the coordinate system perpendicular to the wedge apex are

$$z_2 = \vec{x} \cdot \hat{e}_2 = (x_3 - x_2)/\sqrt{2}$$
$$z_3 = \vec{x} \cdot \hat{e}_3 = (x_2 + x_3 - 2x_1)/\sqrt{6}.$$

Consequently the inclination angle of this initial point with respect to the plane $x_2 = x_3$ in plane-polar coordinates is

$$\lambda = \tan^{-1} \frac{z_2}{z_3}$$
$$= \tan^{-1} \left[\frac{\sqrt{3}(x_3 - x_2)}{x_2 + x_3 - 2x_1} \right]. \tag{8.3.3}$$

From Eq. (7.4.8), the probability that the plane $x_2 = x_3$ is eventually hit, or equivalently, that particle 2 meets particle 3, is simply $1 - \lambda/\Theta$. Therefore this eventual meeting probability is

$$P_{23} = 1 - \frac{3}{\pi} \tan^{-1} \left(\sqrt{3} \frac{1-x}{1+x} \right), \qquad D_1 = D_2 = D_3. \tag{8.3.4a}$$

where we now write $x_1 = 0$, $x_2 = x$, and $x_3 = 1$ without loss of generality.

For general particle diffusivities $D_1 = D_3 = D = \eta D_2$, the counterpart of Eq. (8.3.4a) is

$$P_{23} = 1 - \frac{1}{\cos^{-1}\left(\frac{1}{1+\eta}\right)} \tan^{-1} \left(\sqrt{\eta^2 + 2\eta} \frac{1-x}{1+\eta x} \right). \tag{8.3.4b}$$

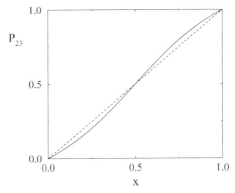

Fig. 8.5. Probability P_{23} that the middle particle, in a group of three particles initially at 0, x, and 1, first meets its right neighbor as a function of x. The diffusivities of the three particles are $D_2 = 0$ and $D_1 = D_3$, where P_{23} exhibits the greatest deviation from linearity. For comparison, a line of unit slope – the meeting probability for the case $D_1 = D_3 = 0$ and D_2 nonzero – is also shown (dashed).

In the limit $\eta \to \infty$, corresponding to the middle particle stationary, this reduces to

$$P_{23} = 1 - \frac{2}{\pi} \tan^{-1} \left(\frac{1-x}{x} \right), \quad D_2 \ll D_1, D_3, \qquad (8.3.4c)$$

whereas, for reference, if $D_2 \gg D_1, D_3$, that is, a diffusing particle in a fixed absorbing interval, the 23 meeting probability is simply

$$P_{23} = x, \quad D_2 \gg D_1, D_3. \qquad (8.3.4d)$$

A plot of the 23 first-meeting probability as a function of x, for the case of a stationary middle particle or, equivalently, infinitely mobile predators ($\eta = \infty$), is shown in Fig. 8.5. This is the case in which P_{23} has the largest deviation from a linear dependence. Even so, this deviation is only barely perceptible. Here $|P_{23} - x|$ attains a maximum value of approximately 0.0453 at $|x - \frac{1}{2}| = 0.262$.

8.3.3. Lead and Order Probabilities

Finally, we study the evolution of the relative particle positions when they are noninteracting. The wedge mapping provides a simple and complete understanding of these probabilities. We treat the case of equal particle diffusivities; the generalization to arbitrary diffusivities is immediate. In the

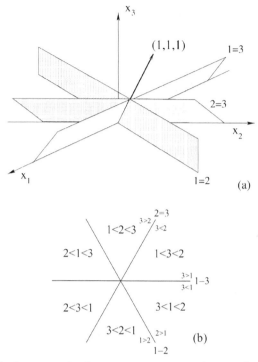

Fig. 8.6. (a) The three constraint planes $x_1 = x_2$, $x_1 = x_3$, and $x_2 = x_3$ for the compound particle. (b) Their projection onto the two-space perpendicular to $(1, 1, 1)$ axis. Each $\pi/3$ wedge corresponds to the specific particle ordering indicated.

three-space defined by the x_i, the constraints $x_i = x_j$ for $i \neq j$ generate the three planes shown in Fig. 8.6. By projecting onto the subspace perpendicular to the $(1, 1, 1)$ axis, the constraint planes symmetrically divide the subspace into six regions, each of which corresponds to the coordinate orderings indicated. From this categorization, the probability that the compound particle remains within a particular wedge corresponds to the particles on the line satisfying a corresponding set of constraints. From this connection, general order probabilities can be deduced easily.

The general result is that the possible wedge angles are $n\pi/3$, with $n = 1, 2, \ldots, 5$, corresponding to survival probabilities $t^{-3/2n}$. All that remains is to ascribe a physical meaning to these different wedge angles in terms of the relative positions of the particle on the line. For example, the probability that a given particle on the line maintains its lead corresponds to the compound particle remaining within a wedge of opening angle $2\pi/3$ (Fig. 8.6). Therefore the single-particle lead probability decays as

$$P_{\text{lead}}(t) \sim t^{-3/4}. \tag{8.3.5a}$$

A specific particle *never* being in the lead is equivalent to the compound particle remaining within a wedge of opening angle $4\pi/3$. Consequently, this no-lead probability decays as

$$P_{\text{no lead}}(t) \sim t^{-3/8}. \tag{8.3.5b}$$

The condition that the initial ordering of the coordinates never gets reversed corresponds to the compound particle remaining within a wedge of opening angle $5\pi/3$. This "no-reversal" probability therefore decays as

$$P_{\text{no reverse}}(t) \sim t^{-3/10}. \tag{8.3.5c}$$

Finally, we can ask for the probability that the particles *always* maintain their relative positions. That is, if initially $x_1 < x_2 < x_3$, what is the probability that this inequality is always maintained up to time t? Another way to state this question is to view the particles as mutually vicious, so that they annihilate whenever any two meet. Then the probability that particles maintain their relative positions is that same as the probability that no vicious annihilation event has occurred. According to Fig. 8.6(b), this is equivalent to the compound particle remaining within a single elemental wedge of opening angle $\pi/3$. Consequently, the order probability decays as

$$P_{\text{order}}(t) \sim t^{-3/2}. \tag{8.3.5d}$$

8.3.4. Extension to Arbitrary Number of Particles

In principle, we can generalize our previous discussion to N particles by mapping the position of each particle in the system onto the coordinates of the trajectory of a diffusing compound particle in N dimensions. Order probabilities now correspond to various constraints of the form $x_i < x_j$ for certain of the i and j. Thus the compound particle is confined to a multidimensional wedge defined by these constraint planes. It still appears to be an open problem to visualize these wedges and thereby determine the time dependence of the ordering probabilities for arbitrary N.

However, one specific ordering probability, namely, the probability that all particles *always* maintain their relative positions, can be computed for arbitrary N. This system has also been appropriately termed vicious random walks, in which two particles annihilate whenever any two meet [Huse & Fisher (1984), and Fisher (1984)]. The probability that all particles maintain their initial order is the same as the probability that all the vicious particles survive up to time t. We can again determine this survival probability by the image method. As in the three-particle system, we first map the coordinates of the N particles on the line to the position of a compound particle in N space, which is subject to boundary conditions that are equivalent to maintenance of the relative particle positions on the line.

To maintain their relative positions, each particle coordinate x_i must satisfy the constraints $x_i < x_{i+1}$ for $i = 1, 2, \ldots, N-1$. The constraint $x_i < x_{i+1}$ can be accounted for in the N-dimensional space defined by $(x_1, x_2, \ldots, x_N)$ by an image anti-Gaussian that is the reflection of the initial particle about the plane $x_i = x_{i+1}$. We determine this mirror image merely by interchanging the coordinates $x_i \leftrightarrow x_{i+1}$. Each of the $N - 1$ constraint planes leads to a "primary" negative image. These images generate "secondary" images that are due to the presence of the other constraint planes, and we must consider the entire set of iterated images. This construction closes after $N!$ particles are generated by all possible transpositions of the components of the initial particle location, $x_i \leftrightarrow x_j$ for $i \neq j$, with odd transpositions leading to a negative image and even transpositions to a positive image.

The probability distribution of the compound particle in N dimensions in the presence of these constraints can be written as

$$
\begin{aligned}
Q^{(N)}(\vec{x}, t; \vec{x}_0) &= \sum_{\mathcal{P}} (-1)^{|\mathcal{P}|} P(\vec{x}, t; \mathcal{P}\vec{x}_0) \\
&= \frac{1}{(4\pi D t)^{d/2}} \sum_{\mathcal{P}} (-1)^{|\mathcal{P}|} e^{-(\vec{x} - \mathcal{P}\vec{x}_0)^2 / 4Dt} \\
&\sim \frac{1}{(4\pi D t)^{d/2}} e^{-x^2/4Dt} \sum_{\mathcal{P}} (-1)^{|\mathcal{P}|} e^{-\vec{x}\cdot(\mathcal{P}\vec{x}_0)/2Dt}.
\end{aligned}
\tag{8.3.6}
$$

Here $\mathcal{P}$ is the permutation operator on the coordinates of $\vec{x}_0$ and $P(\vec{x}, t; \vec{x}_0)$ is the Gaussian probability for free diffusion at $\vec{x}$ at time t, when a particle starts at $\vec{x}_0$. In the long-time limit, the exponential factor in the last line can be expanded in a power series in $1/t$, and the asymptotic survival probability is given by the lowest nonvanishing term in this series. Physically, this is equivalent to determining the lowest nonvanishing electrostatic multipole moment that is due to the set of charges (real and images) generated by the planes $x_i = x_j$.

This expansion of this last exponential factor leads to a homogeneous polynomial in the N variables x_i that is symmetric under the interchange of $\vec{x}$ and $\vec{x}_0$. As a function of $\vec{x}$, this polynomial must also be completely antisymmetric under the interchange of any two components. This condition ensures that the sum over all permutations in the last line of Eq. (8.3.6) is nonzero. The requisite polynomial has the simple form $A(\vec{x}) = \prod_{i<j}(x_i - x_j)$ [Huse & Fisher (1984)] and therefore has degree $N(N-1)/2$.

From this fact, we can easily derive the survival probability. When only the contribution of the lowest nonvanishing term is kept, the dependence of $Q^{(N)}(\vec{x}, t; \vec{x}_0)$ on essential variables is given by

$$Q^{(N)}(\vec{x}, t; \vec{x}_0) \propto \frac{1}{(Dt)^{N/2}} \, e^{-x^2/4Dt} \, \left(\frac{x x_0}{Dt}\right)^{N(N-1)/2} . \tag{8.3.7}$$

We now integrate this concentration over the physical wedge in N space to obtain the survival probability. For asymptotic time dependence, we can replace the Gaussian with a constant and integrate to a distance scale of the order of $\sqrt{Dt}$. This is equivalent to replacing all coordinates x_i with $\sqrt{Dt}$. The volume element cancels the leading factor of $(Dt)^{-N/2}$, and the polynomial factor gives the asymptotic behavior

$$S^{(N)}(t) \sim t^{-N(N-1)/4} \tag{8.3.8}$$

for the probability that N diffusing particles on the line maintain their relative positions up to time t.

8.4. Diffusion-Controlled Reactions

8.4.1. Basic Properties

We now investigate the diffusion-controlled reactions of capture, coalescence, annihilation, and aggregation (Fig. 8.7), primarily in one dimension, where first-passage ideas provide key insights for the kinetics. For the latter three examples, the spatial dimension d is the basic parameter that characterizes the kinetics. For $d > 2$, the transience of diffusion leads to an appreciable probability of reaction by two initially distant particles. This efficient mixing

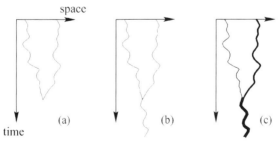

Fig. 8.7. Space–time cartoon of the trajectory of a pair of particles in (a) diffusion-controlled annihilation, $A + A \rightarrow 0$, (b) coalescence, $A + A \rightarrow A$, and (c) aggregation, $A_i + A_j \rightarrow A_{i+j}$. In (c), the thickness of the trajectory indicates the (conserved) aggregate mass.

can be accounted for by a mean-field approximation in which any reactant pair is equally likely to react. Within this approximation, the density $c(t)$ decays as

$$c(t) \sim (kt)^{-1},$$

where k is the reaction rate (see Section 6.5). On the other hand, for $d \leq 2$ the recurrence of diffusion means that reactions are more likely to occur between particles that are initially nearby. In one dimension, in particular, reactions can occur *only* between nearest neighbors. This leads naturally to a description of the reaction kinetics by first-passage processes.

8.4.2. The Capture Reaction, $p + P \rightarrow P$

We begin with the capture reaction, in which a single prey p is "stalked," and ultimately killed on contact, by independent diffusing predators P, as represented by $p + P \rightarrow P$ [Bramson & Griffeath (1991), and Kesten (1992)]. This can be viewed as a variant of the trapping reaction in which both species move and in which there are many traps and one particle. If the prey is immobile, then each prey–predator pair is independent, and it is possible to derive an exact solution for the survival probability in terms of a product of survival probabilities for each prey–predator pair [Blumen, Zumofen, & Klafter (1984)]. If the prey is mobile, then its survival probability can be determined in any dimension by a Lifshitz tail argument [Redner & Kang (1984)]. We focus here on the most interesting situation of a one-dimensional system, with predators all to one side of the prey. By making a correspondence with the survival of a diffusing particle near an advancing, absorbing boundary [Krapivsky & Redner (1996b), and Redner & Krapivsky (1999)], we will show that the prey survival probability in the presence of $N \gg 1$ predators decays as $t^{-\beta(N)}$, where $\beta(N)$ grows logarithmically with N.

It is helpful to begin by categorizing the relative efficiency of the capture process for several basic situations. For spatial dimension $d \geq 2$, $S_N(t)$ asymptotically factorizes [Bramson & Griffeath (1991)], that is, $S_N(t) \sim S_1(t)^N$. From this fact, we can determine the basic time dependence of the reaction for $d \geq 2$. When d is strictly greater than 2, the transience of diffusion means that there is a nonzero probability for the prey to survive to infinite time in the presence of a single predator. This continues to hold for any initial spatial distribution of a finite number of predators. In the terminology of Bramson and Griffeath, the capture reaction is "unsuccessful." For $d = 2$, capture is "successful" and the prey is sure to be killed. However, the average prey lifetime is still infinite! This follows because $S_N(t) \sim (\ln t)^{-N}$, where

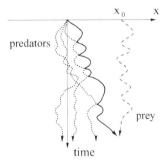

Fig. 8.8. Space–time evolution in one dimension of $N = 4$ diffusing predators (dotted curves) that all start at $x = 0$ and a single diffusing prey (dashed curve) that starts at $x = x_0$. The trajectory of the closest ("last") predator, whose individual identity may change with time, is indicated by the heavy solid path.

$S_1(t) \propto (\ln t)^{-1}$ is the survival probability that is due to a single predator [see approximation (1.5.8)].

In one dimension, there is much less "room" for the prey to escape and its capture is certain. The time dependence of this capture is quite rich. There are two generic situations. When the prey is surrounded by N_ℓ and N_r predators on the left and the right, respectively, the survival probability decays rapidly with time and with N_ℓ and N_r, because the region that remains unvisited by any predators shrinks rapidly. When all N predators are to one side of the prey (Fig. 8.8), the prey has the possibility of running away (as best as a diffusing particle can "run") from the predators. Because a surviving prey typically has a displacement to the right, each additional predator thus has a progressively weaker effect on prey survival. This leads to the prey survival probability decaying as

$$S_N(t) \sim t^{-\beta_N}, \tag{8.4.1}$$

with β_N growing only logarithmically in N for large N.

In principle, this prey–predator system is soluble by mapping the $N + 1$-particle system onto diffusion within an absorbing wedge region in $N + 1$ dimensions. Although much is known about high-dimensional diffusion problems [De Blassie (1987)], there does not appear to be a general approach for the capture reaction. For $N = 1$ and 2, the decay exponents may be deduced from our previous discussions of first passage in few-particle systems. From Section 4.8, $\beta_1 = 1/2$ for a diffusing prey and a single diffusing predator, independent of their diffusion coefficients. From Section 8.3, $\beta_2 = 3/4$ for equally mobile prey and predators, whereas for an arbitrary diffusivity ratio

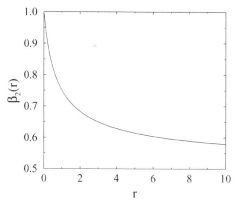

Fig. 8.9. The survival exponent $\beta_2(r)$ given by Eq. (8.4.2) versus diffusivity ratio r.

$r = D_{\text{prey}}/D_{\text{P}}$, Eq. (8.3.2) gives

$$\beta_2(r) = \left(2 - \frac{2}{\pi} \cos^{-1} \frac{r}{1+r}\right)^{-1}. \tag{8.4.2}$$

As $D_{\text{prey}}/D_{\text{P}} \to \infty$, the advantage of two predators versus one is largely negated by their relative inertness (Fig. 8.9).

For arbitrary N, the rightmost predator typically approaches the prey, even though each predator undergoes isotropic diffusion (Fig. 8.10). We can then apply our results from Section 4.8 to determine the prey survival probability

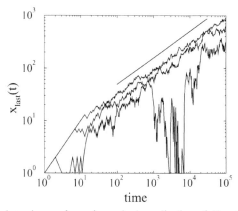

Fig. 8.10. Time dependence of x_{last} for a single realization of $N = 4$, 64, and 1028 nearest-neighbor random walks starting from $x = 0$ (bottom to top). The straight line of slope $1/2$ indicates the expected long-time behavior.

that is due to this approaching absorbing boundary. To determine the location of this last predator, $x_{\text{last}}(t)$, suppose initially that all the predators are at the origin. We estimate $x_{\text{last}}(t)$ by the extreme-statistics condition

$$\int_{x_{\text{last}}(t)}^{\infty} \frac{1}{\sqrt{4\pi D_{\text{P}} t}} e^{-x^2/4D_{\text{P}} t} \, dx = \frac{1}{N}, \qquad (8.4.3)$$

which states that there is one predator from the initial group of N that lies in the range $[x_{\text{last}}(t), \infty]$. The asymptotic behavior of this integral can be expressed in terms of the complementary error function, but we can estimate it directly. Because the integral is nonzero in only a very small range beyond x_{last}, we write $x = x_{\text{last}}(t) + \epsilon$ and expand the integrand for small ϵ to find

$$\int_{x_{\text{last}}}^{\infty} \frac{1}{\sqrt{4\pi D_{\text{P}} t}} e^{-x_{\text{last}}^2/4D_{\text{P}} t} e^{-x_{\text{last}}\epsilon/2D_{\text{P}} t} e^{-\epsilon^2/4D_{\text{P}} t} \, d\epsilon = \frac{1}{N}.$$

Over the range of ϵ for which the second exponential factor is non-negligible, the third exponential factor may be set to one. The resulting integral is then elementary, with the result

$$\frac{1}{\sqrt{4\pi D_{\text{P}} t}} e^{-x_{\text{last}}^2/4D_{\text{P}} t} \frac{2D_{\text{P}} t}{x_{\text{last}}} \approx \frac{1}{N}. \qquad (8.4.4)$$

Defining $y = x_{\text{last}}/\sqrt{4D_{\text{P}} t}$, this condition can be simplified to $y e^{y^2} = N/\sqrt{4\pi}$, with the solution

$$y = \sqrt{\ln N/\sqrt{4\pi}} \left[1 - \frac{1}{4} \frac{\ln(\ln N/\sqrt{4\pi})}{\ln N/\sqrt{4\pi}} + \dots \right].$$

To lowest order, this gives

$$x_{\text{last}}(t) \sim \sqrt{4D_{\text{P}} \ln N \, t} \equiv \sqrt{A_N t} \qquad (8.4.5a)$$

when the number of predators is finite.

For $N = \infty$ it is not sensible to pack all the predators at a single point as $x_{\text{last}}(t)$ would identically equal t. A more suitable initial condition is a uniform concentration c_0 of predators that extends from $-\infty$ to 0. Now only $N \propto \sqrt{c_0^2 D_{\text{P}} t}$ of the predators are "dangerous," that is, within a diffusion distance from the edge of the pack and thus potential killers of the prey. Consequently, for $N \to \infty$, the leading behavior of $x_{\text{last}}(t)$ becomes

$$x_{\text{last}}(t) \sim \sqrt{2D_{\text{P}} \ln \left(c_0^2 D_{\text{P}} t \right) t}. \qquad (8.4.5b)$$

Because the position of the last predator advances toward the prey as $\sqrt{A_N t}$ for N finite and as $\sqrt{2D_{\text{P}} \ln(c_0^2 D_{\text{P}} t) t}$ for $N = \infty$, this suggests that we use the method of Section 4.8 for the survival probability of a diffusing particle near an approaching, absorbing boundary. This method becomes more accurate for large N, as the distribution of x_{last} becomes both smoother and sharper (albeit slowly) as N increases. In the large-N limit, we therefore replace the stochastically moving absorbing boundary induced by the last predator with the smoothly moving boundary whose position is given by Eq. (8.4.5a) or by (Eq. 8.4.5b). Then we may transcribe our results from Section 4.8 to determine the prey survival probability. For finite N, this gives the survival exponent

$$\beta_N(r) \sim A_N/16 D_{\text{prey}} \sim \ln(Nr)/4r, \qquad (8.4.6)$$

whereas for $N \to \infty$, the survival probability decays as

$$S_\infty(t) \sim \exp(-\ln^2 t/8r). \qquad (8.4.7)$$

However different asymptotic behavior should arise in the extreme limits of $r \to 0$ and $r \to \infty$. In the limit of a stationary prey, we know that $S_N \sim t^{-N/2}$, whereas for an infinitely mobile prey, the survival probability should simply decay as $t^{-1/2}$. By matching these with Eq. (8.4.6), we infer that the full dependence of β_N on the diffusivity ratio r is

$$\beta_N(r) = \begin{cases} N/2, & r \ll 1/N \\ \ln(4Nr)/4r, & 1/N \ll r \ll \ln N \\ 1/2, & r \gg \ln N \end{cases}. \qquad (8.4.8)$$

The result for the intermediate regime of $1/N \ll r \ll \ln N$ represents the generalization of the three-particle exponents in Eq. (8.3.2) to arbitrary N.

The actual value of the survival exponent merits comment. As in the approaching cliff problem of Section 4.8, the exponent in Eq. (8.4.8) is not particularly accurate for intermediate values of N, although the correct asymptotic behavior for $N \to \infty$ is reproduced. For example, for the case $r = 1$, estimates for β_N from numerical simulations are $\beta_3 \approx 0.91$, $\beta_4 \approx 1.032 \pm 0.01$, and $\beta_{10} \approx 1.4$ [Bramson & Griffeath (1991)]. In contrast, the asymptotic formula $\beta_N = \frac{1}{4}\ln(N)$ gives $\beta_3 = 0.274\ldots, \beta_4 = 0.346\ldots,$ and $\beta_{10} = 0.575\ldots$.

8.4.3. Coalescence, $A + A \to A$

The kinetics of the coalescence reaction $A + A \to A$ is particularly simple because it can be reduced to an effective one-body system. This stems from the fact that the particles are indistinguishable so that each particle sees only the "cage" formed by its two nearest neighbors. It is immaterial that these

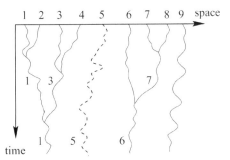

Fig. 8.11. Space–time particle trajectories for $A + A \rightarrow A$ in one dimension showing the propagation of the particle numbers. The trajectories of the nearest neighbors to particle 5 (dashed) form a cage whose boundaries diffuse even as reactions between these neighbors and more distant particles occur. Particle 5 and the two cage particles can be treated independently of the rest of the system.

neighbors may coalesce with more distant particles (Fig. 8.11). Consequently, the kinetics of each particle involves only itself and its two nearest neighbors. By then labeling the particles appropriately after each coalescence event, we can simplify this system still further to an effective one-body problem of a particle and the distance to its nearest neighbor to its left. The following discussion is based on this general approach, following the initial work of ben-Avraham et al. (1990a).

To implement the last reduction to a one-body problem, it is convenient to define a particle as "killed" if it meets its left neighbor, whereas the particle continues to "live" if it meets its right neighbor. This identifies the labels of the surviving particles unambiguously. To determine the fate of any particle, we need consider only a particle and its left neighbor that continues to diffuse even if it experiences additional coalescence events. Suppose that the initial separation of the particle and its left neighbor is x_0. Because both the particle and its neighbor have diffusion coefficients D, the two-particle system is equivalent to a single particle with diffusivity $2D$ that diffuses and is ultimately trapped at a stationary absorbing point.

The probability $\rho(x, t)$ that the separation between the two particles equals x at time t therefore obeys the diffusion equation, subject to an absorbing boundary condition at $x = 0$. The Green's function for this diffusion equation is simply the image solution given in Eq. (3.2.1) but with $D \rightarrow 2D$. Therefore

$$\rho(x, t) = \int_0^\infty G(x, x_0, t)\rho(x_0, 0)\, dx_0, \tag{8.4.9}$$

with

$$G(x, x_0, t) = \frac{1}{\sqrt{8\pi Dt}} \left[e^{-(x-x_0)^2/8Dt} - e^{-(x+x_0)^2/8Dt} \right]$$

$$\sim \frac{1}{\sqrt{8\pi Dt}} \frac{xx_0}{2Dt} e^{-(x^2+x_0^2)/8Dt} \quad t \to \infty. \quad (8.4.10)$$

From the distance distribution $\rho(x, t)$, the particle survival probability $S(t)$ is just the probability that the interparticle distance remains positive at time t. This gives

$$S(t) = \int_0^\infty \rho(x, t)\, dx, \quad (8.4.11)$$

and the particle concentration is simply given by $c(t) = c_0 S(t)$. The distance distribution of the surviving particles at time t, $p(x, t)$, is then given by

$$p(x, t) = \rho(x, t)/S(t). \quad (8.4.12)$$

As long as the probability distribution of initial distances x_0 decays rapidly for large x_0, the long-time behavior of the distance distribution and the particle concentration are *independent* of the initial condition. For a bounded initial distance distribution, we may set $e^{-x_0^2/8Dt} \to 1$ in the Green's function of Eq. (8.4.10) as $t \to \infty$. In this case $\rho(x, t)$ becomes

$$\rho(x, t) \sim \frac{1}{\sqrt{8\pi Dt}} \int_0^\infty \frac{x}{2Dt} e^{-x^2/8Dt} \left[x_0 \rho(x_0, 0) \right] dx_0. \quad (8.4.13)$$

Note the separation of the dependences on x and x_0 and, more importantly, that the integral is just the first moment of the initial distribution of separations. By definition, this is just the inverse of the initial concentration $1/c_0$. We therefore obtain

$$\rho(x, t) \sim \frac{1}{\sqrt{8\pi Dt}} \frac{x}{2Dt} e^{-x^2/8Dt} \frac{1}{c_0}. \quad (8.4.14)$$

Now, integrating over all x, we find that the survival probability is

$$S(t) \sim \frac{1}{\sqrt{2\pi Dt}} \frac{1}{c_0}. \quad (8.4.15)$$

In summary, the concentration at long times is

$$c(t) = c_0 S(t) \sim \frac{1}{\sqrt{2\pi Dt}}, \quad (8.4.16)$$

whereas the probability that two nearest-neighbor particles are separated by a distance x is

$$p(x, t) = \rho(x, t)/S(t) \sim \frac{x}{4Dt} e^{-x^2/8Dt}, \qquad (8.4.17)$$

independent of the initial conditions.

It is worth mentioning that complete results can be derived for a more general single-species coalescence process that includes particle input at a constant rate in space and time, and particle "birth," in which any particle can give birth to an identical offspring that is placed adjacent to the parent, in addition to the basic diffusion and coalescence steps [ben-Avraham et al. (1990a)]. In this case, the solution can be obtained elegantly by the *empty-interval* method, which is based on the fact that the various components of the reaction can be accounted for by an abstract process in which the length of an empty interval undergoes pure diffusion, but subject to an absorbing boundary condition when the interval length reaches zero. Details of this approach can be found both in ben-Avraham et al. (1990a) and in ben-Avraham and Havlin (2000).

8.4.4. Annihilation, $A + A \rightarrow 0$

At first sight, the annihilation reaction $A + A \rightarrow 0$ seems more difficult to treat than coalescence, $A + A \rightarrow A$, as the surrounding cage of a specific particle sometimes diffuses and occasionally increases in size suddenly if one of the cage particles is annihilated (Fig. 8.12). Because of this sporadic motion, the single-particle picture that provided the solution for coalescence is not appropriate for annihilation. Several approaches have been developed to solve annihilation kinetics that are based on a direct attack on the many-body problem. This includes mapping the particle trajectories onto a free fermion problem [Lushnikov (1987)] and the use of the image method [Torney &

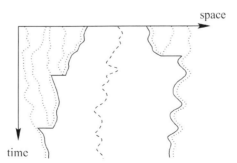

Fig. 8.12. Space–time particle trajectories for $A + A \rightarrow 0$ in one dimension. The boundaries of the "cage" that surrounds the dashed particle trajectory is shown solid.

Fig. 8.13. Equivalence between Ising spins in one dimension and $A + A \rightarrow 0$. Nearest-neighbor antiparallel spins define a domain-wall particle (heavy dot) which hops to a nearest-neighbor site whenever a spin flips. If the target site is already occupied, annihilation occurs.

McConnell (1983a, 1983b)]. Here a different approach is presented that is based on the observation that the motion of reactants in $A + A \rightarrow 0$ is identical to the motion of ferromagnetic domain walls in the kinetic Ising–Glauber model at zero temperature [Rácz (1985), and Amar & Family (1989)]. By this equivalence, we can obtain the solution for the particle concentration in $A + A \rightarrow 0$ merely by transcribing the appropriate results from the Glauber solution of the kinetic Ising model. A notable feature of this approach is that it again reduces the many-body reaction into an effective single-body problem associated with the evolution of the spin correlation function in the Ising–Glauber system.

The starting point for the equivalence between the Ising–Glauber model and annihilation is the identification of domain walls in the former with reactants in the annihilation process. For the Ising model at low temperature, we have seen in Section 4.4 that the system organizes into long domains of aligned spins (Fig. 8.13) that are separated by domain-wall particles. The zero-temperature Glauber kinetics is equivalent to a process in which domain-wall particles diffuse freely and annihilate whenever they meet.

To exploit this correspondence between $A + A \rightarrow 0$ and the Ising–Glauber model and determine the particle density in the former process, we consider the two-spin correlation function in the Ising system (see Section 4.4):

$$c_k = \langle \sigma_i \sigma_{i+k} \rangle,$$

where σ_i is the spin at site i takes the values ± 1 and $\langle \cdots \rangle$ denotes the thermal average. For a translationally invariant system, this correlation function depends on only the separation k between the two spins. We can also use this correlation function to count the number of domain-wall particles. When spins i and $i + 1$ are antiparallel, $(1 - \sigma_i \sigma_{i+1})/2$ equals 1, whereas this quantity equals 0 if the spins are antiparallel. The average number of domain-wall particles in the spin system, or equivalently, the number of A's in the reactive system, is then $A(t) = (1 - \langle \sigma_i \sigma_{i+1} \rangle)/2 = (1 - c_1)/2$. Because the exact behavior of the two-spin correlation function is known for any initial condition, we can transcribe this result to obtain the particle density in $A + A \rightarrow 0$.

From Glauber's original paper [Glauber (1963)], the general solution for the two-spin correlation function at zero temperature is

$$c_k(t) = 1 + e^{-2t} \sum_{m=1}^{\infty} [c_m(0) - 1][I_{k-m}(2t) - I_{k+m}(2t)], \quad (8.4.18)$$

where $I_n(z)$ is the modified Bessel function of the first kind of order n. We now use the well-known asymptotic behavior of the Bessel function to determine the long-time behavior of $A(t)$ for various relevant initial conditions.

For a random initial spin state with $\langle \sigma_i \rangle = m_0$, the initial value of the nearest-neighbor correlation function is $c_1(0) = m_0^2$. Then, using the fact that the modified Bessel function satisfies $I_n(z) = I_{-n}(z)$ for integer n, we find that most of the terms in the infinite series of Eq. (8.4.18) for $c_1(t)$ cancel in pairs and we are left with

$$c_1(t) = 1 - e^{-2t}(1 - m_0^2)[I_0(2t) + I_1(2t)], \quad (8.4.19)$$

from which the density of domain walls is

$$A(t) = (1 - c_1(t))/2 = \frac{1}{2}(1 - m_0^2)e^{-2t}[I_0(2t) + I_1(2t)],$$

$$\sim \frac{1 - m_0^2}{\sqrt{4\pi t}} + \mathcal{O}(t^{-3/2}). \quad (8.4.20)$$

Amusingly, for $m_0 = 0$ the asymptotic density is independent of the initial particle density, whereas for $m_0 \neq 0$, the asymptotic density depends on this initial density. Intuitively, this arises because a nonzero magnetization corresponds to particles in the initial state that are more likely to be arranged in pairs. This is in contrast to $A + A \rightarrow A$, in which the asymptotic behavior of the density is always independent of the initial particle density.

Another useful initial condition is every lattice site initially occupied by a particle, corresponding to an antiferromagnetic initial state in the equivalent spin system. As a consequence, the initial correlation function is $c_k(0) = (-1)^k$, and, substituting this into the general solution of Eq. (8.4.18) for $c_k(t)$, we ultimately obtain the simple result for the particle density,

$$A(t) = e^{-2t} I_0(2t) \sim \frac{1}{\sqrt{4\pi t}} + \mathcal{O}(t^{-3/2}), \quad (8.4.21)$$

which is the same as the asymptotic density for the random initial condition when $m_0 = 0$.

As a final note, the Ising–Glauber model and its correspondence to diffusive annihilation may be extended in a very appealing way to incorporate a temporally and spatially homogeneous source of particles, $0 \to A$, in addition to the reaction, $A + A \to 0$ [Rácz (1985)]. These two processes lead to a steady state whose properties can be inferred easily by the mapping to the Ising–Glauber model. In the framework of Ising spins, the introduction of a particle (a domain wall) between spins i and $i + 1$ is equivalent to flipping *all* the spins from $i + 1$ to infinity. If particle input rate is h, then the corresponding "block" spin–flip process also occurs at this rate.

By augmenting the master equations to account for this block spin–flip process, Rácz (1985) derived the following rate equation for the two-spin correlation function [compare with Eq. (4.4.9)]:

$$\frac{dc_n}{dt} = c_{n-1} - 2(1 + nh)c_n + c_{n+1}. \tag{8.4.22}$$

In the steady state, this difference equation is identical to one of the recursion formulas for the Bessel function [Abramowitz & Stegun (1972)]. By making the appropriate identifications and using the relation $A(t; h) = (1 - c_1(t))/2$, Rácz (1985) found that the steady-state particle density is given by

$$A(t = \infty; h) = -\frac{Ai'(0)}{2^{2/3}Ai(0)} h^{1/3} \approx 0.4593 h^{1/3}, \tag{8.4.23}$$

where Ai is the Airy function. Thus the steady-state density scales as the one-third power of the input rate. For comparison, in the rate-equation approach, the particle density evolves as $\dot{A} = -kA^2 + h$, which leads to $A(\infty; h) \propto h^{1/2}$.

8.4.5. Aggregation, $A_i + A_k \to A_{i+j}$

8.4.5.1. Introduction

Finally, we study the kinetics of irreversible aggregation, $A_i + A_j \to A_k$, with $k = i + j$. Here A_i denotes a cluster of mass i, and in a reaction between an i cluster and a j cluster, the result is a cluster of mass $k = i + j$. Aggregation arises in a wide variety of physical systems such as gelation of jello, curdling of milk, coagulation of blood, etc., and the phenomenology is quite rich [for general discussions, see, e.g., Chandrasekhar (1943), Drake (1972), and Friedlander (1977)]. The basic physical observables in aggregation are the concentrations of clusters of mass k at time t, $c_k(t)$.

Most physical realizations of aggregation occur in three dimensions and also with good mixing of the reactants. Under these conditions, a rate-equation approach – which is the kinetic analog of a mean-field theory – provides an appropriate description. An essential ingredient of these rate equations is the so-called reaction kernel K_{ij} that quantifies the mass dependence of the reaction rate between two clusters with masses i and j. For reaction kernels that correspond to generic types of aggregation processes, the rate equations have been solved and a comprehensive understanding of the behavior of the cluster concentrations $c_k(t)$ has emerged [see, e.g., van Dongen & Ernst (1983)].

Parallel to our treatments of coalescence and annihilation, we again consider aggregation in one dimension, where insights from first-passage phenomena help determine the kinetics. For simplicity, we assume that aggregates are always structureless point clusters [Kang & Redner (1984b)]; thus, by ignoring the cluster masses, aggregation reduces to the coalescence reaction. We also assume that the diffusion coefficients of the aggregates are independent of their mass. Although this latter assumption is generally not true, this idealization allows us to recast aggregation as a classical first-passage problem. The extension to general mass-dependent diffusion and mass-dependent reaction rates represents a challenging open problem.

Quite amazingly, the concentration of k clusters at any time has exactly the same form as the first-passage probability for diffusion in one dimension with absorption at $k = 0$. For example, for the monomer-only initial condition $c_k(t = 0) = \delta_{k,1}$,

$$c_k(t) = \frac{k}{\sqrt{4\pi Dt}} \, e^{-k^2/4Dt}. \tag{8.4.24}$$

This result was apparently first derived by Bramson & Griffeath (1980a) who exploited the equivalence between the time-reversed aggregation process and the Voter-model [Liggett (1985)] and then adapted the known Voter-model solution to determine the cluster concentrations. Subsequently, exact solutions for the cluster concentrations, based on direct approaches, were independently found by Spouge (1988a), Takayasu et al. (1988), and Thompson (1989).

The simple behavior of the cluster concentrations ultimately stems from the fact that, in one dimension, the mass contained inside a fixed-length interval undergoes a diffusion process. This mass can change either by a cluster just inside the interval hopping out, or by a cluster just outside the interval hopping in. The equation of motion that describes the mass evolution in the interval also satisfies an absorbing boundary for intervals of zero length. It is these two attributes that lead to an equivalent first-passage description for $c_k(t)$.

8.4.5.2. Solution for the Cluster Concentrations

Our presentation follows closely that of Thompson (1989). The basic quantity that we study to determine the cluster mass distribution is $A_{i,j}^M$, defined as the set of states in which there is a total mass M in the site interval (i, j) (Fig. 8.14). It is convenient to view the mass as being located at the midpoints

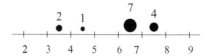

Fig. 8.14. Example configuration that contributes to $A_{3,8}^{14}$ – four clusters within (3, 8) whose total mass is 14. The numbers over the dots denote the individual cluster masses.

between sites so as to include the possibility of a zero-length interval (i, i) that necessarily contains zero mass. In an interval that is part of $A_{i,j}^{M}$, the mass may be distributed arbitrarily among clusters as long as the total mass equals M.

We now enumerate the possibilities that contribute to the change of the mass in this interval. The bookkeeping is somewhat tedious, though straightforward. The crucial point is that in terms of the sets $A_{i,j}^{M}$ a simple diffusion equation emerges whose solution immediately gives the concentration of clusters of a given mass.

- Gain terms: These are illustrated on the left-hand side of Fig. 8.15 from top to bottom.
  - (i) Mass M in $(i, j - 1)$ and an arbitrary-mass k cluster in $(j - 1, j)$ leaves the interval by hopping to the right. Thus mass M exists in $(i, j - 1)$ but not in (i, j) – this is the state set $A_{i,j-1}^{M} - A_{i,j}^{M}$.
  - (ii) M in $(i + 1, j)$ and a k cluster in $(i, i + 1)$ hops to the left – state set $A_{i+1,j}^{M} - A_{i,j}^{M}$.
  - (iii) M in $(i, j + 1)$ and a k cluster in $(j, j + 1)$ hops left to enter (i, j) – state set $A_{i,j+1}^{M} - A_{i,j}^{M}$.
  - (iv) M in $(i - 1, j)$ and a k cluster in $(i - 1, i)$ hops right to enter (i, j) – state set $A_{i-1,j}^{M} - A_{i,j}^{M}$.
- Loss terms: Right-hand side of Fig. 8.15 from top to bottom.
  - (i) M in (i, j) and a k cluster in $(j - 1, j)$ hops right to leave (i, j) –

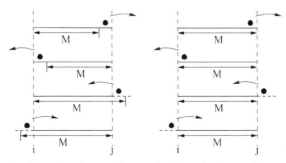

Fig. 8.15. Configurations that contribute to the change in the set of states with mass M in the interval (i, j). The left-hand side shows the four processes that lead to a gain in A_{ij}^{M}, and the right-hand side shows the four loss processes.

state set $A_{i,j}^M - A_{i,j-1}^M$.

(ii) M in (i, j) and a k cluster in $(i, i + 1)$ hops left to leave (i, j) – state set $A_{i,j}^M - A_{i+1,j}^M$.

(iii) M in (i, j) and a k cluster in $(j, j + 1)$ hops left to enter (i, j) – state set $A_{i,j}^M - A_{i,j+1}^M$.

(iv) M in (i, j) and a k cluster in $(i - 1, i)$ hops right to enter (i, j) – state set $A_{i,j}^M - A_{i-1,j}^M$.

The probability for each of these composite configurations can be expressed in terms of elemental configurations. For example, consider the probability $P(A_{i,j-1}^M - A_{i,j}^M)$ to have mass M in $(i, j - 1)$ but not in (i, j). By straightforward counting, the probability for these states to occur equals $P_{i,j-1}^M - P(A_{i,j}^M \mid 0)$. Here the first term is the probability that mass M occupies $(i, j - 1)$ *without* any occupancy restrictions on the boundary and the second term is the probability that mass M occupies (i, j) *and* $(j - 1, j)$ is unoccupied.

By adding all the processes enumerated above, we obtain the diffusionlike equation of motion

$$\frac{\partial P_{i,j}^M}{\partial t} = r\left(P_{i+1,j}^M + P_{i-1,j}^M + P_{i,j+1}^M + P_{i,j-1}^M - 4P_{i,j}^M\right), \quad (8.4.25)$$

where r is the hopping rate for each cluster per unit time. For a translationally invariant system, $P_{i,j}^M$ is just a function of $x = j - i$. We simplify further by taking the continuum limit in which the lattice spacing δx and hopping rate r simultaneously go to zero such that their ratio $D = r(\delta x)^2/2$ remains constant. Then the equation of motion reduces to

$$\frac{\partial P^M(x, t)}{\partial t} = D\frac{\partial^2 P^M(x, t)}{\partial x^2}, \quad (8.4.26)$$

which is valid for all $M \geq 0$.

Thus the interval occupation probabilities are governed by the diffusion equation! Again, this occurs because the mass can change only by local hopping events at the boundary of any interval. The boundary conditions for this diffusion equation are $P^0(0, t) = 1$, $P^M(0, t) = 0$, and $P^M(x, t) \to 0$ as $x \to \infty$. The former two are the continuum limits of the condition that there cannot be any mass in an interval of zero length. The last condition states that the mass in an infinitely long interval cannot remain finite.

Finally, to determine the density of clusters of fixed mass, we start with the fact that $P_{i,i+1}^M$ is the probability that a cluster of mass M occupies $(i, i + 1)$. In the continuum limit, we write $i \to x$, $i + 1 \to x + \delta x$, so that $P_{i,i+1}^M \to P^M(\delta x) = P^M(0) + \delta x \frac{\partial P^M}{\partial x} + \dots$. The zeroth-order term in this expansion is zero by definition. Then, by equating the first derivative to the probability that a cluster of mass M occupies a region of size δx, we obtain

$$c_k(t) = \frac{\partial P^k}{\partial x}\bigg|_{x=0}. \tag{8.4.27}$$

For a lattice system in which each site is initially occupied by a single particle, $P^M(x) = \delta(x - M)$ because there is exactly a mass M inside an interval of length M. With this initial condition and with absorbing boundary conditions, Eq. (8.4.26) is equivalent to the motion of a particle that starts at $x = k$, diffuses on the positive line, and is absorbed whenever $x = 0$ is reached. The solution to this diffusion problem with an absorbing boundary is just our familiar image solution given in Eq. (3.2.1), from which the corresponding flux at the origin is just the cluster concentration quoted in Eq. (8.4.24).

As a postscript, the approach outlined above can be extended to account for a steady monomer source, in addition to the irreversible aggregation. If the monomer input is spatially homogeneous, then the mass inside a fixed-length interval can change both because of the input and because of the hopping of clusters at the interval boundaries. In the continuum limit, the combination of these two processes leads to the equation of motion for the probability that an interval of length x contains mass M [Thompson (1989)]:

$$\frac{\partial P^M(x, t)}{\partial t} = D\frac{\partial^2 P^M(x, t)}{\partial x^2} + x[P^{M-1}(x, t) - P^M(x, t)]. \tag{8.4.28}$$

Clearly, the last two terms account for the creation and the loss of M clusters that are due to the input. In the steady-state solution to this equation, the concentration of clusters of mass k decays asymptotically as $k^{-4/3}$ [Takayasu et al. (1988), Thompson (1989), Doering & ben-Avraham (1989), and Takayasu (1989)]. In contrast, in the mean-field limit, this same steady-state aggregation process has a $k^{-3/2}$ decay for the asymptotic mass distribution [Field & Saslaw (1965), White (1982), and Crump & Seinfeld (1982)].

Finally, for a steady input at a single point $r = 0$, one can generalize the approaches of Spouge (1988a) or ben-Avraham (1990a) to obtain the exact cluster distribution as a function of the distance r from the input [Cheng, Redner, & Leyvraz (1989), Hinrichsen, Rittenberg, & Simon (1997), and Ha, Park, & den Nijs (1999)]. If the source "turns on" at $t = 0$, some of the basic features of aggregation with isotropic diffusion and this steady point monomer input are the following:

- The total density of clusters grows as $\ln t$.
- For $r > (Dt)^{1/2}$, the system is virtually empty, whereas for $r < (Dt)^{1/2}$ the cluster concentrations are close to their steady-state values; that is, $c_k(r, t) \to c_k(r)$.

- For $r \ll k^{1/2}$, $c_k(r) \sim k^{-3/2}$, just as in aggregation with a spatially uniform monomer input.
- For $r \gg k^{1/2}$, $c_k(r) \sim k/r^5$.

These results were further generalized in the above-cited publications to account for a bias in the motion of cluster as well as a steady input at $r = 0$ and a sink at $r = L$.

8.5. Ballistic Annihilation

Our final topic on reactions is the kinetics of ballistically controlled annihilation in one dimension. Here, each particle moves at a constant velocity until it meets another particle and annihilation occurs. This is a purely deterministic system whose evolution depends on only the initial particle velocities and positions. Very rich phenomenology arises for simple and generic velocity distributions. We will focus on two cases for which first-passage ideas help determine the kinetics. These are (i) the two-velocity and (ii) the three-velocity models.

In the former case, the system contains particles with velocities v_1 and v_2 only. Without loss of generality, we redefine these velocities to be $+v$ and $-v$. If the initial concentrations of right-moving and left-moving particles are unequal, there is an exponential decay to a final state in which only one velocity species remains. For equal initial concentrations of the two species, there is a $t^{-1/2}$ asymptotic decay of the density [Elskens & Frisch (1985)]. In the three-velocity model, we consider velocities $\pm v$ and $v = 0$ with respective initial concentrations $c_\pm(0)$ and $c_0(0)$. This is a phenomenologically rich system with many different kinetic regimes, depending on the initial concentrations [Krapivsky et al. (1995)]. For one particular initial state for the densities of the three species, a beautiful exact solution was found in which the particle densities all decay as $t^{-2/3}$ [Droz et al. (1995a, 1995b)]. For the latter reaction, we will focus on situations in which the kinetics can be mapped onto a suitably defined first-passage process. Intriguing kinetics arises in these cases [Krapivsky et al. (1995)].

8.5.1. Two-Velocity Model

For simplicity, suppose that particles are initially distributed according to a Poisson distribution with unit density. With probability p, a given particle has a velocity $+v$, whereas with probability $q = 1 - p$, the particle velocity is $-v$. Other well-behaved density distributions – those in which the mean

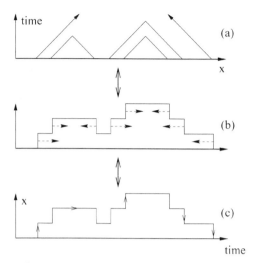

Fig. 8.16. (a) One-dimensional space–time trajectories of particles with velocities $+v$ and $-v$, and (b) a corresponding terraced surface of approaching kinks and antikinks. (c) This surface can also be viewed as the space–time trajectory of a one-dimensional random walk.

separation between nearest neighbors is finite – give qualitatively similar results [Elskens & Frisch (1985)]. We first give a pictorial argument for the concentration decay by mapping the particle trajectories onto an equivalent first-passage process (Fig. 8.16).

To determine the survival of a specific right-moving particle we identify its collision partner. We may accomplish this easily by constructing a geometric equivalence between the particle system and a terraced surface, as indicated in Fig. 8.16(b). A right-moving particle is equivalent to a "kink" that moves to the right at velocity v and a left-moving particle is equivalent to a left-moving "antikink." All particles to the left of the initial particle can be ignored in determining the collision partner. With this mapping, the reaction kinetics is equivalent to the smoothing of a terraced surface of kinks and antikinks that move with respective velocities $+v$ and $-v$. The reaction partner of any kink is simply the closest antikink that is at the same elevation as the initial kink.

We may also view the kinked surface profile as the space–time trajectory of an equivalent random walk in which the horizontal axis is now the "time" and the vertical axis is the displacement [Fig. 8.16(c)]. Here the step lengths of the walk are fixed and the time between hopping events has a Poisson distribution. The meeting of the initial kink and a partner antikink [which must be the $(2n + 1)$th neighbor] is clearly equivalent to the first return of the corresponding random walk to its initial position after $2n + 2$ steps.

We now calculate the survival probability of a right-moving, or positive, particle of velocity v at time t, $S_+(t, v)$ following an approach similar to that of Elskens and Frisch (1985). A positive particle survives to time t if its collision partner is more distant than $2vt$. Thus we may write

$$S_+(t, v) = \sum_{n=0}^{\infty} C_{2n+1} P(x_{2n+1} > 2vt), \qquad (8.5.1)$$

where C_{2n+1} is the probability that the initial particle collides with its $(2n + 1)$th neighbor and the second factor is the probability that their separation is greater than $2vt$. Because C_{2n+1} is the probability that a random walk returns after $2n + 2$ steps, we have (see Section 3.4)

$$C_{2n+1} = 2(pq)^{n+1} \frac{(2n)!}{(n + 1)!n!}. \qquad (8.5.2)$$

To have the $(2n + 1)$th neighbor be more distant than $2vt$, we must have $2n$ particles occupying this interval of length $2vt$. Therefore, for a Poisson distribution or particles,

$$P(x_{2n+1} > 2vt) = \int_{2vt}^{\infty} \frac{z^{2n}}{(2n)!} e^{-z} dz. \qquad (8.5.3)$$

Assembling these results and using the series representation of the modified Bessel function, we obtain

$$S_+(t, v) = \sum_{n=0}^{\infty} C_{2n+1} P(x_{2n+1} > 2vt)$$

$$= \sum_{n=0}^{\infty} \int_{2vt}^{\infty} \frac{z^{2n}}{(2n)!} e^{-z} 2(pq)^{n+1} \frac{(2n)!}{(n + 1)!n!}$$

$$= 1 - \sqrt{4pq} \int_{0}^{2vt} I_1(\sqrt{4pq}z) \frac{e^{-z}}{z} dz. \qquad (8.5.4)$$

For $t \to \infty$, $S_+(t, v) \to p - q$ for $p > q$. For the interesting case of $p = 1/2$, one way to evaluate Eq. (8.5.4) by elementary means is to first use the recursion $I_0(z) - I_2(z) = 2I_1(z)/z$ to eliminate the factor $1/z$ in the integral [Abramowitz & Stegun (1972)]. Then we integrate $\int I_0(z)e^{-z} dz$ by parts twice and use $I_1' = \frac{1}{2}(I_0 + I_2)$ to ultimately give

$$S_+(t, v) = e^{-2vt} [I_0(2vt) + I_1(2vt)]$$

$$\sim \frac{1}{\sqrt{\pi vt}}. \qquad (8.5.5)$$

This power-law decay can also be understood qualitatively in terms of initial density fluctuations. For the random initial condition with equal densities of right- and left-moving particles, there will typically be an imbalance of magnitude $\delta n \simeq \sqrt{c(0)\ell}$ in their numbers within a region of length ℓ. After a time $t \approx \ell/v$, all available collision partners within this region will have reacted and only the local majority particle species will remain. The concentration of this local majority species will be of the order of $c(t) \simeq \delta n/\ell \simeq \sqrt{c(0)/vt}$. Thus the survival probability $S(t) = c(t)/c(0) \propto 1/\sqrt{c(0)vt}$. This reasoning is essentially the same as that used to determine the time dependence of the concentration in diffusion-controlled two-species annihilation [Toussaint & Wilczek (1983), Kang & Redner (1984a, 1985)].

8.5.2. *Three-Velocity Model*

8.5.2.1. *Basic Phenomenology*

A natural and surprisingly rich extension of the two-velocity model is to the case of three velocities, $+v$, $-v$, and 0. It is convenient to represent the system in the composition triangle spanned by the initial concentrations of the three species, c_+, c_-, and c_0 (Fig. 8.17). This triangle is divided into regions where a single species eventually predominates in the long-time limit, as indicated in the figure. More interesting behavior occurs along the phase boundaries that separate the regions where a single species ultimately survives. Between the $+$ and the $-$ phases, the concentration of the minority species $c_0(t)$ decays as t^{-1}, while $c_+(t) = c_-(t) \sim t^{-1/2}$. This behavior persists until a multicritical point at $c_0(0) = 1/4$ [and $c_\pm(0) = 3/8$], where the concentrations of all three species asymptotically decay as $t^{-2/3}$ [Droz et al. (1995a, 1995b)]. On the other hand, along the $+0$ and the -0 boundaries, the minority species

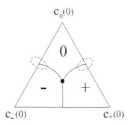

$c_0(0)$

$c_-(0)$ $c_+(0)$

Fig. 8.17. Composition triangle of the one-dimensional three-velocity model. Along the vertical line, $c_\pm(t) \sim t^{-1/2}$, while $c_0(t) \sim t^{-1}$. When $c_\pm(0) = 3/8$, $c_0(0) = 1/4$ (dot), all species decay as $t^{-2/3}$. The region within the dashed ellipses corresponds to the fast-impurity problem. The symbols in the interior of the triangle indicate the surviving species in the long-time limit.

asymptotically decays as $\exp(-\ln^2 t)$ while the majority species decays as $t^{-1/2}$ [Krapivsky et al. (1995)].

We now focus on vanishingly small concentrations of one species, in which the kinetics can be recast as first-passage problems. The two basic cases are

- "slow-impurity" limit – $c_0(0) \ll c_+(0) = c_-(0)$,
- "fast-impurity" limit – $c_\mp(0) \ll c_\pm(0) = c_0(0)$.

The slow-impurity limit applies, in fact, to more general three-velocity systems in which the speed of the slowest species is less than that of the majority. Similarly, the fast-impurity limit applies as long as the speed of the minority species is less than that of the majority when evaluated in the center-of-mass frame of the majority species.

8.5.2.2. Slow-Impurity Limit

Clearly, the survival probability of a single slow impurity factorizes into the product of one-sided survival probabilities; the impurity survives if it does not collide with any particle to its right *and* any particle to its left. Thus the slow-impurity survival probability $S_{SI}(t)$ must decay as t^{-1}. We calculate this survival probability for a slow impurity, which starts at the origin with speed $w < v$, in a sea of majority species that have a Poisson distribution of initial separations with unit density. This survival probability equals the product of one-sided survival probabilities associated with appropriately defined two-velocity problems; that is, $S_{SI}(t) = S_-(t, w)S_+(t, w)$. Because $S_-(t, w) = S_+(t, -w)$, we use Eq. (8.5.5) to obtain

$$S_{SI}(t, w) = e^{-2t}\{I_0[(v + w)t] + I_1[(v + w)t]\}$$
$$\times \{I_0[(v - w)t] + I_1[(v - w)t]\}$$
$$\sim \frac{1}{\pi\sqrt{v^2 - w^2}}\frac{1}{t}. \tag{8.5.6}$$

As expected intuitively, the survival probability of a slow impurity decays as t^{-1}.

8.5.2.3. Fast-Impurity Limit

It is convenient to first transform to a reference frame such that the two majority species move with equal and opposite velocities $\pm v$, while the minority species moves at a faster velocity $w > v$. A Lifshitz tail argument is presented, in conjunction with first-passage ideas, to argue that the density of fast impurities decays faster than any power law, but slower than any stretched exponential function [Krapivsky et al. (1995)].

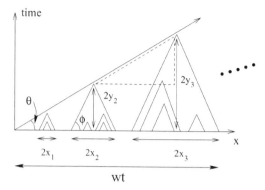

Fig. 8.18. World line of a long-lived fast impurity with velocity w in a background of equal concentrations of background $\pm$ particles with respective velocities $\pm v$. Successive triangles of self-annihilating background particles are indicated. Here $\cot\theta = w$ and $\cot\phi = v$.

The basis of this argument is to identify a subset of configurations that gives the dominant contribution to the survival probability of the impurity, but that is sufficiently simple to evaluate their probability asymptotically. For the impurity to survive to time t, the background $\pm$ particles within a distance wt of the impurity must all annihilate among themselves, with each annihilation occurring before the trajectory of the impurity passes by (Fig. 8.18). On a space–time diagram, these self-annihilation events appear as a sequence of isosceles triangles that must terminate below to the world line of the impurity. The dominant contribution to the impurity survival probability stems from sequences of systematically larger self-annihilation triangles which just miss the impurity world line.

To determine the probability of these configurations, we need the size of each self-annihilation triangle and the number of such triangles up to time t. From the dashed right triangle in Fig. 8.18, the aspect ratio of any self-annihilation triangle $y_n/x_n = \tan\phi = 1/v$, while successive triangle sizes are related by $(y_{n+1} - y_n)/(x_{n+1} - x_n) = \tan\theta = 1/w$. Thus the base of the nth triangle grows as $x_{n+1} = x_n [(w + v)/(w - v)]$ or $x_n \propto [(w + v)/(w - v)]^n \equiv \gamma^n$. The number of triangles in the self-annihilation sequence is determined by the condition that the sum of the base lengths plus the distances between triangles equals the total distance traveled by the impurity. Because each intertriangle distance is of the order of 1, the latter are a negligible correction to the total length. Hence we write $2(x_1 + x_2 + \ldots x_N) \sim wt$ and use $x_n = x_0\gamma^n$ to find $N \sim \ln t / \ln \gamma$.

From our previous discussion of the two-velocity model, a self-annihilation triangle of base x_n corresponds to the first passage of a random walk at time x_n (Fig. 8.16) and thus occurs with probability $x_n^{-3/2}$. The fast-impurity survival probability $S_{FI}(t, w)$ is the product of the occurrence probabilities of an optimal sequence of self-annihilation triangles. This product is

$$S_{FI}(t) \sim \prod_{n=0}^{N} (2x_0\gamma^n)^{-3/2}$$
$$\propto \gamma^{-3N^2/4}$$
$$\sim \exp(-\ln^2 t / \tfrac{4}{3} \ln \gamma). \tag{8.5.7}$$

Thus the survival probability decays faster than any power law in time, but slower than any power of an exponential. It is amusing that this functional form also occurs in the capture reaction from Subsection 8.4.2. Perhaps there is an underlying connection between these two apparently different systems. Finally, because of the faster-than-power-law decay of the survival probability, the mean lifetime of the impurity is finite. From $\langle t \rangle = \int_0^\infty S(t')\,dt'$, a straightforward evaluation of the integral gives the simple result $\langle t \rangle \propto [(w+v)/(w-v)]^{2/3}$, in excellent agreement with numerical simulations [Krapivsky et al. (1995)].

Although ballistic annihilation reactions with two- or three-element discrete-velocity distributions are amenable both to exact analysis [Elskens & Frisch (1985), and Droz et al. (1995a, 1995b)] and to simple first-passage arguments [Krapivsky et al. (1995)], our understanding of systems with continuous-velocity distributions is much more limited [Ben-Naim et al. (1993a)], although current numerical evidence suggests universal kinetic behavior [Rey, Droz, & Piasecki (1998)]. This continuous-velocity model might be an interesting direction for applying first-passage ideas in future studies.

References

Abramowitz, M. & Stegun, I. A. 1972. *Handbook of Mathematical Functions* (Dover, New York).

Agmon, N. & Glasser, M. L. 1986. Complete asymptotic expansion for integrals arising from one-dimensional diffusion with random traps. *Phys. Rev. A* **34**, 656–658.

Alexander, S., Bernasconi, J., Schneider, W. R., & Orbach, R. 1981. Excitation dynamics in random one-dimensional systems. *Rev. Mod. Phys.* **53**, 175–198.

Amar, J. & Family, F. 1989. Diffusion annihilation in one dimension and kinetics of the Ising model at zero temperature. *Phys. Rev. A* **41**, 3258–3262.

Amit, D. J. 1989. *Modeling Brain Function: The World of Attractor Neural Networks* (Cambridge University Press, New York).

Aslangul, C. & Chvosta, P. 1995. Diffusion on a random comb: distribution function of the survival probability. *J. Stat. Phys.* **78**, 1403–1428.

Aslangul, C., Pottier, N., & Chvosta, P. 1994. Analytic study of a model of diffusion on a random comblike structure. *Physica* **203A**, 533–565.

Atkinson, D. & van Steenwijk, F. J. 1999. Infinite resistive lattices. *Am. J. Phys.* **67**, 486–492.

Austin, R. H., Beeson, K. W., Eisenstein, L. Frauenfelder, H., & Gunsalus, I. C. 1975. Dynamics of ligand-binding to myoglobin. *Biochem. J.* **14**, 5355–5373.

Azbel, M. Ya. 1982. Diffusion: a layman's approach and its applications to one-dimensional random systems. *Solid State Commun.* **43**, 515–517.

Bak, P. & Sneppen, K. 1993. Punctuated equilibrium and criticality in a simple model of evolution. *Phys. Rev. Lett.* **71**, 4083–4086.

Balagurov, B. Ya. & Vaks, V. G. 1974. Random walks of a particle on lattices with traps. *Sov. Phys. JETP* **38**, 968–971.

Bak, P. 1996. *How Nature Works: The Science of Self-Organized Criticality* (Copernicus, New York).

Bak, P., Tang, C., & Wiesenfeld, K. 1987. Self-organized criticality: an explanation of the 1/f noise. *Phys. Rev. Lett.* **59**, 381–384.

Ball, R. C. & Brady, R. M. 1985. Large scale lattice effect in diffusion-limited aggregation. *J. Phys. A* **18**, L809–L813.

Ball, R. C., Brady, R. M., Rossi, G., & Thompson, B. R. 1985. Anisotropy and cluster growth by diffusion-limited aggregation. *Phys. Rev. Lett.* **55**, 1406–1409.

Bar-Haim, A. & Klafter J. 1999. Escape from a fluctuating medium: a master equation and trapping approach. *Phys. Rev. E* **60**, 2554–2558.

Barenblatt, G. I. 1996. *Scaling, Self-Similarity, and Intermediate Asymptotics* (Cambridge University Press, New York).

ben-Avraham, D. 1988. Computer simulation methods for diffusion-controlled reactions. *J. Chem. Phys.* **88**, 941–947.

ben-Avraham, D. 1998. Complete exact solution of diffusion-limited coalescence, $A + A \to A$. *Phys. Rev. Lett.* **81**, 4756–4759.

ben-Avraham, D., Burschka, M. A., & Doering, C. R. 1990a. Statics and dynamics of a diffusion-limited reaction: anomalous kinetics, nonequilibrium self-ordering and a dynamics transition. *J. Stat. Phys.* **60**, 695–728.

ben-Avraham, D., Considine, D., Meakin, P., Redner, S., & Takayasu, H. 1990b. Saturation transition in a monomer–monomer model of heterogeneous catalysis. *J. Phys.* A **23**, 4297–4312.

ben-Avraham, D. & Havlin, S. 2000. *The Ant in the Labyrinth: Diffusion and Kinetics of Reactions in Fractals and Disordered Media* (Cambridge University Press, New York).

Ben-Naim, E., Redner, S., & Leyvraz, L. 1993a. Decay kinetics of ballistic annihilation. *Phys. Rev. Lett.* **70**, 1890–1893.

Ben-Naim, E., Redner, S., & Weiss, G. H. 1993b. Partial absorption and "virtual" traps. *J. Stat. Phys.* **71**, 75–88.

Bender, C. M. & Orszag, S. A. 1978. *Advanced Mathematical Methods for Scientists and Engineers* (McGraw-Hill, New York).

Bhattacharya, R. N. & Waymire, E. C. 1990. Stochastic Processes with Applications (Wiley, New York).

Bicout, D. J. & Burkhardt, T. W. 2000. Absorption of a randomly accelerated particle: gambler's ruin in a different game. *J. Phys.* A **33**, 6835–6841.

Bier, M. & Astumian, R. D. 1993. Matching a diffusive and a kinetic approach for escape over a fluctuating barrier. *Phys. Rev. Lett.* **71**, 1649–1652.

Binney, J. J., Dowrick, N. J., Fisher, A. J., & Newman M. E. J. 1992. The Theory of Critical Phenomena: An Introduction to the Renormalization Group (Clarendon, Oxford, U.K.).

Bixon, M. & Zwanzig, R. 1981. Diffusion in a medium with static traps. *J. Chem. Phys.* **75**, 2354–2356.

Blumen, A., Zumofen, G., & Klafter, J. 1984. Target annihilation by random walkers. *Phys. Rev.* B **30**, 5379–5382.

Bonner, R. F., Nossal, R., Havlin, S., & Weiss, G. H. 1987. Model for photon migration in turbid biological media. *J. Opt. Soc. Am.* A **4**, 423–432.

Bouchaud, J.-P. & Georges, A. 1990. Anomalous diffusion in disordered media: statistical mechanisms, models and physical applications. *Phys. Rep.* **195**, 127–293.

Bramson, M. & Griffeath, D. 1980a. Clustering and dispersion rate for some for interacting particle systems on Z^1. *Ann. Prob.* **8**, 183–213.

Bramson, M. & Griffeath, D. 1980b. Asymptotics of interacting particles systems on Z^d. *Z. Wahrscheinlichkeitstheor. Verwandte Geb.* **53**, 183–196.

Bramson, M. & Griffeath, D. 1991. Capture problems for coupled random walks, in *Random Walks, Brownian Motion, and Interacting Particle Systems: a Festschrift in Honor of Frank Spitzer*, R. Durrett & H. Kesten, eds. (Birkhauser, Boston), pp. 153–188.

Breiman, L. 1966. First exit time from the square root boundary, in *Proceedings of the Fifth Berkeley Symposium on Mathematical Statistics and Probability* (University of California, Berkeley, CA), Vol. 2, pp. 9–16.

Brito, V. P. & Gomes, M. A. F. 1995. Block avalanches on a chute. *Phys. Lett. A* **201**, 38–41.

Bulsara, A. R., Elston, T. C., Doering, C. R., Lowen, S. B., & Lindenberg, K. 1996. Cooperative behavior in periodically driven noisy integrate-fire models of neuronal dynamics. *Phys. Rev. E* **53**, 3958–3969.

Bulsara, A. R., Lowen, S. B., & Rees, C. D. 1994. Cooperative behavior in the periodically modulated Wiener process: noise-induced complexity in a model neuron. *Phys. Rev. E* **49**, 4989–5000.

Carslaw, H. S. & Jaeger, J. C. 1959. *Conduction of Heat in Solids* (Oxford University Press, Oxford, U.K.).

Chandrasekhar, S. 1943. Stochastic problems in physics and astronomy. *Rev. Mod. Phys.* **15**, 1–89.

Cheng, Z. & Redner, S. 1990. Kinetics of fragmentation. *J. Phys. A* **23**, 1223–1258.

Cheng, Z., Redner, S., & Leyvraz, F. 1989. Coagulation with a steady point monomer source. *Phys. Rev. Lett.* **62**, 2321–2324.

Chikara, R. & Folks, L. 1989. *The Inverse Gaussian Density* (Dekker, New York).

Clifford, P. & Sudbury, A. 1973. A model for spatial conflict. *Biometrika* **60**, 581–588.

Coats, K. H. & Smith, B. D. 1964. Dead-end pore volume and dispersion in porous media. *Soc. Petrol. Eng. J.* **4**, 73–84.

Considine, D. & Redner, S. 1989. Repulsion of random and self-avoiding walks from excluded points and lines. *J. Phys. A* **22**, 1621–1638.

Cox, T. J. & Griffeath, D. 1986. Diffusive clustering in the two dimensional voter model. *Ann. Prob.* **14**, 347–370.

Cox, D. & Miller, H. 1965. *The Theory of Stochastic Processes* (Chapman & Hall, London, U.K.).

Crank, J. 1987. *Free and Moving Boundary Problems* (Oxford University Press, Oxford, U.K.).

Crump. J. G. & Seinfeld, J. H. 1982. On existence of steady-state solutions to the coagulation equations. *J. Colloid Interface Sci.* **90**, 469–476.

Cserti, J. 2000. Application of the lattice Green's function for calculating the resistance of an infinite networks of resistors. *Am. J. Phys.* **68**, 896–906.

Daniels, H. E. 1969. The minimum of a stationary Markov process superimposed on a U-shaped trend. *J. Appl. Prob.* **6**, 399–408.

Daniels, H. E. 1982. Sequential tests constructed from images. *Ann. Statist.* **10**, 394–400.

Danckwerts, P. V. 1958a. Continuous flow systems: distribution of residence-times. *Chem. Eng. Sci.* **2**, 1–11.

Danckwerts, P. V. 1958b. Local residence-times in continuous-flow systems. *Chem. Eng. Sci.* **9**, 74–77.

Darling, D. A. & Siegert, A. J. F. 1953. The first passage problem for a continuous Markov process. *Ann. Math. Statist.* **24**, 624–639.

De Blassie, R. D. 1987. Exit times from cones in $\mathbf{R}^n$ of Brownian motion. *Z. Wahrscheinlichkeitsthteor. Verwandte Geb.* **74**, 1–29.

de Boer, J., Derrida, B., Flyvbjerg, H., Jackson, A. D., & Wettig, T. 1994. Simple model of self-organized biological evolution. *Phys. Rev. Lett.* **73**, 906–909.

de Boer, J., H., Jackson, A. D., & Wettig, T. 1995. Criticality in simple models of evolution. *Phys. Rev. E* **51**, 1059–1074.

de Gennes, P. G. 1993. Capture of an "ant" by fixed traps on a percolation network *C. R. Acad. Sci. Ser. II* **296**, 881–885.

De Menech, M., Stella, A. L., & Tebaldi, C. 1998. Rare events and breakdown of simple scaling in the Abelian sandpile model. *Phys. Rev. E* **58**, R2677–R2680.

Derrida, B. 1983. Velocity and diffusion constant of a periodic one-dimensional hopping model. *J. Stat. Phys.* **31**, 433–450.

Dhar, D. & Ramaswamy, R. 1989. Exactly solved model of self-organized critical phenomena. *Phys. Rev. Lett.* **63**, 1659–1662.

Dickman, R., Vespignani, A., & Zapperi, S. 1998. Self-organized criticality as an absorbing-state phase transition. *Phys. Rev. E* **57**, 5095–5105.

Doering, C. R. & ben-Avraham, D. 1988. Interparticle distribution functions and rate equations for diffusion-limited reactions. *Phys. Rev. A* **38**, 3035–3041.

Doering, C. R. & ben-Avraham, D. 1989. Diffusion-limited coagulation in the presence of particle input: exact results in one dimension. *Phys. Rev. Lett.* **62**, 2563–2566.

Doering, C. R. & Gadoua, J. C. 1992. Resonant activation over a fluctuating barrier. *Phys. Rev. Lett.* **69**, 2318–2321.

Donsker, M., & Varadhan, S. R. S. 1975. Asymptotic for the Wiener sausage. *Commun. Pure Appl. Math.* **28**, 525–565.

Donsker, M., & Varadhan, S. R. S. 1979. On the number of distinct sites visited by a random walk. *Commun. Pure Appl. Math.* **32**, 721–747.

Doyle, P. G. & Snell, J. L. 1984. Random Walks and Electric Networks, Carus Mathematical Monographs #22 (Mathematical Association of America, Oberlin, OH).

Drake, R. L. 1972. A general mathematical servey of the coagulation equation, in *Topics in Current Aerosol Research*, G. M. Hidy and J. R. Brock, eds. (Pergamon, Oxford, U.K.), Vol. III, Part 2, pp. 201–376.

Droz, M., Rey, P. A., Frachebourg, L., & Piasecki, J. 1995a. New analytic approach to multivelocity annihilation in the kinetic theory of reactions. *Phys. Rev. Lett.* **75**, 160–163.

Droz, M., Rey, P. A., Frachebourg, L., & Piasecki, J. 1995b. Ballistic-annihilation kinetics for a multivelocity one-dimensional ideal gas. *Phys. Rev. E* **51**, 5541–5548.

Dynkin, E. B. & Yushkevich, A. A. 1969. *Markov Processes: Theorems and Problems* (Plenum, New York).

Eizenberg, N. & Klafter, J. 1995. Molecular motion under stochastic gating. *Chem. Phys. Lett.* **243**, 9–14.

Elskens, Y. & Frisch, H. L. 1985. Annihilation kinetics in the one-dimensional ideal gas. *Phys. Rev. A* **31**, 3812–3816.

Ernst, M. H. 1985. Kinetic Theory of Clustering, in *Fundamental Problems in Statistical Physics*, VI, E. G. D. Cohen, ed. (Elsevier, New York), pp. 329–364.

Faber, T. E. 1995. *Fluid Dynamics for Physicists* (Cambridge University Press, Cambridge, U.K.).

Family, F. & Landau, D. P. 1984. *Kinetics of Aggregation and Gelation* (North-Holland, Amsterdam).

Feller, W. 1968. *An Introduction to Probability Theory and Its Applications* (Wiley, New York).

Feynman, R. P., Leighton, R. B., & Matthew Sands, M. 1963. *The Feynman Lectures on Physics* (Addison-Wesley, Reading, MA), Vol. II, Chap. 41.

Field, G. B. & Saslaw, W. C. 1965. A statistical model of the formation of stars and interstellar clouds. *Astrophys. J.* **142**, 568–583.

Fienberg, S. E. 1974. Stochastic models for single neuron firing trains: a survey. *Biometrics* **30**, 399–427.

Fisher, M. E. 1984. Walks, walls, wetting, and melting. *J. Stat. Phys.* **34**, 667–729.

Fisher, M. E. 1988. Diffusion from an entrance to an exit. *IBM J. Res. Dev.* **32**, 76–81.

Fisher, M. E. & Gelfand, M. P. 1988. The reunions of three dissimilar vicious walkers. *J. Stat. Phys.* **53**, 175–189.

Fletcher, J. E., Havlin, S., & Weiss, G. H. 1988. First passage time problems in time-dependent fields. *J. Stat. Phys.* **51**, 215–232.

Flyvbjerg, H. 1996. Simplest possible self-organized critical system. *Phys. Rev. Lett.* **76**, 940–943.

Flyvbjerg, H., Sneppen, K., & Bak, P. 1993. Mean field theory of a simple model of evolution. *Phys. Rev. Lett.* **71**, 4087–4090.

Frachebourg, L. & Krapivsky, P. 1996. Exact results for kinetics of catalytic reactions. *Phys. Rev. E* **53**, R3009–R3012.

Frachebourg, L., Krapivsky, P. L., & Redner, S. 1998. Alternating kinetics of annihilating random walks near a free interface. *J. Phys. A* **31**, 2791–2799.

Franklin, J. N. & Rodemich, E. R. 1968. Numerical analysis of an elliptic-parabolic partial differential equation. *SIAM J. Numer. Anal.* **5**, 680–716.

Friedlander, S. K. 1977. *Smoke, Dust and Haze: Fundamentals of Aerosol Behavior* (Wiley, New York).

Frisch, H. L., Privman, V., Nicolis, C., & Nicolis, G. 1990. Exact solution of the diffusion in a bistable piecewise linear potential. *J. Phys. A* **23**, L1147–L1153.

Galambos, J. 1987. *The Asymptotic Theory of Extreme Order Statistics* (Krieger, Malabar, FL).

Gálfi, L. & Rácz, Z. 1988. Properties of the reaction front in an $A + B \to C$ type reaction–diffusion process. *Phys. Rev. A* **38**, 3151–3154.

Gardiner, C. W. 1985. *Handbook of Stochastic Methods* (Springer-Verlag, New York).

Gerstein, G. L. & Mandelbrot, B. B. 1964. Random walk models for the spike activity of a single neuron. *Biophys. J.* **4**, 41–68.

Ghez, R. 1988. *A Primer on Diffusion Problems* (Wiley, New York).

Gillespie, D. T. 1992. *Markov Processes: An Introduction for Physical Scientists* (Academic, San Diego, CA).

Glauber, R. J. 1963. Time-dependent statistics of the Ising model. *J. Math. Phys.* **4**, 294–307.

Gnedenko, B. V. & Kolmogorov, S. N. 1954. *Limit Distributions for Sums of Independent Random Variables* (Addison-Wesley, Reading, MA).

Goldenfeld, N. 1992. *Lectures on Phase Transitions and the Renormalization Group* (Addison-Wesley, Reading, MA).

Goldhirsch, I. & Gefen, Y. 1986. Analytic method for calculating properties of random walks on networks. *Phys. Rev. A* **33**, 2583–2594.

Grassberger, P. & Procaccia, I. 1982a. The long time properties of diffusion in a medium with static traps. *J. Chem. Phys.* **77**, 6281–6284.

Grassberger, P. & Procaccia, I. 1982b. Diffusion and drift in a medium with randomly distributed traps. *Phys. Rev. A* **26**, 3686–3688.

Ha, M., Park, H., & den Nijs, M. 1999. Particle dynamics in a mass-conserving coalescence process. *J. Phys. A* **32**, L495–L502.

Hagan, P. S., Doering, C. R., & Levermore, C. D. 1989a. The distribution of exit times for weakly colored noise. *J. Stat. Phys.* **54** 1321–1352.

Hagan, P. S., Doering, C. R., & Levermore, C. D. 1989b. Mean exit times for particles driven by weakly colored noise. *SIAM J. Appl. Math.* **49,** 1480–1513.

Hänggi, P., Talkner, P., & Borkovec, M. 1990. Reaction rate theory: fifty years after Kramers. *Rev. Mod. Phys.* **62,** 251–341.

Hardy, G. H. 1947. *Divergent Series* (Oxford University Press, Oxford, U.K.).

Havlin, S. & ben-Avraham, D. 1987. Diffusion in disordered media. *Adv. Phys.* **36,** 695–798.

Havlin, S., Dishon, M., Kiefer, J. E., & Weiss, G. H. 1984. Trapping of random walks in two and three dimensions. *Phys. Rev. Lett.* **53,** 407–410.

Havlin, S., Kiefer, J. E., & Weiss, G. H. 1987. Anomalous diffusion on a random comblike structure. *Phys. Rev. A* **36,** 1403–1408.

Havlin, S., Larralde, H., Kopelman, R., & Weiss, G. H. 1990. Statistical properties of the distance between a trapping center and a uniform density of diffusing particles in two dimensions. *Physica* **169 A,** 337–341.

Hinch, E. J. 1991. *Perturbation Methods* (Cambridge University Press, Cambridge, U.K.).

Hinrichsen, H., Rittenberg, V., & Simon, H. 1997. Universality properties of the stationary states in the one-dimensional coagulation-diffusion model with external particle input. *J. Stat. Phys.* **86,** 1203–1235.

Hoffmann, K. H. & Sibani, P. 1988. Diffusion in hierarchies. *Phys. Rev. A* **38,** 4261–4270.

Holley, R. & Liggett, T. M. 1975. Ergodic theorems for weakly interacting infinite systems and the voter model. *Ann. Prob.* **3,** 643–663.

Huberman, B. A. & Kerszberg, M. 1985. Ultradiffusion: the relaxation of hierarchical systems. *J. Phys. A* **18,** L331–L336.

Hughes, B. D. 1995. *Random Walks and Random Environments* (Oxford University Press, New York).

Hughes, B. D. & Sahimi, M. 1982. Random walks on the Bethe lattice. *J. Stat. Phys.* **29,** 781–794.

Huse, D. A. & Fisher, M. E. 1984. Commensurate melting, domains walls, and dislocations. *Phys. Rev. B* **29,** 239–270.

Huang, K. 1987. *Statistical Mechanics,* 2nd ed. (Wiley, New York).

Iglói, F. 1992. Directed polymer inside a parabola: exact solution. *Phys. Rev. A* **45,** 7024–7029.

Ishimaru, A. 1978. *Wave Propagation and Scattering in Random Media* (Academic, New York).

Jackson, J. D. 1999. *Classical Electrodynamics,* 3rd ed. (Wiley, New York).

Jain, N. S. & Orey, S. 1968. On the range of random walk. *Isr. J. Math.* **6,** 373–380

Kahng, B. & Redner, S. 1989. Scaling of the first passage time and survival probability on exact and quasi self-similar structures *J. Phys. A* **22,** 887–902.

Kang, K. & Redner, S. 1984a. Scaling approach for the kinetics of recombination processes. *Phys. Rev. Lett.* **52,** 955–958.

Kang, K. & Redner, S. 1984b. Fluctuation effects in Smoluchowski reaction kinetics. *Phys. Rev. A* **30,** 2833–2836.

Kang, K. & Redner, S. 1984c. Novel behavior of biased correlated walks in one dimension. *J. Chem. Phys.* **80,** 2752–2755.

Kang, K. & Redner, S. 1985. Fluctuation-dominated behavior of diffusion-controlled reactions. *Phys. Rev. A* **32**, 435–447.

Karlin, S. & Taylor, H. M. 1975. *A First Course in Stochastic Processes*, 2nd ed. (Academic, New York).

Kayser, R. F. & Hubbard, J. B. 1983. Diffusion in a medium with a random distribution of static traps. *Phys. Rev. Lett.* **51**, 79–82.

Kehr, K. W. & Murthy, K. P. N. 1990. Distribution of mean first-passage times in random chains due to disorder. *Phys. Rev. A* **41**, 5728–5730.

Keirstead, W. P. & Huberman, B. A. 1987. Dynamical singularities in ultradiffusion. *Phys. Rev. A* **36**, 5392–5400.

Kemperman, J. H. B. 1961. *The Passage Problem for a Stationary Markov Chain* (University of Chicago Press, Chicago, IL).

Kesten, K. 1992. An absorption problem for several Brownian motions, in *Seminar on Stochastic Processes, 1991*, E. Cinlar, K. L. Chung, & M. J. Sharpe, eds. (Birkhäuser, Boston).

Khintchine, A. 1924. Über einen Satz der Wahrscheinlichkeitsrechnung. *Fund. Math.* **6**, 9–20.

Koplik, J., Redner, S., & Hinch, E. J. 1994. Tracer dispersion in planar multipole flows. *Phys. Rev. E* **50**, 4650–4671.

Koplik, J., Redner, S., & Wilkinson, D. 1988. Transport and dispersion in random networks with percolation disorder. *Phys. Rev. A* **37**, 2619–2636.

Kramers, H. A. 1940. Brownian motion in field of force and diffusion model of chemical reactions. *Physica* **7**, 284–304.

Krapivsky, P. L. 1992a. Kinetics of monomer–monomer surface catalytic reactions. *Phys. Rev. A* **45**, 1067–1072.

Krapivsky, P. L. 1992b. Kinetics of a monomer–monomer model of heterogeneous catalysis. *J. Phys. A* **25**, 5831–5840.

Krapivsky, K. L., Redner, S., & Leyvraz, F. 1995. Ballistic annihilation kinetics: the case of discrete velocity distributions. *Phys. Rev. E* **51**, 3977–3987.

Krapivsky, P. L. & Redner, S. 1996a. Life and death in an expanding cage and at the edge of a receding cliff. *Am. J. Phys.* **64**, 546–552.

Krapivsky, P. L. & Redner, S. 1996b. Kinetics of a diffusion capture process: lamb besieged by a pride of lions. *J. Phys. A* **29**, 5347–5357.

Ktitarev, D. V., Lubeck, S., Grassberger, P., & Priezzhev, V. B. 2000. Scaling of waves in the Bak–Tang–Wiesenfeld sandpile model. *Phys. Rev. E* **61**, 81–92.

Le Doussal, P. 1989. First-passage time for random walks in random environments. *Phys. Rev. Lett.* **62**, 3097–3097.

Le Doussal, P., Monthus, C., & Fisher, D. S. 1999. Random walkers in 1-d random environments: exact renormalization group analysis. *Phys. Rev. E* **59**, 4795–4840.

Lerche, H. R. 1986. *Boundary Crossing of Brownian Motion* (Springer-Verlag, Berlin).

Levenspiel, O. & Smith, W. K. 1957. Notes on the diffusion-type model for the longitudinal mixing of fluids in flow. *Chem. Eng. Sci.* **6**, 227–233.

Lifshitz, I. M., Gredeskul, S. A., & Pastur, L. A. 1988. *Introduction to the Theory of Disordered Systems* (Wiley, New York).

Lifshitz, I. M. & Slyozov, V. V. 1961. The kinetics of precipitation for supersaturated solid solutions. *J. Phys. Chem. Solids* **19**, 35–50.

Liggett, T. M. 1985. *Interacting Particle Systems* (Springer-Verlag, New York).

Lima, A. R., Moukarzel, C. F., Grosse, I., & Penna, T. J. P. 2000. Sliding blocks with random friction and absorbing random walks. *Phys. Rev. E* **61**, 2267–2271.

Lushnikov, A. A. 1987. Binary reaction $1 + 1 \to 0$ in one dimension. *Sov. Phys. JETP* **64**, 811–815.

Maritan, A. & Stella, A. 1986. Exact renormalisation group approach to ultradiffusion in a hierarchical structure. *J. Phys. A* **19**, L269–L273.

Marshall, T. W. & Watson, E. J. 1985. A drop of ink falls from my pen . . . . It comes to earth, I know not when. *J. Phys. A* **18**, 3531–3559.

Marshall, T. W. & Watson, E. J. 1987. The analytical solutions of some boundary layer problems in the theory of Brownian motion. *J. Phys. A* **20**, 1345–1354.

Maslov, S. & Zhang, Y.-C. 1995. Exactly solved model of self-organized criticality. *Phys. Rev. Lett.* **75**, 1550–1553.

Masoliver, J. & Porrà, J. M. 1995. Exact solution to the mean exit time problem for free inertial processes driven by Gaussian white noise. *Phys. Rev. Lett.* **75**, 189–192.

Masoliver, J. & Porrà, J. M. 1996. Exact solution to the exit-time problem for an undamped free particle driven by Gaussian white noise. *Phys. Rev. E* **53**, 2243–2256.

Meakin, P. 1985. The structure of two-dimensional Witten–Sander aggregates. *J. Phys. A* **18**, L661–L666.

Meakin, P. 1986. Universality, nonuniversality, and the effects of anisotropy on diffusion-limited aggregation. *Phys. Rev. A*, 3371–3382.

Moffat, H. K. & Duffy, B. R. 1979. Local similarity solutions and their limitations. *J. Fluid Mech.* **96**, 299–313.

Monthus, C. & Texier, C. 1996. Random walk on the Bethe lattice and hyperbolic motion. *J. Phys. A* **29**, 2399–2409.

Montroll, E. 1965. Random walks on lattices, in *Proceedings of the Symposium on Applied Mathematics* (American Mathematical Society, Providence, RI), Vol. 16, pp. 193–230.

Montroll, E. & Weiss, G. H. 1965. Random walks on lattices. II. *J. Math. Phys.* **6**, 167–181.

Movaghar, B., Sauer, G. W., & Würtz, D. 1982. Time decay of excitations in the one-dimensional trapping problem. *J. Stat. Phys.* **27**, 472–485.

Murthy, K. P. N. & Kehr, K. W. 1989. Mean first-passage time of random walk on a random lattice. *Phys. Rev. A* **40**, 2082–2087 (Erratum, **41**, 1160).

Nagel, K. & Paczuski, M. 1995. Emergent traffic jams. *Phys. Rev. E* **51**, 2909–2918.

Nagel, K. & Schreckenberg, M. 1992. A cellular automaton model for freeway traffic. *J. Phys. I (Paris)* **2**, 2221–2229.

Nørrelykke, S. F. & Bak, P. 2000. Self-organized criticality in a transient system. *Phys. Rev. Lett.* (to be published).

Noskowicz, S. H. & Goldhirsch, I. 1988. Average versus typical mean first-passage time in a random random walk. *Phys. Rev. Lett.* **61**, 500–502.

Paczuski, M. & Boettcher, S. 1997. Avalanches and waves in the Abelian sandpile model. *Phys. Rev. E* **56**, R3745–R3748.

Pólya, G. 1921. Über eine Aufgabe der Wahrscheinlichkeitsrechnung betreffend die Irrfahrt im Strassennetz. *Mathematische Annalen* **84**, 149–160.

Prigogine, I. & Herman, R. 1971. *Kinetic Theory of Vehicular Traffic* (American Elsevier, New York).

Privman, V. 1997. *Nonequilibrium Statistical Mechanics in One Dimension* (Cambridge University Press, New York).

Privman, V. & Frisch, H. L. 1991. Exact solution of the diffusion problem for a piecewise linear barrier *J. Chem. Phys.* **94**, 8216–8219.

Rácz, Z. 1985. Diffusion-controlled annihilation in the presence of particle sources: exact results in one dimension. *Phys. Rev. Lett.* **55**, 1707–1710.

Raykin, M. 1993. First passage probability of a random walk on a disordered one-dimensional lattice. *J. Phys. A* **26**, 449–466.

Redner, S. 1997. Survival in a random velocity field. *Phys. Rev. E* **56** 4967–4972.

Redner, S. & ben-Avraham, D. 1990. Nearest-neighbor distances of diffusing particles from a single trap. *J. Phys. A* **23**, L1169–L1173.

Redner, S. & Kang, K. 1984. Kinetics of the scavenger reaction. *J. Phys. A* **17**, L451–L455.

Redner, S. & Krapivsky, P. L. 1996. Diffusive escape in a nonlinear shear flow: life and death at the edge of a windy cliff. *J. Stat. Phys.* **82**, 999–1014.

Redner, S. & Krapivsky, P. L. 1999. Capture of the lamb: diffusing predators seeking a diffusing prey. *Am. J. Phys.* **67**, 1277–1283.

Redner, S., Koplik, J., & Wilkinson, D. 1987. Hydrodynamic dispersion on self-similar structures: the convective limit. *J. Phys. A* **20**, 1543–1555.

Reif, F. 1965. *Fundamentals of Statistical and Thermal Physics* (McGraw-Hill, New York).

Reiss, H., Patel, J. R., & Jackson, K. A. 1977. Approximate analytical solutions of diffusional boundary-value problems by the method of finite zone continuity. *J. Appl. Phys.* **48**, 5274–5278.

Rey, P.-A., Droz, M., & Piasecki, J. 1998. Search for universality in one-dimensional ballistic annihilation kinetics. *Phys. Rev. E* **57**, 138–145.

Rice, S. A. 1985. *Diffusion-Limited Reactions* (Elsevier, Amsterdam).

Risken, H. 1988. *The Fokker–Planck equation: Methods of Solution and Applications* (Springer-Verlag, New York).

Rosenstock, H. B. 1969. Luminescent emission from an organic solid with traps. *Phys. Rev.* **187**, 1166–1168.

Saffman, P. G. 1959. A theory of dispersion in a porous medium. *J. Fluid Mech.* **6**, 321–349.

Saffman, P. G. 1960. Dispersion due to molecular diffusion and macroscopic mixing in flow through a network of capillaries. *J. Fluid Mech.* **7**, 194–208.

Salminen, P. & Borodin, A. N. 1996. *Handbook of Brownian Motion: Facts and Formulae* (Birkhäuser-Verlag, Basel, Switzerland).

Salminen, P. 1988. On the hitting time and the exit time for a Brownian motion to/from a moving boundary. *Adv. Appl. Prob.* **20**, 411–426.

Sander, E., Sander, L. M., & Ziff, R. M. 1994. Fractals and fractal correlations. *Comput. Phys.* **8**, 420–425.

Scheucher, M. & Spohn, H. 1988. A soluble kinetic model for spinodal decomposition. *J. Stat. Phys.* **53**, 279–294.

Schiff, L. I. 1968. *Quantum Mechanics* (McGraw-Hill, New York).

Seiferheld, U., Bässler, H., & Movaghar, B. 1983. Electric-field-dependent charge-carrier trapping in a one-dimensional organic solid. *Phys. Rev. Lett.* **51**, 813–816.

Sinai, Ya. G. 1992. Distribution of some functionals of the integral of a random walk. *Theor. Math. Phys.* **90**, 219–241.

Spalding, D. B. 1958. A note on mean residence-times in steady flows of arbitrary complexity. *Chem. Eng. Sci.* **9**, 74–77.

Sparre Andersen, E. 1953. On the fluctuations of sums of random variables. *Math. Scand.* **1**, 263–285; **2**, 195–223.

Spitzer, F. 1976. *Principles of Random walk*, 2nd ed. (Springer-Verlag, New York).

Spitzer, F. 1981. Infinite systems with locally interacting components. *Ann. Prob.* **9**, 349–364.

Spouge, J. L. 1988a. Exact solutions for a diffusion-reaction process in one dimension. *Phys. Rev. Lett.* **60**, 871–874.

Spouge, J. L. 1988b. Exact solutions for a diffusion-reaction process in one dimension. *J. Phys. A* **21**, 4183–4200.

Spouge, J. L., Szabo, A., & Weiss, G. H. 1996. Single-particle survival in gated trapping. *Phys. Rev. E* **54**, 2248–2255.

Stehfest, H. 1970. Numerical inversion of Laplace transforms. *Commun. ACM* **13**, 47–49.

Stanley, H. E. 1971. *Introduction to Phase transitions and Critical Phenomena* (Oxford University Press, New York).

Syski, R. 1992. *Passage times for Markov chains* (IOS, Amsterdam).

Takayasu, H. 1989. Steady-state distribution of generalized aggregation system with injection. *Phys. Rev. Lett.* **63**, 2563–2565.

Takayasu, H., Nishikawa, I., & Tasaki, H. 1988. Power-law mass distribution of aggregation systems with injection. *Phys. Rev. A* **37**, 3110–3117.

Teitel, S., Kutasov, D., & Domany, E. 1987. Diffusion and dynamical transition in hierarchical systems. *Phys. Rev. B* **36**, 684–698.

Thompson, B. R. 1989. Exact solution for a steady-state aggregation model in one dimension. *J. Phys. A* **22**, 879–886.

Titschmarsh, E. C. 1945. *The Theory of Functions* (Oxford University Press, Oxford, U.K.).

Torney, D. C. & McConnell, H. M. 1983a. Diffusion-limited reactions in one dimension. *Proc. R. Soc. Lond. A* **387**, 147–170.

Torney, D. C. & McConnell, H. M. 1983b. Diffusion-limited reaction rate theory for two-dimensional systems. *J. Phys. Chem.* **87**, 1941–1951.

Toussaint, D. & Wilczek, F. 1983. Particle-antiparticle annihilation in diffusive motion. *J. Chem. Phys.* **78**, 2642–2647.

Tuckwell, H. C. 1988. *Introduction to Theoretical Neurobiology* (Cambridge University Press, Cambridge, U.K.), Vol. 2.

Tuckwell, H. C. 1989. *Stochastic Processes in the Neurosciences* (Society for Industrial and Applied Mathematics, Philadelphia, PA).

Turban, L. 1992. Anisotropic critical phenomena in parabolic geometries: The directed self-avoiding walk. *J. Phys. A* **25**, L127–L134.

Turkevich, L. A. & Scher, H. 1985. Occupancy-probability scaling in diffusion-limited aggregation. *Phys. Rev. Lett.* **55**, 1026–1029.

Uchiyama, K. 1980. Brownian first exit from sojourn over one sided moving boundary and application. *Z. Wahrscheinlichkeitsthteor. Verwandte Geb.* **54**, 75–116.

Uhlenbeck, G. E. & Ornstein, L. S. 1930. On the theory of the Brownian motion. *Phys. Rev.* **36**, 823–841.

Van den Broeck, C. 1989. Renormalization of first-passage times for random walks on deterministic fractals. *Phys. Rev. A* **40**, 7334–7345.

Van den Broeck, C. 1993. Simple stochastic model for resonant activation. *Phys. Rev. E* **47**, 4579–4580.

van der Pol, B. & Bremmer, H. 1955. *Operational Calculus Based on the Two-Sided Laplace Integral* (Cambridge University Press, Cambridge, U.K.).

van Dongen, P. G. J. & Ernst, M. H. 1983. Pre- and post-gel size distributions in (ir)reversible polymerization. *J. Phys. A* **16**, L327–L332.

van Kampen, N. G. 1997. *Stochastic Processes in Physics and Chemistry* (revised edition) (North-Holland, Amsterdam).

Vergeles, M., Maritan, A., & Banavar, J. R. 1998. Mean-field theory of sandpiles. *Phys. Rev. E* **55**, 1998–2000.

von Smoluchowski, M. 1917. Versuch einer mathematischen Theorie der koagulationskinetik kolloider Lösungen. *Z. Phys. Chem.* **92**, 129–168.

Wald, A. 1947. *Sequential Analysis* (Wiley, New York).

Wang, M. C. & Uhlenbeck, G. E. 1945. On the theory of Brownian motion II. *Rev. Mod. Phys.* **17**, 232–342.

Watson, G. N. 1952. *Theory of Bessel Functions*, 2nd ed. (Cambridge University Press, Cambridge, U.K.).

Weiss, G. H. 1994. *Aspects and Applications of the Random Walk* (North-Holland, Amsterdam).

Weiss, G. H. & Havlin, S. 1986. Some properties of a random walk on a comb structure. *Physica* **134A**, 474–482.

Weiss, G. H. & Havlin, S. 1987. Use of comb-like models to mimic anomalous diffusion on fractal structures. *Philos. Mag. B* **56**, 941–947.

Weiss, G. H., Kopelman, R., & Havlin, S. 1989. Density of nearest-neighbor distances in diffusion-controlled reactions at a single trap. *Phys. Rev. A* **39**, 466–469.

Weiss, G. H. & Rubin, R. J. 1983. Random walks: theory and selected applications. *Adv. Chem. Phys.* **52**, 363–505.

White, W. H. 1982. On the form of steady-state solutions to the coagulation equations. *J. Colloid Interface Sci.* **87**, 204–208.

Wolf, D. E. & Schreckenberg, M. 1998. *Traffic and Granular Flow II* (Springer-Verlag, Heidelberg, Germany).

Wolfram, S. 1991. *Mathematica: A System for Doing Mathematics by Computer* (Addison-Wesley, Reading, MA).

Zumofen, G. & Blumen, A. 1981. Survival probabilities in three-dimensional random walks. *Chem. Phys. Lett.* **83**, 372–375.

Zürcher, U. & Doering, C. R. 1993. Thermally activated escape over fluctuating barriers. *Phys. Rev. E* **47**, 3862–3869.

Zwanzig, R. 1990. Rate processes with dynamical disorder. *Acc. Chem. Res.* **23**, 148–152.

Zwanzig, R. 1992. Dynamical disorder: passage through a fluctuating bottleneck. *J. Chem. Phys.* **97**, 3587–1589.

Index

57712196R00197

Made in the USA
Lexington, KY
06 December 2016